A Modern Guide to Labour and the Platform Economy

ELGAR MODERN GUIDES

Elgar Modern Guides offer a carefully curated review of a selected topic, edited or authored by a leading scholar in the field. They survey the significant trends and issues of contemporary research for both advanced students and academic researchers.

The books provide an invaluable appraisal and stimulating guide to the current research landscape, offering state-of-the-art discussions and selective overviews covering the critical matters of interest alongside recent developments. Combining incisive insight with a rigorous and thoughtful perspective on the essential issues, the books are designed to offer an inspiring introduction and unique guide to the diversity of modern debates.

Elgar Modern Guides will become an essential go-to companion for researchers and graduate students but will also prove stimulating for a wider academic audience interested in the subject matter. They will be invaluable to anyone who wants to understand as well as simply learn.

Titles in the series include:

A Modern Guide to Public Policy
Edited by Michael Howlett and Giliberto Capano

A Modern Guide to the Economics of Happiness
Edited by Luigino Bruni, Alessandra Smerilli and Dalila De Rosa

A Modern Guide to Economic Sociology
Edited by Milan Zafirovski

A Modern Guide to National Urban Policies in Europe
Edited by Karsten Zimmermann and Valeria Fedeli

A Modern Guide to Wellbeing Research
Edited by Beverley A. Searle, Jessica Pykett and Maria Jesus Alfaro-Simmonds

A Modern Guide to Philosophy of Economics
Edited by Harold Kincaid and Don Ross

A Modern Guide to the Urban Sharing Economy
Edited by Thomas Sigler and Jonathan Corcoran

A Modern Guide to the Digitization of Infrastructure
Edited by Juan Montero and Matthias Finger

A Modern Guide to Sports Economics
Edited by Ruud H. Koning and Stefan Késenne

A Modern Guide to Labour and the Platform Economy
Edited by Jan Drahokoupil and Kurt Vandaele

A Modern Guide to Labour and the Platform Economy

Edited by

Jan Drahokoupil

Senior Researcher, European Trade Union Institute, Brussels, Belgium

Kurt Vandaele

Senior Researcher, European Trade Union Institute, Brussels, Belgium

Elgar Modern Guides

Cheltenham, UK • Northampton, MA, USA

Published by
Edward Elgar Publishing Limited
The Lypiatts
15 Lansdown Road
Cheltenham
Glos GL50 2JA
UK

Edward Elgar Publishing, Inc.
William Pratt House
9 Dewey Court
Northampton
Massachusetts 01060
USA

Paperback edition 2023

A catalogue record for this book
is available from the British Library

Library of Congress Control Number: 2021943470

This book is available electronically in the **Elgar**online
Political Science and Public Policy subject collection
http://dx.doi.org/10.4337/9781788975100

ISBN 978 1 78897 509 4 (cased)
ISBN 978 1 78897 510 0 (eBook)
ISBN 978 1 0353 1210 8 (paperback)

Printed and bound by CPI Group (UK) Ltd, Croydon, CR0 4YY

Contents

PART II REGULATING PLATFORM WORK

PART III CASE STUDIES ACROSS THE GLOBE: ONLINE LABOUR PLATFORMS

PART IV CASE STUDIES ACROSS THE GLOBE: LOCATION-BASED LABOUR PLATFORMS

Figures

Tables

Contributors

Mehtap Akgüç is Senior Researcher at the European Trade Union Institute in Brussels, Belgium. Her research interests include labour markets, inequalities, migration and mobility, the green and circular economy and economic development. She has contributed to various projects on employment and social policies, social dialogue, migration, the platform economy, social innovation and trade. Mehtap has a PhD in economics from Toulouse School of Economics in France.

Mariya Aleksynska is Economist at the OECD Development Centre. Mariya holds a PhD in economics from Bocconi University. She is a former ILO officer, a former CEPII researcher and a former lecturer. Her main areas of work include labour migration, labour market institutions and work in the platform economy.

Adam Badger is Researcher for the Fairwork project. His research focuses on the lived experience of work for those in the platform economy and on the manifestations of algorithmic management in everyday working life. He explores the role of technology and other components of contemporary work in addition to the ways in which workers are beginning to resist through, and beyond, their work practice. He is an active member of the European Network on Digital Labour.

Helena Barnard is Director of the doctoral programme at the Gordon Institute of Business Science at the University of Pretoria. She studies how knowledge moves between more and less developed countries, particularly in Africa, with a particular interest in internet-enabled organizational practices and innovation.

Pierre Bérastégui is Researcher at the European Trade Union Institute. He holds a PhD in cognitive ergonomics and an MSc in industrial and organizational psychology. He is also Lecturer at the University of Liège where he teaches ergonomics. His fields of research include psychosocial risks, musculoskeletal disorders, fatigue-related risk management and digitalization. He is also a member of the Tools and Awareness Raising Advisory Group.

Janine Berg is Senior Economist with the Conditions of Work and Equality Department of the International Labour Office in Geneva, Switzerland. She is the author of several books and numerous articles on employment, labour market institutions and the digital transformation of work.

Julie (Yujie) Chen is Assistant Professor in the Institute of Communication, Culture, Information and Technology and holds a graduate appointment at the Faculty of Information at the University of Toronto, Canada. She is interested in digital labour studies, the political economy of media and communications and critical data studies. She is the lead author of *Super-sticky WeChat and Chinese society* (2018) and the co-author of *Media and management* (2021).

Pauline de Becdelièvre holds a PhD in industrial relations. She is Assistant Professor at École normale supérieure Paris-Saclay where her research is on independent workers. She has published several papers on trade unions and careers and has been a visiting scholar at Florida State University to study unions.

Valerio De Stefano is BOFZAP Research Professor of Labour Law at KU Leuven, Belgium, where he does research on non-standard employment, work in the platform economy, technology and fundamental labour rights. He obtained his PhD at Bocconi University where, after his doctorate, he received a postdoctoral fellowship for four years (2011–2014). From 2014 to 2017, Valerio worked as an officer of the International Labour Office in Geneva. At KU Leuven, he leads a team of young researchers working on labour and technology issues.

Jan Drahokoupil is Senior Researcher at the European Trade Union Institute where he coordinates research on digitalization and the future of work. His publications include 'The limits of foreign-led growth: demand for skills by foreign and domestic firms' (with Brian Fabo, 2020) and 'The challenge of digital transformation in the automotive industry: jobs, upgrading and the prospects for development' (2020).

Darcy du Toit is Co-Investigator on the Fairwork project. He is also Emeritus Professor at the University of the Western Cape, and former dean of law, and currently coordinates a new niche area in the Faculty of Law, Labour Law 4.0. His research has focused on the impact of digitalization on labour rights, changing forms of work, discrimination in the workplace, workplace democratization, employment equity, strike law and domestic work.

Sai Englert is Lecturer in the Institute for Area Studies, Leiden University, where he teaches political economy of the Middle East. His research focuses on the labour movement in Israel, settler colonialism, accumulation by dispossession, the transformation of work and anti-Semitism. He is on the editorial board of *Historical Materialism* and is Editor at *Notes from Below*.

Maria Figueroa is Dean of the Harry Van Arsdale Jr. School of Labor Studies at the State University of New York. Her work has involved applied research and technical assistance in the area of non-standard work arrangements, including digitally mediated employment. She has worked on several projects commissioned by unions, employer groups, government agencies and private foundations. Maria has more than 25 years of experience in labour and industry research, having worked as a senior analyst and researcher for national and international organizations, including the Teamsters union and the United Nations.

Jörg Flecker is Professor of Sociology in the Department of Sociology of the University of Vienna, Austria. His research areas are labour process analysis, industrial relations, digitalization, social change and the political far right. He is the editor of *Space, place and global digital work* (2016).

Sandra Fredman is Professor of Law at the University of Oxford. She is a Fellow of the British Academy and Queen's Council (*honoris causa*). She has published widely and has numerous peer-reviewed publications in the fields of gender equality, labour law and human rights. She is Co-Investigator on the Fairwork project in South Africa.

Sacha Garben is Professor of European Union Law at the Legal Studies Department of the College of Europe. She is furthermore an officer in the European Commission, currently on special leave to be full time at the College of Europe, and a replacement judge at the Amsterdam Court of Appeal. She has published widely on a range of constitutional and substantive aspects of European Union law and policy.

Raoul Gebert is Assistant Professor at *Université de Sherbrooke* in Quebec, Canada, and holds a PhD in industrial relations from *Université de Montréal*. Between 2012 and 2015 he served as chief of staff to the leader of the official opposition in Canada's parliament and then worked for the Friedrich-Ebert Foundation in Washington, DC. He is affiliated to the Interuniversity Research Centre on Globalization and Work (*Centre de recherche interuniversitaire sur la mondialisation et le travail*) as co-researcher.

Nora Gobel is Junior Research Officer within the Research Department of the International Labour Office in Geneva, Switzerland. She is currently working on issues related to the digital economy.

Mark Graham is Professor of Internet Geography at the Oxford Internet Institute and Director of the Fairwork Foundation. Together with Jamie

Woodcock he is the author of *The gig economy: a critical introduction* (2019). The full list of his publications is available at www.markgraham.space.

Richard Heeks is Chair in Digital Development at the Global Development Institute, University of Manchester, as well as Director of the Centre for Digital Development and Researcher with Fairwork. He has been consulting and researching on digital issues and development for more than 30 years with research interests encompassing digital labour, digital authoritarianism and data-intensive development.

Benjamin Herr is a doctoral student, funded by the Austrian Academy of Sciences, at the Department of Sociology of the University of Vienna, Austria.

Isis Hjorth is Research Associate at the Oxford Internet Institute. She is a cultural sociologist who specializes in analysing emerging practices associated with networked technologies. She completed her PhD at the Oxford Internet Institute in 2014. Her thesis 'Networked cultural production: filmmaking in the Wreckamovie community' was an ethnographic study of four crowdsourced feature films tackling the emergence of distributed collaborative production models spanning the boundaries between non-market and market-oriented production. Grounded in critical sociological theory, it examined the division of labour and theorized the dynamics of the various forms of capital enabling the realization of these novel forms of cultural goods.

Kristin Jesnes is Researcher at the Fafo Institute for Labour and Social Research in Norway where her main areas of research are industrial relations and non-standard forms of work including platform work. She is also Doctoral Research Fellow at the University of Gothenburg in Sweden, exploring in her PhD how platform companies influence work in different labour market models.

Hannah Johnston is a geographer and Postdoctoral Fellow at Northeastern University in Boston, MA, where she researches algorithmic management and digital labour platforms.

Simon Joyce is Research Fellow at the Digital Futures at Work Research Centre, University of Leeds. His research interest centres on the processes of change in employment relations, examining how shifts in state policy, political economy, employer strategy and management systems affect the everyday experience of work and how workers' responses to those changes generate resistance to and the reshaping of management approaches. He is currently researching platform work and the platform economy.

Vili Lehdonvirta is Professor of Economic Sociology and Digital Social Research at the Oxford Internet Institute, University of Oxford, and a Fellow of the Alan Turing Institute. He is an economic sociologist whose research examines how digital technologies are shaping the organization of economic activities in society and with what implications for workers, consumers, business and policy. He is currently the principal investigator of iLabour, a major research project on the online platform economy funded by the European Research Council. His publications include *Virtual economies: design and analysis* (2014, with Edward Castronova).

Jack Linchuan Qiu is Professor at the Department of Communications and New Media of the National University of Singapore. He works on digital labour, development and media policy in Asian contexts. He is also President of the Chinese Communication Association.

Pamela Meil is a sociologist of work and Senior Research Fellow at the Institute for Social Science Research, Munich. Her current research includes the impact of work organization and skill on hybrid value-added systems, the development of a European perspective on work and employment conditions for slash workers and the effects of Industry 4.0 and digital transformation on value chains.

Denis Neumann is Researcher at the Centre for Employment Relations, Innovation and Change at Leeds University Business School. He is currently holding an Economic and Social Research Council scholarship within the Digital Futures at Work Research Centre that is led by both the University of Sussex Business School and Leeds University Business School. His main research interests include industrial relations in platform-mediated food delivery and grassroots trade unionism.

David Peetz is Professor Emeritus of Employment Relations at Griffith University and Co-Researcher at the Interuniversity Research Centre on Globalization and Work (*Centre de recherche interuniversitaire sur la mondialisation et le travail*) in Canada. He has been a Distinguished Visiting Fellow at the Advanced Research Collaborative in the Graduate Center of the City University of New York, a manager in the Senior Executive Service of the Australian government's Department of Industrial Relations and has undertaken work for trade unions, employers, the International Labour Organization, the Organisation for Economic Co-operation and Development and governments of various political persuasions in and outside Australia.

Susanne Pernicka is Professor of Economic Sociology and Chair of the Department of Economic and Organizational Sociology at

Johannes-Kepler-University Linz. Her research interests include employment relations in international comparative perspective and sociological theory.

Agnieszka Piasna is Senior Researcher at the European Trade Union Institute (ETUI). She holds a PhD in sociology from the University of Cambridge. Her research interests lie in the areas of job quality, working time, labour market regulation, gender and digitalization. She currently carries out research on the platform economy in Europe and she coordinates the ETUI Internet and Platform Work Survey.

Carlos Piñeyro Nelson is a doctoral candidate in sociology at the New School for Social Research in New York City. He currently teaches in the Humanities Department at *Universidad Iberoamericana* in Puebla, Mexico. His dissertation examines how the National Domestic Workers Alliance organizes low-wage domestic workers, why and how emotions are tied to organizing this workforce and how affective ties between employees and employers influence these efforts.

Uma Rani is Senior Economist with the Research Department of the International Labour Office, Geneva, Switzerland. Her research focuses on the informal economy, minimum wages and digital transformations in the world of work.

Andrea Santiago Páramo has a BA and MA in philosophy from the *Universidad Nacional Autónoma de México* and is a PhD candidate in anthropological sciences at the *Universidad Autónoma Metropolitana.* She is Executive Director of *Nosotrxs* and a member of *Red de Investigación sobre Trabajo del Hogar en América Latina.* She has coordinated and directed initiatives and projects for the promotion and defence of the rights of domestic workers and digital platform delivery workers.

Philip Schörpf is Senior Researcher at FORBA, the Working Life Research Centre in Vienna. In recent years his research interests have included online platforms, virtual work, crowdwork, outsourcing, offshoring and digitalization in the service sector.

Andrey Shevchuk is Associate Professor and Senior Researcher at the Laboratory for Studies in Economic Sociology at the National Research University Higher School of Economics, Moscow, Russia. His recent work primarily examines the development of freelance contracting and online labour markets in Russia.

David Peter Simon is an ethnographer specializing in informational spaces and experience design. He has conducted fieldwork with populations as varied as small business merchants in the Middle East, public servants in western Europe, community health workers in east Africa and platform workers in south-east Asia. He currently leads research at Shopify Inc.

Denis Strebkov is Associate Professor and Deputy Head at the School of Sociology and Senior Researcher at the Laboratory for Studies in Economic Sociology at the National Research University Higher School of Economics, Moscow, Russia. His recent work primarily examines the development of freelance contracting and online labour markets in Russia.

Mark Stuart is Montague Burton Professor of Human Resource Management and Employment at Leeds University Business School, University of Leeds, and Co-Director of the Economic and Social Research Council Digital Futures at Work Research Centre. A past president of the British Universities Industrial Relations Association, Mark has published widely on trade union change, the modernization of employment relations, skills and organizational restructuring. His current research is exploring the differential impacts of new digital technologies on the changing world of work. He is a Fellow of the Academy of Social Sciences.

Ping Sun is Assistant Professor at the Institute of Journalism and Communication of the Chinese Academy of Social Sciences. She is interested in platform economy, digital labour, gender and critical media studies.

Vera Trappmann is Professor of Comparative Employment Relations at Leeds University Business School. Her main areas of research are the precarization of work, including platform work, restructuring and organized labour, as well as climate change and just transition. She is involved in the Leeds Index of Platform Labour Protest.

Gérard Valenduc is Associate Researcher to the Foresight Unit of the European Trade Union Institute and at Chair Labour-University, UCLouvain. He holds a PhD in informatics. Until his retirement in 2017, he was an invited professor at the Universities of Louvain-la-Neuve and Namur and a director of the Research Centre of *Fondation Travail-Université* in Namur, Belgium.

Jean-Paul Van Belle is Professor in the Department of Information Systems at the University of Cape Town and Director of the Centre for IT and National Development in Africa. His research areas are the adoption and use of emerging technologies in developing world contexts including mobile, cloud computing, the platform economy, the Fourth Industrial Revolution, artificial intelligence, open and big data. He has over 200 peer-reviewed publications including 25 chapters in books and around 40 refereed journal articles.

Kurt Vandaele is Senior Researcher at the European Trade Union Institute. He holds a PhD in political science from the University of Ghent. His research interests include trade union revitalization, the workers' repertoire of collective action, the platform economy and the political economy of Belgium and the Netherlands.

Alex J. Wood is Lecturer at the University of Bristol and Research Associate at the Oxford Internet Institute, University of Oxford. His research focuses on the future of work and the economy and he is currently researching how platforms transform labour relations. His book *Despotism on demand: how power operates in the flexible workplace* (2020) investigates the impact of precarious scheduling on workplace power relations. His research has appeared in academic journals such as *Sociology*, *Work, Employment and Society* and *Human Relations*.

Mathias Wouters is a PhD candidate at KU Leuven, Belgium where he works for the Institute for Labour Law. His research primarily focuses on the linkages between platform work and triangular employment relationships, home-based work and domestic work.

Acknowledgements

Just as it takes a village to raise a child, publishing an edited volume is not solely the work of the editors. First of all, we are grateful to our employer, the European Trade Union Institute, based in Brussels, for creating a stimulating intellectual environment for doing this kind of work and for providing encouragement and support. We hope that this book can contribute to the debate on the platform economy and especially on the role trade unions and the labour movement in general can play in this type of economy.

Two major events happened during the course of the book's adventure. The decision of the United Kingdom in 2016 to leave the European Union brought legal uncertainty that made us postpone the launch of this project. It took a while to sort this out before the book could finally go ahead with the despatch of invitations to possible authors in summer 2019. In this process, we aimed for a gender balance and almost succeeded, particularly if we were to consider only the lead authors.

The second event is of course the Covid-19 pandemic which, from 2019, has disrupted the lives and work of many. This book project was not immune either. A few weeks before the first lockdown in Belgium, in February 2020, an authors' meeting took place in Brussels. We could not provide a comprehensive assessment of the impacts of the crisis, but some of it has been taken into account in some of the chapters.

We have also very much appreciated the engagement and enthusiasm of all the contributors, from various disciplines, during these remarkable times. We are very grateful to our colleague Kristel Vergeylen for organizing the meeting. Also, we owe many thanks to our language editor, Calvin Allen, who always hit the nail on the head, and to our colleague Giovanna Corda of the Documentation Centre of the European Trade Union Institute for thoroughly checking the references. Last but not least, we are grateful to Stephanie Hartley and Matthew Pitman of Edward Elgar for having offered us this opportunity and then remaining patient in the course of the Brexit shenanigans.

Brussels, February 2020
Jan Drahokoupil and Kurt Vandaele

Abbreviations

AB5	Assembly Bill 5
AI	Artificial intelligence
AMT	Amazon Mechanical Turk
B2B	Business to business
BATX	Baidu, Alibaba, Tencent, Xiaomi
BPO	Business process outsourcing
BZÖ	Bündnis Zukunft Österreich (Alliance for the Future of Austria)
CGT	Confédération Générale du Travail (General Confederation of Labour)
CJEU	European Court of Justice
CLAP	Collectif des livreurs autonomes de Paris (Independent Couriers Collective of Paris)
CUPW	Canadian Union of Postal Workers
CWS	Contingent Worker Supplement
DCCA	Danish Competition and Consumer Authority
DIHK	Deutscher Industrie- und Handelskammertag (Association of German Chambers of Industry and Commerce)
ECJ	European Court of Justice
ETUC	European Trade Union Confederation
ETUI	European Trade Union Institute
EU	European Union
FAU	Freie Arbeiterinnen- und Arbeiter-Union (Free Workers' Union)
FNV	Federatie Nederlandse Vakbeweging (Federation of Dutch Trade Unions)
FPÖ	Freiheitliche Partei Österreichs (Freedom Party of Austria)
FWC	Fair Work Commission

GAFA	Google, Apple, Facebook, Amazon
GDPR	General Data Protection Regulation
GE	General Electric
GSM	Global System for Mobile Communications
GWR	Gig Workers Rising
ICT	Information and communications technologies
IDG	Independent Drivers Guild
IHK	Industrie- und Handelskammer (German chamber of commerce)
ILO	International Labour Organization, International Labour Office
ISCO	International Standard Classification of Occupations
IT	Information technology
IWGB	International Workers Union of Great Britain
LA	Los Angeles
LATWA	Los Angeles Taxi Workers Alliance
MLC	Maritime Labour Convention
MTBOT	Metropolitan Taxicab Board of Trade
NACE	Nomenclature Statistique des activités économiques dans la Communauté européenne (Statistical Classification of Economic Activities in the European Community)
NASA	National Aeronautics and Space Administration
NDWA	National Domestic Workers Alliance
Neos	Das Neue Österreich und Liberales Forum (New Austria and Liberal Forum)
NGG	Gewerkschaft Nahrung-Genuss-Gaststätten (Food, Beverages and Catering Union)
NTF	Norsk Transportarbeiderforbund (Norwegian Transport Workers' Union)
NYC	New York City
NYTWA	New York Taxi Workers Alliance
OHS	Occupational Health and Safety
OLI	Online Labour Index

ÖVP	Österreichische Volkspartei (Austrian People's Party)
PUC	Public Utilities Commission
SEIU	Service Employees International Union
SEO	Search engine optimization
SMart	Société Mutuelle pour artistes (Mutual Society for Artists)
SNCF	Société nationale des chemins de fer français (National Company of French Railways)
SPÖ	Sozialdemokratische Partei Österreichs (Social Democratic Party of Austria)
TFEU	Treaty on the Functioning of the European Union
TLC	Taxi and Limousine Commission
TPAC	Taxicab and Paratransit Association of California
TWA	Temporary work agency
UFCW	United Food and Commercial Workers Union
UK	United Kingdom
US	United States
USD	United States dollar
USSR	Union of Soviet Socialist Republics
Ver.di	Vereinte Dienstleistungsgewerkschaft (United Services Trade Union)

1. Introduction: Janus meets Proteus in the platform economy

Jan Drahokoupil and Kurt Vandaele

1.1 THE ORGANIZATION OF WORK IN THE PLATFORM ECONOMY

Platform companies are market makers. They unleash market forces from the constraints of hierarchies and long-term contracts. Digital technology and the internet have allowed platform companies to run digital marketplaces that can link clients across the globe, with cost structures that allow mediation of the cheapest services through sophisticated matching mechanisms. A search on the internet triggers an auction in which one's attention is sold to the advertisers before the results are displayed. That spare room might not need to stay idle for the night. Platforms also make matching less arduous. A dating app saves one an awkward visit to a dodgy bar, or the need to sign up for a dance class.

The world of work has not been immune to restructuring by platforms. Digital labour platforms organize marketplaces that match supply and demand for paid work. Long-term engagements are thus replaced by one-off transactions. Jobs are transformed into tradeable tasks. Employment agencies have long been in the business of matching workers with their clients, but only labour platforms promise to get away from the hierarchies and long-term contracts that tie workers in the employment relationship. Digital labour platforms thereby transform labour markets from markets for jobs to markets for tasks. Although the platform workforce is apparently relatively small in terms of employment, the economic transformation symbolized by labour platforms has attracted considerable attention, triggering imaginations about the new flexible future for the world of work. While some have welcomed, if not celebrated, the emancipation, freedom and autonomy this gives to workers, others have fretted about the loss of social protection for the most vulnerable workers and the prospect of algorithmic enslavement and race-to-the-bottom terms and conditions of work.

Combining Janus and Proteus – figures from Roman and Greek mythology, respectively – might be an appropriate fit to describe the current and complex state of digital labour platforms. First, regarding the temporal dimension, labour platforms have a Janus face: they are combining the 'new' with the 'old'. The 'new', or discontinuity, refers to the platforms' use of digital technology, the algorithmic management approach and the 'production and commodification of data' (Grabher and Köning 2020, p. 106), but it relates as well, perhaps more importantly, to their narrative ability to produce a discourse of novelty and disruptiveness which, at the same time, 'function[s] to detract attention from underlying continuities and constitutive contexts' (Peck and Philipps 2020, p. 82). The 'old' concerns the types of employment practices and intermediation that are reminiscent of informal and triangular work arrangements during earlier phases within capitalist development in the Global North, although this still exists in certain industries today. Thus, in practice, the borderline between platform work and non-standard and casual employment is blurred so that the employment status of platform workers is often contested. Second, concerning the spatial dimension, just like the prophetic sea god Proteus, labour platforms can take numerous forms and shapes influencing workers' experiences and potentially their resistance to, for instance, algorithmic control. One could distinguish between the scale of tasks mediated, ranging from micro to large-scale project work, the nature of the skills involved and the system of task distribution such as the direct allocation of offers, auctions or contests (see also Florisson and Mandl 2018). Moreover, as so often, context matters: institutions and policies shape the prevalence of platform work and the conditions within it.

One of the key distinctions in the Proteus-like heterogeneity of platform work relates to the format of service provision and to that of the geography of the respective market. Location-based platforms offer services that are delivered and performed in or around the locality while the intermediation occurs online. They include ride-sharing, food delivery and domestic services like care work, cleaning and repair and maintenance. Matching supply and demand in local markets for physically delivered services, location-based platforms are often run by international operators that might be considered transnational corporations. The limited scalability of their markets leaves space also for companies operating locally, however. A different case is what we call online platforms. The latter, including, for example, Amazon Mechanical Turk, TaskRabbit or Upwork, organize digitally delivered services that can, in principle, be delivered from anywhere in the world. Online platforms thus have the potential to operate on a global scale. Yet, in practice, they are characterized by regional divisions and distinct geographies of operation. The distinction between location-based and online platforms, and hence between location-based and online platform work, is related to the main differences in opportunities for worker organizing and alliance-building, regulatory issues and challenges and that of business strategies. According to the International

Labour Organization (ILO), the number of location-based platforms increased almost ten-fold in the period between 2010 and 2020 while the number of online platforms tripled (ILO 2021).

This key distinction between location-based and online labour platforms informs many of the comparative chapters in this book and it structures our organization of the material. The book aims to provide a guide to the diverse landscape of platform work. It delivers a comprehensive overview of the key issues relating to labour in the platform economy, focusing in particular on the challenges for labour agency, including the role of institutions and policies. Designed to provide a stimulating guide to the current research landscape, it offers comparative overviews of the key topics followed by case studies of platform work from across the globe. The case studies cover working conditions in major segments of platform work and focus on the strategies, struggles and resistance of the workers involved. Covering the diversity of platform work in a global comparative perspective, the experience of different world regions is intersected with the variety of the platform economy.

Part I of the book puts digital labour platforms in their historical, economic and geographical context. It also considers the implications of platform work for working conditions and worker health and tackles the challenges of measuring the extent and nature of platform work. The rise of labour platforms is thus situated in the context of the range of forces that are driving change in the organization of work and that act towards undermining the 'standard employment relationship' (Bosch 2004). The employment relationship has played a key role in regulating labour relations and organizing protection for workers, including social insurance. Such a relationship has been less prominent in developing countries, however, where large parts of the workforce earn their living in the informal economy without any access to social insurance and other forms of social protection. The emergence of platform work thus has different implications in such contexts, possibly offering a way of formalizing the labour market.

Labour platforms are therefore part and parcel of longer-term processes of work externalization and value chain fragmentation. Driven by the search for allocation efficiency, cost savings and regulatory arbitrage, these processes can be observed, in different forms and with different implications, across the world. Digital technology is a key enabler of this process but the outcomes are not determined by it. Policies and institutions not only shape the outcomes but also enable the rise of the platform economy and platform work. Drawing on comparative evidence, the chapters in Part I also help to understand the diversity of labour platforms and the types of work mediated through them.

Part II addresses the regulatory response to platform work and the role of institutions in shaping the outcomes. The chapters are motivated by the question of what regulatory framework is needed to support good working conditions for platform workers and guarantee their fundamental rights, taking into account self-regulation by the platforms, regulation via collective bargaining and legal regulation. It considers in particular the role of the conventions of the ILO and regulation at European level. Above all, the question of the employment status of platform workers dominates discussion of the regulatory responses. The combination of the high degree of worker control by the platform and working time arrangements that allow flexibility and short-term engagements has made some forms of platform work sit uneasily within many regulatory frameworks. Another source of legal uncertainty is the question whether platform workers, working on a self-employed basis, can conclude collective agreements without breaching competition law. The list of regulatory issues is much longer, however. For instance, there is the issue of the fees charged by some platforms to workers; there is the lack of redress in the event of refusal of payment for completed work or the suspension or closure of worker accounts; the automated systems for rating and allocating work are often opaque and error-prone; and clients or platform operators, or both, often do not communicate with workers to resolve issues (Silberman and Johnston 2020).

In the case of location-based platform work, the key regulatory challenge is to make the existing regulations fit for purpose. Online platforms are more difficult to regulate as a larger number of jurisdictions are typically involved and organizing workers collectively is challenging, where even possible. ILO conventions may represent a way of addressing some of these issues (see ILO 2021). In practice, however, the international labour movement may need to rely on voluntary codes of conduct, alliance-building with non-governmental consumer organizations and informal worker coordination in digital online communities, although the forums and other types of online interaction are often not independent of the platforms themselves (see, for example, Bucher et al. 2020; Gegenhuber et al. 2020; Gerber 2020; Gerber and Krzywdzinski 2019; Panteli et al. 2020; Wood and Lehdonvirta 2019; Wood et al. 2018).

The remainder of the book investigates how these general trends play out in the individual types of platform work across the globe. Part III encompasses three case studies of online labour platforms while Part IV consists of six case studies of location-based platforms either in private or public settings. The studies indeed underline the diversity of experience of platform work. For some platform workers, particularly those working on online platforms in lower-income countries, it provides good pay, often higher than alternatives in the offline economy. At the same time, remuneration on online platforms tends to be characterized by large inequalities between platform workers in a single place. Even poorly paid online work, such as microwork, can be attractive for other reasons, however. These include the lack of a commute or the possibility of combining it with other work or care obligations. Relatively low remuneration tends to characterize location-based labour platforms that are able to tap into a pool of low-skilled workers, often migrants, who tend to be abundant on the labour market or who are willing to take a poorly paid job for limited access to the regular labour market. In such a context, adequate pay can be achieved by guaranteeing platform workers a minimum wage or coverage by a collective agreement.

The case studies also identify the problems associated with platform work in terms of insecurity, poor work–life balance and discrimination. Online platform workers, in particular, suffer also from social isolation. A general problem experienced by both location-based and online platform workers is a lack of skills development and training opportunities in platform work and hence their low employability outside of this low-skilled work. Despite its potential to formalize activities that would otherwise happen in the informal economy, much of platform work is, in fact, not declared and workers are not covered by social protection. This also creates a fiscal problem for the states that, among others, finance the education of these workers but do not recuperate this investment through taxation. Finally, location-based platform work often entails dangerous health and safety practices. This is most evident in the case of food delivery in which algorithmic management often puts workers under considerable pressure for speedy delivery.

Working time flexibility and – more generally – the ability to work autonomously as 'your own boss' is often seen as a key benefit of platform work. It is also the reason why workers take up, and value, platform work (Adams-Prassl and Berg 2017; Lehdonvirta 2018; Pesole et al. 2018). Platform work is, however, characterized by diverse experiences in this respect, too. Many platform workers benefit from flexible work arrangements and, for some, it is the reason why they are willing to accept lower remuneration. At the same time, our case studies contribute to the body of evidence that shows that, for many platform workers, working time flexibility is only notional (Drahokoupil and Piasna 2019; Schor et al. 2020; Piasna and Drahokoupil 2021). The nature of work, or the need to maximize work to earn sufficient income, often limits the scope for exercising control over working time. Many workers need to work long hours to earn sufficient income, putting in hours that substantially exceed the standard in regular employment. Platform workers often spend a lot of unpaid time waiting and searching for adequate work while facing costs related to fluctuations in demand and the loss of income when demand is low (Berg 2016; Kuhn and Maleki 2017; Wood et al. 2019). Our evidence from food delivery also shows incentive structures that encourage long hours. Moreover, algorithmic management often actually imposes authority structures that leave workers with little autonomy (Aloisi 2020). Again, as shown in the case studies, food delivery is a case in point: workers experience their work as a complete subordination to the algorithms of the company's platform (see also Cant 2018).

Finally, the evidence presented in the case studies also contributes to the debate on worker preferences vis-à-vis working time flexibility and security triggered by the rise of platform work. The debate is often centred around the alleged trade-off between the schedule flexibility of self-employment on the one hand and the protection offered by employment status on the other (Hall and Krueger 2018; Berger et al. 2019). In fact, worker protection and flexible work arrangements can both be understood as the means towards a single end: enhancing worker autonomy and control over the conditions of work. Interested in exercising autonomy and control, workers might, in fact, have ambiguous preferences towards employment status. They are informed, among others, by the specific trade-offs in a given regulatory framework and shaped by the uncertainty about alternative arrangements (Dubal 2020). Income security, in particular, is often valued by workers but guaranteeing a minimum income (especially a minimum hourly wage) may require a more regulated schedule that restricts extreme forms of working time flexibility. The devil is in the detail. Employee status may entail considerable schedule flexibility (see also Berg et al. 2004). Self-employment, in turn, may provide a degree of autonomy and control if incomes are predictable and workers can influence their conditions of work (for example, Dubal 2017).

1.2 KEY ISSUES OF PLATFORM WORK AND ITS CONTEXT: CHAPTERS 2 TO 7

Platforms make and organize markets: they enable their customers to interact with each other, coordinate demand on both sides of the market they serve and lower the transaction costs involved. Market making by platforms has particularly significant implications in the world of work in which it shifts the provision of labour from employment relations, characterized by longer-term engagements and worker protection, to one-off transactions on open markets.

The economic role of platforms is discussed by Jan Drahokoupil in Chapter 2. It distinguishes between the intermediary role of labour platforms, which has specific implications as far as the market for labour is concerned, and the infrastructure role. The latter allows the platforms to structure interactions between clients through its regulatory power and, less directly, by behavioural design. These roles are associated with three distinct business models: regulatory arbitrage; an intermediation business relying on network effects; and a surveillance business associated with infrastructural power.

In practice, labour platforms pursue regulatory arbitrage and the intermediation model. Turning either of them into a successful business strategy is fraught with difficulties, however.

Regulatory arbitrage exploits the cost advantage of replacing employment relations with one-off market transactions via the use of 'independent contractors'. As many aspects of managerial prerogative are built into algorithmic management and performance and monitoring systems, working with independent contractors is a highly ambiguous arrangement, albeit one that is (legally) contested and thus uncertain. It also means facing collective organizing efforts and worker resistance. At the same time there are cases, particularly in Nordic Europe (see Jesnes and Nordli Oppegaard 2020), where the operation of labour platforms goes hand in hand with a genuine employment relationship, although the arrangements are somewhat unusual featuring contracts of a rather non-standard nature. Even outside Nordic Europe, it should be added that some labour platforms have not made use of 'independent contractor' status in the first instance but only returned to it later on.

Furthermore, a notable case has also been the arrangement set up by the labour market intermediary and member-owned and governed cooperative SMart (Société mutuelle pour artistes) for platform-based food delivery couriers in Belgium (Charles et al. 2020; Drahokoupil and Piasna 2019). This wage portage arrangement put the couriers in a genuine employment relationship, providing them with an extra guarantee that legal minimum employment rights would be respected by the platform. Trade unions considered the arrangement as the second-best option in anticipation of the regulatory employment classification of the couriers (Vandaele 2020; see also Xhauflair et al. 2018). The SMart arrangement ended in early 2018, though, when Deliveroo unilaterally decided to work only with ad hoc contracts largely devoid of minimum standards or social protection.

The intermediation model of labour platforms allows the generation of income from matching without the need to invest in underlying assets. Network effects promise monopoly rents. Market dominance is not guaranteed, however, and the costs of sustaining the network effects can be considerable. It is the latter in particular that puts question marks over location-based platforms as an unstoppable juggernaut. They have to create and sustain network effects in price-sensitive, fragmented markets, in each and every location, while upscaling network effects is not possible as, by definition, location-based platforms are restricted to the location. In general, while labour platforms save costs on capital equipment and its maintenance, they also need a critical mass of users and clients or customers in order to gain liquidity and become self-sustaining, while both users and clients have the possibility of using multiple platforms so that loyalty to the platform is undermined. Sustaining the network effects therefore implies the need for continued investment. Also, there might be possibilities for users and clients or customers simply to circumvent the platform as an intermediary.

Given all these difficulties, is it any wonder that the profits of labour platforms, at least the publicly listed ones, and location-based platforms in particular, are close to non-existent so far? Still, the prospect of monopoly rents can seduce investors while the ability of some platforms to capture and utilize data about transactions, users and customers seems also of importance in investment decisions.

The platforms' control of workers' economic assets obtained from user-generated data – their ratings and their status they themselves have earned – is a key question for the engagement of regulatory institutions since it concerns, ultimately, the ownership of data. Within the workplace there are important debates concerning the use of surveillance software, not least with the rise of remote working for office workers under Covid-19, and the data which platform workers generate while going about their work fit precisely into that set of debates. At national level, it is clear that the commissions and agencies responsible for information privacy need to be both better engaged in those debates and more understanding of and responsive to the types of issues being raised – and the same is true of regulatory agencies at international level where there is a degree of policy coordination. It seems that such agencies have been somewhat behind the curve when it comes to the workplace; and that platforms have latched onto this rather unfortunate lack of market intelligence.

Much of the discussion and research focuses on consumer-facing platforms and less is known about business-to-business platforms (see also Grabher and van Tuijl 2020). The extent to which companies in the conventional economy are utilizing digital labour platforms, and the forms this takes, is addressed by Pamela Meil and Mehtap Akgüç. In Chapter 3, they explore the linkage between digital labour platforms and conventional labour markets, situating the platform model in the context of global production networks and long-term value-chain restructuring.

Data collection on the specific types of activities performed on labour platforms is difficult: either this belongs to the internal operation of companies, while the list of business clients on the websites of some labour platforms can simply be considered as marketing, or it might hide complex processes of outsourcing within the value chain. Nevertheless, based on the distribution of work, the authors discern three outsourcing patterns of interaction between labour platforms and companies across a diverse set of industries in the conventional economy. Two of these patterns are in accordance with the two types of the online platform economy: jobs and tasks that can be considered microwork, also known as 'ghost work' (Gray and Suri 2019), which can be reintegrated into the company without much effort; and (stand-alone) macrowork being similar to freelance work and which needs little reintegration as the tasks are mainly separated from (day-to-day) company processes. The third pattern relates to platforms as an intermediary coordinator and manager of dispersed activities, making their status as labour platforms, rather than service suppliers, contested. Such companies split complex tasks, organize their sourcing and execution, and then integrate them as a solution for the client.

While those three patterns are dominant, this does not exclude, however, that companies could make use of location-based platforms, like cleaning services, although this use of (personal) services by companies seems of secondary importance. Future in-depth case studies are needed here for further deepening and juxtaposing of the different patterns. Even so, outsourcing via labour platforms is a continuation of the long-term trend of the externalizing of activities within companies across expanding value chains. This process affects the working conditions and pay of workers in client companies who find themselves negotiating with their employer in the shadow of the conditions of the externalized workforce (for example, Drahokoupil 2015).

While it can be assumed that the connection between labour platforms and companies is intensifying, its amplitude is uncertain. Simultaneously, the relationship between labour platforms and the conventional economy has a certain Faustian overtone, at least in the Global North. Platforms are 'free-riding' on the conventional economy and taking their gains, ignoring the consequences for the latter (see also Schor et al. 2020). This clearly holds true if it concerns platform workers for whom the income retrieved from platform work is supplementary either because they have another (main) job or they are entitled to social benefits. Put differently, labour platforms are hinging upon existing labour markets in the conventional economy and social protection systems, whereas their 'creative destruction' tends to undermine existing labour market regulations and tax-financed social protection.

Chapter 4 by Agnieszka Piasna critically spells out the pitfalls of measuring the size and scope of the platform economy. Surprisingly, despite its digital character and recording of the employment relationship in this type of economy, not much is known about its prevalence. Digital labour platforms are mostly not obliged to be transparent about their workforce, while platform workers are a hard-to-reach population as platform work is largely 'invisible' (see also Gruszka and Böhm 2020; Mateescu and Ticona 2021) or underreported by the workers themselves for tax or other reasons. Dedicated surveys by private or official statistical agencies are still in their infancy and thus of an explorative nature. They usually suffer from methodological difficulties like fuzzy taxonomy and measurement bias, inadequate data cleaning or self-selecting samples. Depending on the sample method used, estimates of the frequency and intensity of platform work diverge substantially between surveys, calling for caution in interpreting their results. Also, longitudinal data are commonly missing so that possible employment dynamics and trends are largely obscured and difficult to examine. Multi-apping – platform workers combining different platform tasks – adds further to the complication.

Three major remarks can nevertheless be made based on survey results on the size of the platform economy. First, the platform economy is still marginal in quantitative terms if offset against the total working population – yet its importance lies in the implications for employment relations, labour markets and business models in certain industries in the conventional economy and the financing of social security systems. In other words, the platform economy warrants specific study as it is relatively distinctive from other transformations within capitalist development given its specific features like algorithmic management and its use of a contractor relationship to lessen labour costs and maximize labour control (Silver 2003; Vandaele 2021). That being said, as turnover among platform workers tends to be high, exposure to platform work might be quite sizable in absolute terms, with a considerable number of workers experiencing it at least for a short period.

Second, it is necessary to distinguish between different levels of economic dependency or attachment to the platform economy which can be measured by the extent the income level retrieved from platform work is able to 'pay basic expenses' (Schor et al. 2020, p. 833). A first but simple bifurcation can be made between those workers for whom platform work is only supplemental to other (offline) earnings and the (smaller) group of workers who are platform dependent with platform work being their predominant source of income. The economic dependency of the latter group opens up possibilities for labour platforms to discipline their workers. A more fine-grained classification of economic dependence on platform remuneration can be achieved via introducing thresholds based on income retrieved from the conventional economy. Put differently, platform workers are not a homogeneous group – neither from the perspective of their attachment to the platform economy and the degree of algorithmic control nor from other perspectives like the specific occupation or industry they are working in or their level of education. Nevertheless, research from Europe has found that platform workers tend, on average, to be younger, although older workers are also involved (Piasna and Drahokoupil 2019; Urzì Brancati et al. 2020). Also, the proportion of workers with a migrant background is generally higher among platform workers than offline workers, and this is even more true where the income from platform work is at least 50 per cent (Urzì Brancati et al. 2020).

Third, while access in itself in the platform economy might be less marked by discrimination for certain worker categories than in traditional labour markets (van Doorn 2017), it is highly questionable if certain types of platform work, such as in transportation (including delivery), act as a stepping stone to jobs with better employment terms and conditions. Compared to workers not using digital labour platforms to find paid work, the employment trajectory of platform workers is marked by greater instability and fragmentation, with a higher incidence of non-standard contracts than workers solely employed offline (Piasna and Drahokoupil 2019; see also Borchert et al. 2018; van Doorn 2020).

The rise of organizational work practices associated with digital labour platforms should be appreciated within a longitudinal perspective of capitalist development. There are historical comparisons with similar types of piecework, informal and triangular work arrangements and decentralized and on-demand production systems in proto-industrialization areas of the Global North in the nineteenth century (Finkin 2016; Stanford 2017). Although such organizational arrangements still exist in certain non-digital occupations and industries today, their prevalence diminished with the advancement of the standard employment relationship in the twentieth century. While this kind of employment relationship remains dominant in the Global North, contingent forms of employment have nevertheless synchronously developed alongside it from the 1970s onwards (Crouch 2019; Herod and Lambert 2016; Hyman 2018).

In Chapter 5, Gérard Valenduc contributes to our understanding of the long-term trends that have given rise to platform work by analysing how developments in information and communications technologies and policies fostering labour market deregulation and liberalization have enabled and stimulated flexible work arrangements from the mid-1980s onwards. Call centres have been a quintessential example of the application of such arrangements that pertain to flexible working time and work location, atypical employment contracts or blurred subordination links.

Showing continuity with the past, the advancement of the platform economy after the crisis of the finance-led accumulation regime in 2007–2008 fits within this broader trend of contingent employment relationships and fissured workplaces (Weil 2014). Thus, while informality and precariousness have never been away in the Global South, the platform economy denotes a return to a (dystopian) past for workers in the Global North (van der Linden and Breman 2019). At the same time, making use of various innovations in digital technologies, digital labour platforms have a disruptive impact on existing markets and embody a (new) business model with two-sided markets (see Chapter 2). Valenduc also optimistically suggests, however, that the current digital transition can incorporate new types of work, like 'co-creation' (see also Seppänen et al. 2020) and 'open work', which can potentially lead to a withdrawal from the confining straitjacket of work arrangements associated with platforms.

Pierre Bérastégui and Sacha Garben explore further the theme of contingent employment in Chapter 6: they focus on the potential risks of platform-mediated work for the occupational health and safety of workers. The authors discern how the business model of digital labour platforms fits into, and is a profound deepening of, three ongoing and long-term trends in employment relations that are engrained in various industries in the conventional economy.

First, the authors put algorithmic management and digital surveillance – rooted in managerial approaches towards engineering processes and not necessarily confined to the platform economy – central to understanding the monitoring, controlling and disciplining of workers. While the degree of algorithmic management and digital surveillance differs from platform to platform, in almost all instances their neutrality as merely technology-innovating companies is illusory. Second, the on-demand character of platform-mediated work implies that income volatility and feelings of job insecurity are omnipresent so that health and safety risks are augmented. Third, increased workplace fragmentation and the isolation of platform workers adds to the tendencies of mutual competition and undermines the building of professional identities.

These three trends associated with the business model of platforms have no uniform impact but are contingent on the type of platform work. Taking this platform variety into account, Bérastégui and Garben end on a positive note by hinting at the possibilities for labour agency and voice in certain platform types for applying and strengthening occupational health and safety regulations.

The diversity of the platform economy and labour agency is further analytically explored from a labour geography perspective in Chapter 7, written by Benjamin Herr, Philip Schörpf and Jörg Flecker. Applying the concepts of place and space reveal that the geographical appearance of digital labour platforms is, in fact, a *trompe l'oeil*. Thus, while online platforms seem 'placeless', the work is actually performed and territorially embedded somewhere, revealing regional patterns linked, for instance, to colonialization, cultural affinities and linguistic resemblances. Equally, whereas location-based platforms are not only subjected to local regulation, they are also a part of spatial structures, in particular where platform companies are operating at international level. Effective labour agency in the platform economy therefore entails the development of trade union strategies at interlinked, multiple geographical entities or spatial scales, which is a condition not at all different to the conventional economy. In this context, it is revealing that the headquarters of platform companies are geographically concentrated in a small number of countries (ILO 2021).

Herr and his colleagues use a local, national and transnational lens to examine labour agency in the platform economy. Two major conclusions can be drawn from this. First, it is clear that the opportunities for and obstacles to labour agency are unevenly spread across the types of platform, which is linked to their 'spatial fix'; that is, their geographical mobility of production for facilitating profitability and controlling labour (Johnson 2020; Silver 2003; Vandaele 2018, 2021). Workers particularly in location-based platform work, and specifically in food delivery, are commonly able to meet physically as the platforms themselves divide cities into certain areas where restaurants are clustered, which acts as a starting zone for the couriers. Couriers' understanding of and experience with those shared log-in zones not only give rise to individual practices of resistance, but they also provide a spatial context fostering face-to-face contact and the building of collective identities and solidarities (Heiland 2021). Second, as a consequence, although union strategies are ideally multi-scalar from the outset, the emphasis on a specific scale in labour agency differs along with platform type. Thus, whereas labour agency in online platform work operates, almost by default, on a transnational scale, this is less evident in location-based platform work which is more strongly embedded in local and national regulatory contexts. Still, both main types of platform work experience the absence of a physical, shared workplace.

A bottom-up response to workers' isolation and spatial fragmentation is the emergence and development of digital, online communities of platform workers. Whether such communities are simply confined to the virtual world or can be considered prototypes of 'platform unions' remains to be seen. Their development, whether solely virtual or not, entails opening the lens to other collective organizations in the labour movement, like (platform) cooperatives, than simply (and rather myopically) analysing (long-established) trade unions (Atzeni 2020; Vandaele 2021). Even so, as the state has noticeably refrained from legal action so far, the combination of bottom-up approaches with regulatory creativity promises to be fruitful in the building of countervailing power vis-à-vis the platforms.

1.3 REGULATING PLATFORM WORK: CHAPTERS 8 TO 11

The four chapters in Part II of this volume engage with debates on the regulation of digital labour platforms. There are various regulatory ways of addressing the working conditions of platform workers and enforcing their fundamental rights, contingent on the type and features of platform work (see also Frenken et al. 2020; Koutsimpogiorgos et al. 2020).

In Chapter 8, Valerio De Stefano and Mathias Wouters propound the embedding of platforms in contemporary regulatory frameworks. Here, uncertainty about the employment status of platform workers is the key issue to be addressed. The authors review court cases across the globe to illustrate their arguments and point to inconsistencies in the classification of platform workers both between countries as well as within them. They consider two regulatory possibilities. First, some platforms can simply be treated as employers. This implies a proposal for a reversal of the usual burden of proof, such that platforms take on the responsibility for producing evidence that they do not act like an employer. Second, other platforms show similarities with private employment agencies, leading the authors to lean on two conventions of the ILO – the Private Employment Agencies Convention 1997 (No. 181) and the

Home Work Convention 1996 (No. 177) – in making this interpretation of labour market intermediation and in proposing an international governance system for online platforms.

Chapter 9 by Sacha Garben addresses the regulatory options in the European context, addressing in particular legal developments at the level of the European Union (EU). Compared to organizational work arrangements which are based on non-standard types of employment in the informal or shadow economy, the author points out that the intentional use of such arrangements by platforms makes them more detectable and formal such that they may be made subject to regulatory frameworks.

In reality, applying and enforcing existing employment regulation to the triangular, or other, arrangements of platforms is challenging. Case-by-case determinations indeed show mixed results across EU member states, so the outcome is unpredictable. A reliance on existing regulations may not imply a static and passive response, however. The court may adapt the existing definitions of employment to the specific features of platform work. An alternative option is to narrow the group of people that will be considered self-employed. This can be done by introducing an intermediate category beyond the employed/self-employed dichotomy to extend some of the protection that applies to employees. However, this may prove an equally unsatisfactory solution as it still leaves room for interpretation. It may thus be preferable to narrow the self-employment category by introducing a presumption of employee status to platform workers. The approach of the European Trade Union Confederation (ETUC) follows this option, advocating the presumption of an employee relationship along with other measures to reduce the incentive to hire workers as self-employed.[1] Finally, Garben also considers the option of providing specific protection for platform workers regardless of employment status through dedicated regulation.

Turning ultimately to the European level, following the developments in the relevant judgments of the EU Court of Justice shows a pattern that is not quite crystal clear. While the Court considered the intermediation service provided by Uber to be a transportation service, and thus that it was not a merely technology-innovating company, the judgment did not address the employment status of platform workers. Even so, the judgment could be a basis for further regulation in this regard. At the same time, the Court's older *FNV* judgment did in fact introduce legal uncertainty about the collective bargaining rights of platform and other self-employed workers. Interpretations of this judgment differ on whether platform workers who are considered self-employed are able to conclude collective agreements to set minimum employment terms and conditions or whether this might, alternatively, be at odds with European competition law. At the time of writing, the European Commission launched a first-stage consultation with the social partners – representatives of employers and trade union organizations – on how to protect people working on platforms. It remains to be seen how, what and to what extent such a regulatory initiative will effectively reconcile the status of self-employment, within and beyond the platform economy, with collective bargaining rights.

The promise of self-regulation by platforms and their advocates can be considered one of the reasons why the state has been absent from regulating the platform economy. This narrative, rooted in a neoliberal ideology, is a falsehood, however, as Sai Englert, Mark Graham, Sandra Fredman, Darcy du Toit, Adam Badger, Richard Heeks and Jean Paul Van Belle show in Chapter 10. The authors discern two steps in how platforms are deliberately moulding regulations so that they fit with their dominant business model based upon algorithmic management and 'independent contractors'. First, the parlance of self-regulation by platforms is often a cover for a strategy based on simply rule-ignoring or rule-evading or interpreting existing regulations in an unconventional way to their advantage. States are, then, seemingly confronted with a *fait accompli*. Second, platforms try to perpetuate this deregulation or re-regulation or to install new regulatory frameworks. They do this via their lobbying power (see, for example, Corporate Europe Observatory and AK Europe 2019) or via their campaigning power in shaping public opinion to their regulatory preferences (see also Chapter 19).

Still, as several case studies in Part III of this volume demonstrate, the embeddedness of platforms in local and national regulatory contexts means that they intersect with existing regulations so that their actual operation is marked by contextual variation. Crucial in understanding this variation is also the resistance of platform workers, trade unions or their supporters via, for instance, industrial action and litigation. Other initiatives like that of the Fairwork Foundation of the Oxford Internet Institute, encouraging platforms to improve their employment terms and conditions by mapping their compliance with international minimum labour standards, have a role to play here, too.[2]

Given the deficiencies in self-regulation and the lack or insufficiency so far of an adequate state response, it is no wonder that trade unions have stepped in to guarantee minimum employment terms and conditions and social protection for platform workers and to assure their right to collective bargaining. Pursuing a logic-of-influence approach, mainstream unions in particular have initiated several cases of litigation over the past few years, related chiefly to the misclassification of workers in location-based platform companies as 'independent contractors' and to enforce the application to them of the existing regulations. However, the jury is still out on the success of such litigation by mainstream and, occasionally also, grassroots unions. Certainly, and the extent of union resources is an issue here, it can take some time before national court rulings are issued and, while they are setting legal precedents, they sometimes concern only individuals or small groups of platform workers.

That litigation is an avenue with uneven outcomes also partly informs Chapter 11, by Simon Joyce and Mark Stuart. Mainly focusing on location-based platforms in the Global North, the authors supplement, in addition, the issue of the so far ambivalent results of the collective agreements which have been negotiated by mainstream unions. Indeed, while such agreements aim to incorporate platform workers, this is hardly the case in practice since most platform companies do not consider themselves as employers and thus do not adhere to collective agreements.

Pointing to the importance of the institutional context, some recent agreements regarding platform-based food delivery couriers in Nordic Europe do seem to be an exception, but they relate to couriers in a genuine employment relationship. Even in the case of *Hilfr*, a Danish platform company providing cleaning services in households, and which has been willing to negotiate with the union in question, the collective agreement of 2018 was thrown into doubt two years later by the Danish Competition and Consumer Authority which, following the European Court in the *FNV* case, considered some aspects of it to be at odds with competition law.

The *Hilfr* agreement also introduced a kind of third employment status, offering better employment terms and conditions than 'independent contractor' status would allow but poorer than in genuine employment. This has been criticized as mainstream unions being too pragmatic and lenient in the face of platform reality but, on the other hand, talk is cheap and it might be a first step in the direction that unions ideally strive for. Much of all of this can be seen as 'institutional experimentation' or regulatory frameworks 'in the making' from a historical and optimistic perspective (Degryse 2020).

The starting point and mainstay of collective bargaining strategies by mainstream unions remains the genuine employment relationship. As long as the status of 'independent contractor' is not adequately tackled by enforcing existing or new regulations, the effectiveness of the current strategies of mainstream unions is somewhat limited. Bottom-up self-organization among platform workers and grassroots unions is, almost by default, leading them to opt for a logic-of-membership approach; that is, a strategy of (app-based and online) mobilizing, organizing and direct action and in which formal union membership seems a secondary consideration (see also Vandaele et al. 2019). Designed to catch attention from the press and media, labelled as 'discursive power', old tactics are thereby combined with creative tactics bred by and adapted to the platform economy and therefore considered 'new'. So far, however, concessions by the platform companies seem limited whereas there is a danger as well of a certain 'mobilization fatigue' in adverse algorithmic-heavy circumstances.

Three additional points can be made here. First, platform companies are still a relatively new phenomenon, but mainstream unions have somewhat caught up after initially being puzzled about their rapid dynamics. There is, however, still room for learning from self-organization and grassroots unions. Second, the distinction between logic-of-influence and logic-of-membership approaches, linked respectively to mainstream unions and to self-organized and grassroots unions, is sometimes more complex in reality. Also, while historical, ideological or other tensions between grassroots and mainstream unions impedes any rapprochement, this might be less the case for self-organized platform workers, especially where grassroots unions are absent (see also Vandaele 2021). Finally, apart from some initiatives by, for instance, *IG Metall*, the largest trade union in Germany, little is known so far about union action in the online platform economy, although here it is imaginable that a logic-of-influence approach by mainstream unions might be a better fit.

1.4 CASE STUDIES ON ONLINE PLATFORM WORK: CHAPTERS 12 TO 14

The remainder of the book explores the issues in greater detail through case studies of platform work. Part III includes case studies of online labour platforms while location-based platforms are covered in Part IV. The case studies are based on a range of research designs and methods for collecting and analysing data. Focusing on the actors, strategies and struggles involved in individual types of platform work, the research designs are, among others, informed by concepts of job quality and work-related capabilities and insights drawn from (historical) institutionalism, mobilization theory and the power research approach. Methods include ethnographic fieldwork such as informal conversations and participant observation; face-to-face (semi-structured) interviews with platform workers themselves or selected key persons like union officials, representatives of platforms, non-governmental officials and policy-makers; exploring platform transaction data; conducting survey-based research; and using descriptive and inferential statistics. Quantitative and qualitative methods are sometimes combined in a multi-method approach.

Many of the case studies included here share a(n) (implicit) comparative dimension, either by comparing the platform economy with similar developments or industries within the conventional economy via hypothetico-deductive reasoning or by making a comparison between well-defined geographical areas via an inductive approach. In geographical terms, the focus ranges from cities to regions within countries and to countries and country groups. Thus, the following cities and regions are covered in Part IV of this book: Beijing; Berlin; Cologne; Los Angeles; New York City; Oslo; Paris; Toronto; Vienna; and the state of Queensland in Australia – all demonstrating the highly city-based character of location-based platform work, in particular in transportation. Furthermore, platform workers in India, in particular its south-east region

Tamil Nadu, as well as in Kenya, Malaysia, Mexico, Nigeria, the Philippines, Russia, South Africa, Ukraine, the United States (US) and Vietnam are studied.

Part III includes three chapters covering online platform work in low- to middle-income countries, covering both routine-like microtasks and specific skills-required macrotasks, and all three engage with the extent to which digital labour platforms stimulate economic and human development. Platform work is indeed often promoted and supported by policy-makers in lower-income countries via investment in digital architecture and training programmes.

Chapter 12 by Mark Graham, Vili Lehdonvirta, Alex J. Wood, Helena Barnard, Isis Hjorth and David Peter Simon analyses the conditions of online platform workers in the Global South. Here, such workers benefit from better remuneration than could be earned on local labour markets while they also appreciate the autonomy and flexibility of platform work. Yet, importantly, the benefits come at a cost. The authors report on the risks of platform workers in sub-Saharan Africa (Kenya, Nigeria and South Africa) and south-east Asia (Philippines, Malaysia and Vietnam). For instance, there are notable differences in remuneration between the aforementioned regions and platform workers performing micro- and macrotasks. Remuneration is also not equally distributed among platform workers, which can be explained by differences in reputations, while there is also a downwards pressure on platform-based remuneration due to overt oversupply. Job insecurity, due to a lack of employment protection, opacity (as clients are often unknown), discrimination and social isolation and a poor work–life balance (due to overwork and work intensification) are other features of platform work which negatively influence job quality. By way of at least mitigating these drawbacks, and for generating tax revenues, the authors propose to embed platforms in local labour market and welfare institutions.

In Chapter 13, with a strong focus on the gender dimension, Janine Berg, Uma Rani and Nora Gobel come to similar conclusions based on their survey findings of microtasks performed by platform workers in India. The three main incentives to engage in this type of platform work is to work from home, thus avoiding commuting, especially among women (which, therefore, reinforces gender segmentation on the labour market); to supplement income from other jobs; or to do it as a form of leisure. The authors focus in particular on remuneration and training opportunities and put this in comparison with India's industries involved in business process outsourcing: online platform work can be considered a type of offshore outsourcing, but to individuals instead of companies. The profile of the workers demonstrates indeed some similarities: both platform workers and workers in industries linked to business process outsourcing are young, well educated and they often have an information technology background. Also, work scheduling is comparable as both are performed during the evening or at night. Finally, the tendency to monitor, control and supervise platform work is, to a certain extent, similar to call centre work, although in the latter industry there is still room for human interaction instead of the anonymous algorithmic management associated with platform work.

As with Chapter 12, the remuneration from platform work in India is highly unevenly distributed and this deteriorates when unpaid work is taken into account. Indeed, on average, perhaps astonishingly, more than one-fifth of platform work is unpaid work involving activities such as searching for tasks or taking unpaid qualification tests. Including also unpaid work, the remuneration of microwork is lower on average when compared to similar work in the conventional economy although proportionately fewer women are employed in online platform work. In other words, skill arbitrage for platform-based microwork is quite unlikely as such workers are usually not in a position to set a higher price for their labour via the platforms than with local employers. Unlike in call centres, increases in remuneration due to tenure over time are non-existent on platforms. This can be explained by the constant influx of new labour on platforms and by the preferences of clients in the Global North to select fewer workers from the Global South for the more demanding tasks which are associated with higher remuneration. Furthermore, while work in call centres improves employability through skill upgrading, the (de-skilling) nature of platform-based microtasks does not allow much room for skill development and learning which can be used in other jobs outside the platform economy. The authors thus question governmental efforts to promote online platform work in India and the Global South in general.

Chapter 14 by Mariya Aleksynska, Andrey Shevchuk and Denis Strebkov is an antidote to the narrative of borderless, global labour markets through online platform work, at least when it concerns macrotasks. The authors review how a sole focus on English-language online platforms can be misleading: it underestimates the global size of the platform economy and it provides a false picture of the real geographical presence of digital labour platforms. There is a variety in the geographical scale of platforms. Online platforms might be part of global value chains but can also (additionally) operate in domestic and regional online labour markets. Reasons for the latter are manifold. First, a common language and culture eases communication and fosters trust between workers and clients, while geographic proximity means people are working in the same time zone. Second, depending on their degree of transferability and recognition, workers' skills can be valued differently in different markets which can give rise to domestic and regional online labour markets.

The authors illustrate the 'stickiness' of such markets by reference to domestic and regional online digital labour platforms in Russia and Ukraine, in which the Russian language delineates the regional dimension of the platforms (which are located in both countries). This multi-scalarity of the platforms should be taken into consideration. First, survey results show not only that some socio-demographic and other characteristics of platform workers differ along with the market, but also that remuneration and working hours are likely to be lower and less irregular, respectively, in domestic markets than in regional or international ones. Second, it also represents a case for the domestic and regional regulation of online platform work in addition to international regulation – not least because much of this type of work remains informal in Russia and Ukraine.

1.5 CASE STUDIES ON LOCATION-BASED PLATFORM WORK: CHAPTERS 15 TO 20

Almost all the chapters in Part IV cover location-based platform work in public settings, in particular within transportation. Food delivery and ride-sharing are the two industries that dominate the public discourse and have also attracted considerable attention from researchers. Chapter 15 by Andrea Santiago Páramo and Carlos Piñeyro Nelson, however, focuses on location-based platform work in a private setting: domestic cleaning. At the intersection between informality, precariousness and platform work, location-based platform work in private settings is under-researched (Ticona and Mateescu 2018; Mateescu and Ticona 2021).

Santiago Páramo and Piñeyro Nelson contrast a for-profit platform in Mexico and, as an ethical alternative, a non-profit platform in the US. Similarities in the workforce are manifold in both countries: it is a job overwhelmingly done by women, more than a quarter of whom are of indigenous origin in Mexico while most of those in the US have immigrant status, although undocumented. They are in informal employment arrangements, often facing discrimination, while they work in isolation and their job is considered of low status.

The analysis makes clear how the managerial prerogative which is built into algorithmic management does little to improve employment terms and conditions in the case of the for-profit platform in Mexico. In fact, the platform tends to reproduce cultural prejudices and stereotypes about cleaning and reinforces the existing dimensions of precarious work such as job insecurity, despite its choice of narrative centred around professionalizing domestic cleaning.

In contrast, the National Domestic Workers Alliance, an organization stimulating capacity-building among domestic workers and which campaigns for better rights for them, has established a non-profit platform for domestic cleaning in the US offering workers basic social protection. This 'community logic' (Frenken et al. 2020) demonstrates a future, countervailing avenue for organized labour, combining platform cooperativism with unionism or other types of workers' collective organization and representation (see also Scholz 2016; Schneider 2018), which is also reminiscent of the creative past of the labour movement. Based on a model of democratic ownership and governance, platform cooperatives could set improved minimum employment terms and conditions, embrace data transparency and incorporate environmental concerns as well. Before such an alternative can be successful and competitive vis-à-vis the (international) platform corporations, however, this necessitates a number of requirements such as sufficient financial investment (for instance via crowdfunding), regulatory enabling frameworks at local or higher policy levels and adequate scalability and price-setting.

The remaining five chapters put the focus on an often much more 'visible' type of platform work – transportation – with three chapters on food delivery and two on ride-sharing. The geographical lens in all the chapters is directed towards cities and urban contexts. These contexts, as then highlighted by Maria Figueroa in Chapter 21 – the concluding chapter – seem to offer a fertile ground for 'institutional experimentation' by the social actors.

Turning first to platform-based food delivery, and probably also other industries within the platform economy (see, for example, Bucher et al. 2020), worker resistance – where it exists at all – often has an individualized character (Barratt et al. 2020; Heiland 2021; Shanahan and Smith 2021). Individualized coping strategies have been explained by food delivery couriers' mainly 'entrepreneurial' interpretations of their employment terms and conditions, being manipulated by algorithmic management and their own internalization of the platforms' neoliberal discourse so that the organizational work arrangements associated with the platforms are sustained and reinforced rather than altered. While this has given rise to pessimistic accounts of labour agency in the (food delivery) platforms, the three case studies here offer a more optimistic view of the possibilities of collective resistance and counter-discourses among platform-based food delivery couriers, as well as of their self-organization and unionization.

Chapter 16 by Kristin Jesnes, Denis Neumann, Vera Trappmann and Pauline de Becdelièvre studies the mobilization process of platform-based food delivery couriers in Cologne in Germany, Oslo in Norway and Paris in France, as well as the variety in the couriers' collective (proto-)organization, coalition-building with trade unions and the effectiveness of social protest. The authors accentuate the similarities in the mobilization process of the couriers between the three cities, irrespective of the specific contractual arrangement, while simultaneously allowing for subjective understandings of trade unionism among those couriers who took a leadership role and the contextual differences in the prevailing national system of industrial relations.

As documented elsewhere for other cities as well (for example, Vandaele 2020), some couriers have taken up a leadership role, driving the mobilization process. Thus, a critical mass of couriers has been instrumental in identifying how algorithmic management, combined with non-standard contractual arrangements, negatively affects employment terms and conditions, especially regarding remuneration. They have made fellow couriers aware of this 'algorithmic injustice', attributing it to human management within the food delivery platforms. They have also fostered group identification in online digital communities and offline physical gatherings at meeting points and defended social protest, sometimes including industrial action, as an effective means of mitigating or undoing the feelings of 'algorithmic injustice'. The forging of coalitions with trade unions and the selection of specific unions have been (partially) dependent on the subjectivities and norms of those key couriers, however. In turn, those city-dependent configurations of courier unions – the extent to which couriers are formally working together with unions – have oriented the action repertoire and the aims and effectiveness of social protest. As a result, the outcomes of protests have varied, much in line with the dominant logic of the national industrial relations system.

The study on organizing food delivery bike couriers in Toronto, Canada, by Raoul Gebert in Chapter 17, with an equal focus on labour agency, runs fairly parallel with Chapter 16. Yet, the author shifts the lens from the couriers to the trade unions in this case study, with the couriers' grassroots association being seen as a pre-existing structure from the viewpoint of the union (Vandaele 2020, p. 13).

While platform-based food delivery can be considered a fairly novel industry, challenging prevailing regulatory and institutional arrangements for unionization and collective bargaining, unions have of course previously been confronted with new industries or new groups of workers within existing industries. Union responses towards new groups of (precarious) workers have been diverse (Heery 2009; Keune and Pedaci 2020) and the advent of digital labour platforms is no different. Gebert understands the union response from the Canadian Union of Postal Workers as 'institutional experimentation'. First, the union legally endeavoured to establish 'dependent contractor' status for the couriers, instead of the traditional 'salaried employee' status, which resulted in precedent being set in Canadian labour jurisprudence in February 2020. In this way, platform-based food delivery could act as leverage for regulating other industries within the platform economy. Second, under the influence of its construction of a relationship with the union, 'Foodsters United', initially a grassroots association of bike couriers who shared a strong collective identity, shows features of the 'whole-worker approach' (McAlevey 2016) and community unionism today; that is, a presence going beyond a sole focus on collective bargaining. Although the organizing drive of the union was successful, Foodora, the platform company in this case, left the Canadian market in spring 2020, demonstrating the evident geographical binarity behind location-based platforms: they are either present and operating in a given locality or they are not (Vandaele 2021).

Chapter 18 by Jack Linchuan Qiu, Ping Sun and Julie Chen engages with the food delivery platforms themselves in Beijing, and especially how algorithmic management is manipulating employment terms and conditions. A short (labour) market analysis deciphers that platform-based food delivery is ultra-competitive and a very significant market in China in terms of revenue and employment. Similar to food delivery in other countries (in the Global North), low remuneration and a high level of turnover mark the labour market. Interestingly, the industry in China has been characterized by a de-flexibilization, with incentive structures and algorithmic management encouraging full-time engagement with very little working time flexibility. Couriers are young, they have a migratory background (coming from rural areas in the Chinese context) and employment relationships are complex, due also to the involvement of temporary staffing agencies, and are often initially informal. Couriers perform emotional labour when dealing with customers, influencing their ratings and thus remuneration, while the platforms try to discipline them via gamification and managing their working time in order to maximize orders and minimize delivery time.

As in the conventional economy, however, labour control through algorithmic management is not absolute: courier resistance is, individually and collectively, via the avenues of deceiving and subverting the algorithms, social protest and industrial action. The value of platform-based food delivery has, in particular, been highlighted during the Covid-19 pandemic within urban lockdown areas and local authorities have put pressure on platforms to smooth the edges of algorithmic management so that couriers' employment terms and conditions are improved.

The authors conclude by considering platform-based food delivery a 'digital utility' – that is, a public service delivered by platforms but driven by public values embedded in solid regulatory frameworks – since it has increased its relevance in everyday life in Chinese cities, becoming an essential part of the urban service infrastructure.

Chapter 19 by Hannah Johnston and Susanne Pernicka turns to ride-sharing – another very visible industry within the location-based platform economy. Their account reflects the historical, pendular tendency of regulation, deregulation and re-regulation within the taxi sector in urban contexts. The authors add a constructive, subjective dimension into their analysis as they put the perceived legitimacy of power resources, or symbolic power, central to an understanding of the variation in resistance and regulatory outcomes vis-à-vis ride-sharing in the cities of Berlin, Los Angeles, New York City and Vienna. To explain the responses of platform workers and unions towards ride-sharing platforms, they develop a power resources perspective that takes into account the interplay of local, national and transnational scales.

Given their strong institutional power and symbolic recognition of this power resource, trade unions in Vienna oriented themselves to a successful logic-of-influence approach so that it looks like ride-sharing platforms such as Uber will be embedded into regulatory frameworks and institutions, at least at national level. Workers' institutional power has diminished over time in Berlin, while their associational power is underdeveloped so that regulation tends to favour the new ride-sharing platform companies.

Institutional power and the legitimacy thereof are comparatively weaker in the US, although with considerable differences between states and, thus, the two cities studied here, Los Angeles and New York City. While a logic-of-membership approach is pursued in both cities, (platform) workers in Los Angeles have, so far, not been able to build sufficient associational power. Their efforts are largely oriented to the city level while platform companies have been able to deregulate the taxi sector at state level in California, especially as the symbolic struggle is currently also bending towards them. Thus, political authorities and the public in general are inclined towards an interpretation of platform companies in the taxi sector as a technological app-based solution to transportation needs. This is in contrast to New York City, where (platform) workers in ride-sharing have been engaging in a symbolic struggle about the meaning of platform work and where they have been able to build associational power which has been instrumental in the building of institutional power at local regulatory level.

Interestingly, an initiative has been recently taken by three labour activists and progressives to launch a driver-owned ride-sharing app-based platform, The Drivers Cooperative, as an alternative to platform companies (Nolan 2020). Its crowdfunding campaign for developing this initiative quickly reached its goal which hints at how information asymmetry and a lack of trust between funders and investors can be transcended by platform cooperativism and how this stimulates interaction between the future driver-owners and potential users (Talonen et al. 2020).

Finally, in Chapter 20, David Peetz analyses regulatory responses towards the health and safety risk – injuries or fatalities due to accidents – associated with ride-sharing in the state of Queensland, Australia. While it looks like there have been far more motor vehicle fatalities since the arrival of Uber, it is clear that vehicle passenger transport is an industry which presents above-average risk. Platform workers in ride-sharing in Australia are, however, not covered by injury compensation insurance as they have so far been considered as 'independent contractors'.

In 2018, the Queensland government reviewed the injury compensation system and the extent to which this could be extended to workers in location-based platforms like ride-sharing. The review made the innovative proposal to redefine insurance coverage so that workers under agency arrangements would be included and to require the payment of injury compensation premiums by intermediaries or agencies, in this case by digital labour platforms in the taxi sector. The premiums, which would also have an experience-rated character, would be set as a proportion of the commission received by the platform from the customer. Feedback to the review's recommendations varied, with the ride-sharing platform companies themselves either rejecting the proposals or being lukewarm about them.

After the publication of the review, the Covid-19 pandemic broke out, impeding further steps as the policy focus shifted towards handling the pandemic. Needless to say, the pandemic has increased the health risks for platform workers in ride-sharing while their income from this kind of work has both dropped and become even more insecure.[3]

1.6 PROSPECTS

Writing amidst the Covid-19 pandemic, it is still rather uncertain whether the economic consequences of the pandemic will stimulate the growth of the platform economy. While the complexity of labour platforms can be characterized by their Janus and Proteus tendencies, Proteus's prophetic ability is not of much help here with evidence about the impact of the pandemic on platform workers being scant and patchy. As far as it concerns location-based platforms, the economic effect is, in all probability, uneven, with some industries (like platform-based food and grocery delivery) booming, while others (such as ride-sharing) suffer declining demand. Online platforms saw a drop in demand across the board in the second quarter of 2020 but that was followed by a recovery with demand for information technology-related services growing rapidly (cf. Stephany et al. 2020). Also, certain platforms might benefit from shifting consumer preferences, new business strategies and an acceleration in remote work or from states' post-Covid-19 policy responses regarding platform work.

Despite the self-regulating efforts of certain platforms to offer some health insurance schemes or support funds, there is no doubt that most platform workers, especially where it concerns their main income, are essentially bearing the risks during the pandemic (Fairwork 2020; ILO 2021). At the same time this exposure might push the need for a regulatory framework cushioning such risks rather higher up the policy agenda.

Furthermore, as employment via digital labour platforms or the conventional economy looks like compensating vessels (Schor et al. 2020), at least in the Global North, taking into account developments within conventional labour markets provides some guidance for the future of labour platforms. Increased unemployment due to restructuring as a consequence of the pandemic will undoubtedly feed a fresh reservoir of on-demand labour. All of this will demand a post-pandemic regulatory embeddedness of digital labour platforms at different policy levels.

As to what that regulatory embeddedness might look like, Chapter 21 by Maria Figueroa helpfully prompts us towards the potential contents of a policy framework within which a response might be made to the challenges laid down by the activity of location-based and online platforms.

The fluidity of the context in which platforms operate, even ignoring the effects of Covid-19, indicates that institutional experimentation is, as Figueroa also identifies, likely to be a key part of that policy framework for some time to come as the social actors, and regulatory agencies, get to grips with the meaning and implications of the business model developed by the platforms. This is fertile ground for experimentation – but, at the same time, it is also uncertain ground in the sense that there are question marks over the sustainability of that model given the sizable year-on-year financial losses being recorded by some of the platforms and which, as the court rulings mount up, are heading in one direction only as the competitive effects of their regulatory advantage are eaten away.

The bulk of the call for institutional experimentation might seem to fall on labour organizations – and, as many of the contributions to this volume point out, there is actually quite a bit going on there already, both among mainstream trade unions and grassroot organizations. More could, inevitably, always be done and there remain some crucial barriers at the organizational level and institutional level, as a result not least of the understanding gap, but none of these contributions call into question the continuing logic of collective action. Above that principle, however, the debate is a broad and lively one.

NOTES

1. See ETUC Resolution on the protection of the rights of non-standard workers and workers in platform companies (including the self-employed), adopted 28 October 2020, available at www.etuc.org/en/document/etuc-resolution-protection-rights-non-standard-workers-and-workers-platform-companies.
2. On the Fairwork Foundation, see https://fair.work/en/fw/homepage/.
3. Meanwhile, in that other type of location-based platform work, namely food delivery, Deliveroo couriers have elected seven health and safety representatives from their Sydney ranks after five fatalities occurred between September and October 2020 (*Australian Financial Review*, 4 January 2021). The representatives are paid by Deliveroo for the time spent dealing with safety issues.

REFERENCES

Adams-Prassl, A. and J. Berg (2017), *When home affects pay: an analysis of the gender pay gap among crowdworkers*, 6 October, accessed 15 February 2021 at https://papers.ssrn.com/abstract=3048711.

Aloisi, A. (2020), 'Hierarchies without firms? Vertical disintegration, outsourcing and the nature of the platform', *Quaderni 8th Giorgio Rota Best Paper Award*, 11–32.
Atzeni, M. (2020), 'Worker organisation in precarious times: abandoning trade union fetishism, rediscovering class', *Global Labour Journal*, **11** (3), 311–314.
Barratt T., C. Goods and A. Veen (2020), '"I'm my own boss ...": active intermediation and "entrepreneurial" worker agency in the Australian gig-economy', *Environment and Planning A: Economy and Space*, **52** (8), 1643–1661.
Berg, J. (2016), 'Income security in the on-demand economy: findings and policy lessons from a survey of crowdworkers', Conditions of Work and Employment Series 74, Geneva: ILO.
Berg, P., E. Appelbaum, T. Bailey and A. L. Kalleberg (2004), 'Contesting time: international comparisons of employee control of working time', *ILR Review*, **57** (3), 331–349.
Berger, T., C. B. Frey, G. Levin and S. R. Danda (2019), 'Uber happy? Work and well-being in the "gig economy"', *Economic Policy*, **34** (99), 429–477.
Borchert, K., M. Hirth, M. Kummer, U. Laltenberger, O. Slivko and S. Viete (2018), 'Unemployment and online labor', ZEW Discussion Paper 18-023, Mannheim: ZEW.
Bosch, G. (2004), 'Towards a new standard employment relationship in western Europe', *British Journal of Industrial Relations*, **42** (4), 617–636.
Bucher, E. L., P. Kalum Schou and M. Waldkirch (2020), 'Pacifying the algorithm: anticipatory compliance in the face of algorithmic management in the gig economy', *Organization*, **28** (1), 44–67.
Cant, C. (2018), 'The warehouse without walls: a workers' inquiry at Deliveroo', *Ephemera*, **20** (4), 131–161.
Charles, J., I. Ferreras and A. Lamine (2020), 'A freelancers' cooperative as a case of democratic institutional experimentation for better work: a case study of SMart-Belgium', *Transfer*, **26** (2), 157–174.
Corporate Europe Observatory and AK Europe (2019), 'Über-influential? How the gig economy's lobbyists undermine social and workers rights', Brussels: Corporate Europe Observatory and AK Europa.
Crouch, C. (2019), *Will the gig economy prevail?*, Cambridge: Polity Press.
Degryse, C. (2020), 'Du flexible au liquide: le travail dans l'économie de plateforme', *Relations Industrielles/Industrial Relations*, **74** (4), 660–683.
Drahokoupil, J. (ed.) (2015), *The outsourcing challenge: organizing workers across fragmented production networks*, Brussels: ETUI.
Drahokoupil, J. and A. Piasna (2019), 'Work in the platform economy: Deliveroo riders in Belgium and the SMart arrangement', Working Paper 2019.01, Brussels: ETUI.
Dubal, V. B. (2017), 'The drive to precarity: a political history of work, regulation, and labor advocacy in San Francisco's taxi and Uber economies', *Berkeley Journal of Employment and Labor Law*, **38** (1), 73–135.
Dubal, V. B. (2020), 'An Uber ambivalence: employee status, worker perspectives, and regulation in the gig economy', in D. Das Acevedo (ed.), *Beyond the algorithm: qualitative insights for gig work regulation*, Cambridge: Cambridge University Press, pp. 33–56.
Fairwork (2020), *The gig economy and Covid-19: looking ahead. September 2020*, Oxford: The Fairwork Project.
Finkin, M. (2016), 'Beclouded work, beclouded workers in historical perspective', *Comparative Labor Law and Policy Journal*, **37** (3), 603–618.

Florisson, R. and I. Mandl (2018), *Platform work: types and implications for work and employment – literature review*, Dublin: Eurofound.

Frenken, K., T. Vaskelainen, L. Fünfschilling and L. Piscicelli (2020), 'An institutional logics perspective on the gig economy', in I. Maurer, J. Mair and A. Oberg (eds), *Theorizing the sharing economy: variety and trajectories of new forms of organizing*, Bingley: Emerald Publishing, pp. 83–105.

Gegenhuber, T., M. Ellmer and E. Schüßler (2020), 'Microphones, not megaphones: functional crowdworker voice regimes on digital work platforms', *Human Relations*, accessed 15 February 2021 at https://doi.org/10.1177/0018726720915761.

Gerber, C. (2020), 'Community building on crowdwork platforms: autonomy and control of online workers?', *Competition and Change*, accessed 15 February 2021 at https://doi.org/10.1177/1024529420914472.

Gerber, C. and M. Krzywdzinski (2019), 'Brave new digital work? New forms of performance control in crowdwork', in S. Vallas and A. Kovalainen (eds), *Work and labor in the digital age*, Bingley: Emerald Publishing, pp. 121–143.

Grabher, G. and J. Köning (2020), 'Disruption, embedded: a Polanyian framing of the platform economy', *Sociologica*, **14** (1), 95–118.

Grabher, G. and E. van Tuijl (2020), 'Uber-production: from global networks to digital platforms', *Environment and Planning A: Economy and Space*, **52** (5), 1005–1016.

Gray, M. L. and S. Suri (2019), *Ghost work: how to stop Silicon Valley from building a new global underclass*, Boston, MA: Houghton Mifflin Harcourt.

Gruszka, K. and M. Böhm (2020), 'Out of sight, out of mind? (In)visibility of/in platform-mediated work', *New Media and Society*, accessed 15 February 2021 at https://doi.org/10.1177/1461444820977209.

Hall, J. V. and A. B. Krueger (2018), 'An analysis of the labor market for Uber's driver-partners in the United States', *ILR Review*, **71** (3), 705–732.

Heery, E. (2009), 'Trade unions and contingent labour: scale and method', *Cambridge Journal of Regions, Economy and Society*, **2** (3), 429–442.

Heiland, H. (2021), 'Controlling space, controlling labour? Contested space in food delivery gig work', *New Technology, Work and Employment*, accessed 15 February 2021 at https://doi.org/10.1111/ntwe.12183.

Herod, A. and R. Lambert (2016), 'Neoliberalism, precarious work and remaking the geography of global capitalism', in R. Lambert and A. Herod (eds), *Neoliberal capitalism and precarious work: ethnographies of accommodation and resistance*, Cheltenham, UK and Northampton, MA, USA: Edward Elgar Publishing, pp. 1–35.

Hyman, L. (2018), *Temp: how American work, American business, and the American dream became temporary*, New York: Viking Penguin.

ILO (2021), *World employment and social outlook: the role of digital labour platforms in transforming the world of work*, Geneva: International Labour Office.

Jesnes, K. and S. M. Nordli Oppegaard (eds) (2020), *Platform work in the Nordic models: issues, cases and responses*, Copenhagen: Nordic Council of Ministers.

Johnson, H. (2020), 'Labour geographies of the platform economy: understanding collective organizing strategies in the context of digitally mediated work', *International Labour Review*, **159** (1), 25–45.

Keune, M. and M. Pedaci (2020), 'Trade union strategies against precarious work: common trends and sectoral divergence in the EU', *European Journal of Industrial Relations*, **26** (2), 139–155.

Koutsimpogiorgos, N., J. van Slageren, A. M. Herrmann and K. Frenken (2020), 'Conceptualizing the gig economy and its regulatory problems', *Policy and Internet*, **12** (4), 525–545.

Kuhn, K. M. and A. Maleki (2017), 'Micro-entrepreneurs, dependent contractors, and instaserfs: understanding online labor platform workforces', *Academy of Management Perspectives*, **31** (3), 183–200.

Lehdonvirta, V. (2018), 'Flexibility in the gig economy: managing time on three online piecework platforms', *New Technology, Work and Employment*, **33** (1), 13–29.

Mateescu A. and J. Ticona (2021), 'Invisible work, visible workers: visibility regimes in online platforms for domestic work', in D. Das Acevedo (ed.), *Beyond the algorithm: qualitative insights for gig work regulation*, Cambridge: Cambridge University Press, pp. 57–81.

McAlevey, J. (2016), *No shortcuts: organizing for power in the new gilded age*, Oxford: Oxford University Press.

Nolan, H. (2020), 'New York City drivers cooperative aims to smash Uber's exploitative model', *In These Times blog*, 10 December, accessed 16 February 2021 at https://inthesetimes.com/article/new-york-city-drivers-cooperative-uber-lyft.

Panteli, N., A. Rapti and D. Scholarios (2020), '"If he just knew who we were": microworkers' emerging bonds of attachment in a fragmented employment relationship', *Work, Employment and Society*, **34** (3), 476–494.

Peck, J. and R. Phillips (2020), 'The platform conjuncture', *Sociologica*, **14** (3), 73–99.

Pesole, A., M. C. Urzì Brancati, E. Fernández-Macías, F. Biagi and I. González Vázquez (2018), *Platform workers in Europe evidence from the COLLEEM survey*, Luxembourg: Publications Office of the European Union.

Piasna, A. and J. Drahokoupil (2019), 'Digital labour in central and eastern Europe: evidence from the ETUI Internet and Platform work survey', Working Paper 2019.12, Brussels: ETUI.

Piasna, A. and J. Drahokoupil (2021), 'Flexibility Unbound: Understanding the Heterogeneity of Preferences among Food Delivery Platform Workers', *Socio-Economic Review*, https://doi.org/10.1093/ser/mwab029.

Schneider, N. (2018), *Everything for everyone: the radical tradition that is shaping the next economy*, New York: Nation Books.

Scholz, T. (2016), *Platform cooperativism: challenging the corporate sharing economy*, New York: Rosa Luxemburg Stiftung.

Schor, J. B., W. Attwood-Charles, M. Cansoy, I. Ladegaard and R. Wengronowitz (2020), 'Dependence and precarity in the platform economy', *Theory and Society*, **49** (5–6), 833–861.

Seppänen, L., C. Spinuzzi, S. Poutanen and T. Alasoini (2020), 'Co-creation in macrotask knowledge work on online labor platforms', *Nordic Journal of Working Life Studies*, accessed 15 February 2021 at https://doi.org/10.18291/njwls.123166.

Shanahan, G. and M. Smith (2021), 'Fair's fair: psychological contracts and power in platform work', *International Journal of Human Resource Management*, accessed 15 February 2021 at https://doi.org/10.1080/09585192.2020.1867615.

Silberman, M. 'Six' and H. Johnston (2020), *Using GDPR to improve legal clarity and working conditions on digital labour platforms*, Working Paper 2020.05, Brussels: ETUI.

Silver, B. (2003), *Forces of labor: workers' movements and globalization since 1870*, Cambridge: Cambridge University Press.

Stanford, J. (2017), 'The resurgence of gig work: historical and theoretical perspectives', *Economic and Labour Relations Review*, **28** (3), 382–401.

Stephany, F., M. Dunn, S. Sawyer and V. Lehdonvirta (2020), *Distancing bonus or downscaling loss? The changing livelihood of US online workers in times of*

Covid-19, preprint, SocArXiv, 20 April, accessed at https://doi.org/10.31235/osf.io/vmg34.

Talonen, A., J. Pasanen and O.-P. Ruuskanen (2020), 'Exploring the co-operative form's potential in crowdfunding: a non-monetary perspective', *FIIB Business Review*, accessed 15 February 2021 at doi.org/10.1177/2319714520920798.

Ticona, J. and A. Mateescu (2018), 'Trusted strangers: carework platforms' cultural entrepreneurship in the on-demand economy', *New Media and Society*, **20** (11), 4384–4404.

Urzì Brancati, M. C., A. Pesole and E. Fernández-Macías (2020), *New evidence on platform workers in Europe: results from the second COLLEEM survey*, Luxembourg: Publications Office of the European Union.

van der Linden, M. and J. Breman (2019), 'Informalizing the economy: the return of the social question on a world scale', in M. van der Linden (ed.), *The global history of work: critical readings, Vol. III: Labour market*, London: Bloomsbury Publishing, pp. 254–271.

van Doorn, N. (2017), 'Platform labor: on the gendered and racialized exploitation of low-income service work in the "on-demand" economy', *Information, Communication and Society*, **20** (6), 898–914.

van Doorn, N. (2020), 'Stepping stone or dead end? The ambiguities of platform mediated domestic work under conditions of austerity. Comparative landscapes of austerity and the gig economy: New York and Berlin', in D. Baines and I. Cunningham (eds), *Working in the context of austerity: challenges and struggles*, Bristol: Bristol University Press, pp. 49–69.

Vandaele, K. (2018), 'Will trade unions survive in the platform economy? Emerging patterns of platform workers' collective representation and voice', Working Paper 2018.05, Brussels: ETUI.

Vandaele, K. (2020), 'From street protest to improvisational unionism: platform-based food delivery couriers in Belgium and the Netherlands', Trade Unions in Transformation 4.0, Berlin: Friedrich-Ebert-Stiftung.

Vandaele, K. (2021), 'Collective resistance and organizational creativity amongst Europe's platform workers: a new power in the labour movement?', in J. Haidar and M. Keune (eds), *Work and labour relations in global platform capitalism*, Cheltenham, UK and Northampton, MA, USA: Edward Elgar Publishing and Geneva: ILO.

Vandaele, K., A. Piasna and J. Drahokoupil (2019), '"Algorithm breakers" are not a different "species": attitudes towards trade unions of Deliveroo riders in Belgium', Working Paper 2019.06, Brussels, ETUI.

Weil, D. (2014), *The fissured workplace: why work became so bad for so many and what can be done to improve it*, Cambridge, MA: Harvard University Press.

Wood, A. and V. Lehdonvirta (2019), 'Platform labour and structured antagonism: understanding the origins of protest in the gig economy', accessed 16 February 2021 at https://ssrn.com/abstract=3357804.

Wood, A., V. Lehdonvirta and M. Graham (2018), 'Workers of the internet unite? Online freelancer organisation among remote gig economy workers in six Asian and African countries', *New Technology, Work and Employment*, **33** (2), 95–112.

Wood, A., M. Graham, V. Lehdonvirta and I. Hjorth (2019), 'Networked but commodified: the (dis)embeddedness of digital labour in the gig economy', *Sociology*, **53** (5), 931–950.

Xhauflair, V., B. Huybrechts and F. Pichault (2018), 'How can new players establish themselves in highly institutionalized labour markets? A Belgian case study in the area of project-based work', *British Journal of Industrial Relations*, **56** (2), 370–394.

PART I

Context and issues

2. The business models of labour platforms: Creating an uncertain future

Jan Drahokoupil

2.1 INTRODUCTION

Platform companies have come to assume a central economic role since digital technologies and the internet facilitated the development of two-sided markets. The leading platforms include the GAFA (Google, Apple, Facebook, Amazon) and, in China, the BATX (Baidu, Alibaba, Tencent, Xiaomi) technology companies. These critical intermediaries and market makers have risen to dominate the business landscape. This can be attributed, in large part, to their ability to take advantage of the data they collect about their users. Interestingly, though, it is the much smaller Uber that is often seen as emblematic of the business form to dominate twenty-first-century capitalism (e.g. Rahman and Thelen 2019). Being a labour platform, its business model has potentially the most comprehensive impact as it also affects the way companies organize relations with their workforce.

This chapter addresses the transformatory potential of labour platforms by investigating their economic role and business strategies. In order to do so, it builds on insights from the economics of two-sided markets as well as from the broader literature on the political economy of the platform business model.

Platform businesses are market makers, enabling their customers to interact with each other and coordinate demand on both sides of the market they serve. As far as labour platforms are concerned, it is possible to identify three types of economic roles they may take, each associated with a distinct business strategy. First, platforms are intermediaries involved in organizing markets through matchmaking, audience building and minimizing transaction costs (Evans and Schmalensee 2007). The business of intermediation is based on the rents that come from matching. In this context the network effects associated with demand coordination play a key role. Second, market making by platforms has a specific role in labour markets where transaction costs provide the economic rationale for hiring workers on a longer-term basis and as employees. Some labour platforms are thus in the business of regulatory arbitrage, exploiting the cost advantage of hiring workers as self-employed rather than as employees. The third business strategy to consider is the one associated with the infrastructural role of platforms as organizers of the marketplace. The infrastructural role, including the ability to structure interactions, is emphasized in the political economy literature. One can indeed distinguish between digital intermediaries that specialize in simple matching and platforms that coordinate network effects to create and capture additional value (e.g. Langley and Leyshon 2017). The ability to capture and utilize data on users is key to the ability to monetize the infrastructural role and generate value beyond matching.

While labour platforms indeed enjoy considerable infrastructural power in the markets they organize, this chapter argues that they are able to use it primarily in support of their intermediation business. The labour platform business model is thus limited to the business of matching and that of regulatory arbitrage. While the viability of the latter is contingent on the regulatory response, the viability of the business of matching needs to be considered together with the costs of sustaining the network effects. These are relatively high in the case of local platforms, making it challenging to transform them into viable business models.

2.2 THE MATCHING BUSINESS: THE COST OF NETWORK EFFECTS

Platform businesses serve 'two-sided', or 'multi-sided', markets (Rochet and Tirole 2003). They are thus defined as intermediaries serving two, or more, different groups of customers (Evans 2003). Put differently, platforms are in the business of organizing a marketplace by coordinating demand among distinct groups of customers who benefit from trading or interacting with each other. They do this by offering their customers access to each other and enabling them to realize gains from trade, or other interactions, by reducing the transaction costs of finding each other and of interacting. This can be done by providing real or virtual marketplaces for their customers.

Demand coordination is an essential aspect of organizing a market, allowing effective matching. Platforms use the price system to regulate the volume of transactions and participation on each side of the market (Rochet and Tirole 2006). A critical mass of users on each side is required for a market to gain sufficient liquidity and become self-sustaining (Evans 2009). A ridesharing platform thus needs to attract sufficient drivers in a neighbourhood so that users can obtain a ride with a tolerable waiting time. In turn, a sufficient number of users is needed to give drivers a satisfactory occupancy rate. Platforms may need to offer various discounts or subsidies to attract the required number of users on both sides of the market.

Demand coordination is thus associated with network effects whereby the value of a service increases with the number of its users. On a ridesharing platform, a higher number of drivers gives users a lower waiting time. In turn, a higher number of users reduces the time that drivers spend idle. Importantly, the network effects might give advantage to first movers (Arthur 1989; Katz and Shapiro 1994): a platform that obtains a lead tends to widen it as a result of the positive feedback effects and loops in which additional users are attracted by the larger audience that already uses the platform. In the absence of countervailing forces, a leader is destined to win the race for the market. The possibility of achieving such a monopoly position makes platform start-ups attractive to investors (Shapiro and Varian 1998; Langley and Leyshon 2017; van Doorn and Badger 2020). The latter are thus often willing to subsidize market expansion with a view to monopolizing rents.

The business of matching can thus appear an attractive business proposition. Platform business is an asset-light model where value is generated from mediating transactions. The cost of capital equipment and that of its maintenance is borne by one side of the market. Uber drivers not only provide their labour; they also need to finance the purchase of the car and then take care of its maintenance. Similarly, delivery couriers and workers on online platforms use their own equipment, be it bicycle, mobile phone or computer software and hardware. In this way, platforms transform the fixed costs of a business into variable costs (Muehlberger 2005). Income can therefore be generated without deploying large amounts of fixed capital. Moreover, while the costs of maintaining a platform are relatively low, there are significant economies of scale. Market growth thus does not require investment in fixed capital while the cost of serving an additional user is close to zero. Moreover, once they establish a critical mass of users, platforms can rely on organic growth through the network effects established as a result of additional users being attracted by the existing user base.

Countervailing forces are common, however (Evans and Schmalensee 2007). In such a context, the presence of network effects does not guarantee market dominance to the first movers (Evans 2003). Industries characterized by indirect network effects often rely on several platforms competing for fragmented markets. These countervailing forces include congestion and search optimization in which search and transaction costs increase with the number of users exceeding an optimal level. These are relevant mainly in offline contexts but online platforms are not immune to differentiation and, crucially, the possibility of multi-homing. Differentiation makes it viable for competing platforms to divide markets into niches. Multi-homing then refers to the situation in which users find it possible, and advantageous, to use several platforms at the same time. This can increase the costs of audience building.

Audience building and sustaining the network effects can thus be costly, even if financed through variable costs. Building a network effect may require considerable investment in incentives. The act of participation in, for instance, a ridesharing platform is costly for drivers as they may need to drive even without a customer. They might also expect the depreciation costs of the car to be covered. What is more, even once the critical thresholds for market liquidity are exceeded, multi-homing may make it costly to maintain the network effects. User promiscuity makes it expensive to defend market position in the face of competing platforms. Competition is an inherent feature of labour platform businesses characterized by low barriers to entry: the investment required to launch such a platform is relatively low and, once acquired, the technology is transferrable to different markets.

Location-based platforms in particular suffer from the high costs related to generating the demand which is required to maintain the network effects: by definition, they cannot scale network effects but need to build and sustain them in each market separately. Market dominance in one city does not give a ridesharing platform any particular advantage in other markets they serve. Ridesharing and food delivery, dominant among location-based platforms, are in a particularly difficult position. To start with, they face high liquidity thresholds: they need a high number of users logged in at one point in one place to ensure instant and local delivery.

Ridesharing platforms also observe a lack of loyalty and price sensitivity on both sides of the market, with the driver side characterized by particularly promiscuous behaviour. It is common for drivers to use two ridesharing platforms at the same time, reacting to the incentives that the platforms offer to retain them on their network. Similarly, competition among food delivery platforms prompts them to continue using incentives to attract customers who can often choose a different platform to order from the very same restaurant. As an alternative, platforms can rely on partnerships with sought-after restaurant chains. That, however, gives negotiating power to these chains which are then able to negotiate lower provisions for food delivery platforms.

Further, ridesharing and food delivery platforms are characterized by high worker churn (e.g. Drahokoupil and Piasna 2019). High labour turnover represents another reason that makes it costly to sustain market liquidity. Turnover can be related to poor remuneration, the actual level of which may become apparent to workers only over time, particularly in the case of ridesharing platforms where drivers' expenses are an important factor (Mishel 2018). As discussed above, the labour management system allows platforms to deal with labour turnover at operational level, but the recruitment of workers can require the payment of incentives, such as bonuses to existing workers for recruiting new ones, something that is common on ridesharing platforms.

The economic sustainability of location-based platforms is also put in question by the possibility to order from suppliers directly. Customers can use food delivery platforms as a source of information and then order food directly from the restaurant.[1] Similarly, platforms mediating location-based work in private settings, such as care and repair services, can be used simply to obtain initial contact with the worker.

High network maintenance costs put additional pressure on these platforms to keep labour costs low. The cost pressure is especially high in the food delivery business which is particularly price sensitive – delivery is added on top of the price of food or cuts into the margin of the restaurant, or both. At the same time, as discussed in Chapter 7, location-based platforms are more likely to face pressures from worker organizing (see also Vandaele 2021).

Ridesharing and food delivery businesses can thus be characterized by low barriers to entry but high barriers to success (MBI 2020). In contrast, online platforms benefit from scalable network effects with lower maintenance costs. Linguistic and other boundaries may limit the scalability of their markets, though, particularly as far as tasks requiring a good command of language or skill recognition are concerned (see Chapter 14). Advances in machine translation make linguistic and other boundaries less relevant for a range of tasks, however. In any case, it is much easier for online platforms to build networks with sufficient liquidity, because they also face relatively less pressure for instant delivery. Moreover, the importance to workers of non-transferable rankings based on past performance complicates multi-homing and limits the ability of workers to change platforms: they are less likely to be able to vote with their feet if another platform offers somewhat more attractive conditions. This, in turn, reduces the need to offer incentives to attract workers and also raises the barriers to entry for other labour platforms, helping to entrench the position of incumbents or first movers.

2.3 LOWERING TRANSACTION COSTS (FOR SOME)

Apart from audience building and matching, platforms make markets by lowering transaction costs. This makes the exchange, or market interactions, more efficient. High transaction costs may also be the reason why some assets stay idle and why some tasks are not traded at all. Transaction costs explain why some transactions are better governed inside hierarchically organized firms than in open markets (Coase 1937). Platforms allow markets to function in contexts where high transaction costs would outweigh any gains from trade. By using digital technology, online platforms have reduced the costs related to matching and audience building to such an extent that it makes trade possible even if small amounts are involved. Online labour platforms, such as Amazon Mechanical Turk, thus mediate trade in microtasks worth as little as a few cents.

The economics of two-sided markets focus on the transaction costs involved in settling the externalities related to audience building and matching. Platforms also address other types of externalities and transaction costs, however (Drahokoupil and Piasna 2017). These include the costs related to obtaining information, bargaining the terms and conditions of the relevant contract, and monitoring and enforcing the agreements (Williamson 1981). For instance, a key problem faced by platforms and their customers is that interactions with strangers are risky: people might opt to do business with a vetted taxi driver rather than with a stranger moonlighting on Uber Pop. Platforms reduce these information asymmetries through reputation systems. Star rankings identify drivers and other providers that delivered service of satisfying quality in the past. Platforms may also internalize the transaction costs involved in enforcing agreements by offering insurance, dispute resolution mechanisms and other services to reduce the costs related to opportunistic behaviour and fraud.

However, platforms may have little incentive to lower transaction costs in a symmetric way, for both sides of the market. In many markets, the potential labour supply far exceeds demand. Online labour platforms mediating low-skilled work face virtually no limits when it comes to audience building on the worker side (see Chapter 10). Under excess labour supply, platforms can expect to attract a sufficient number of workers even if these workers need to internalize the transaction costs. To sustain an optimal volume of transactions, platforms only need to address the market inefficiencies that affect primarily clients or customers. Matching mechanisms on online platforms often provide effective ways for clients or customers to find a worker, but the latter often spend large parts of their work time searching for jobs by going through unsuitable tasks offered to them by the platforms (Berg 2016; Cant 2018; Wood et al. 2019). Further, platforms commonly include reputation mechanisms to provide information on the past performance of workers, but they often lack the mechanisms that would address the risks of cheating by clients. In practice, platforms also lack effective dispute resolution mechanisms between workers and clients, as the risks related to fraudulent behaviour on the side of the client or customer can simply be borne by workers who may not be paid at all if a client finds the quality of work inadequate (see Silberman and Johnston 2020).

2.4 FROM TRANSACTION COSTS TO REGULATORY ARBITRAGE

It is the presence of high transaction costs that underpins the economic rationale for mediating access to labour through (longer-term) employment relationships rather than through (one-off) market transactions (e.g. Simon 1951).

Mediating work through market exchange, as in self-employment, requires complete contracts that fully specify the work to be done or that are based on a piece rate specifying payment per unit of output. However, these types of contracts are often impractical and, in fact, incompatible with the nature of work. To start with, the need to specify all the job requirements and rates for each task entails high transaction costs given the variety of tasks present in many jobs. Further, work often involves cooperation and coordination that would imply the constant redefinition of the requirements of the worker and a reworking of the respective task description. Cooperation also complicates, if not makes impossible, any breaking down of the input, or value added, to the level of the individual, which may be necessary for payment systems based on output. Moreover, given the information asymmetries, it is often difficult to assess what work has been done.

In such a context, work can be more efficiently mediated through incomplete contracts in which the employer brings work inside the firm by purchasing the labour time, rather than the actual labour or tasks to be performed. This can be described as an arrangement of command where the employee, in return for remuneration per hours worked, agrees to subject herself to the direction of the employer (Bowles et al. 2005). Soliciting labour through the employment relationship addresses the transaction costs involved in market transactions, but it comes at a price to the employer: the employment contract gives the worker a level of protection that effectively transfers part of the costs of managing market and life-cycle risk to the employer. Relying on the employment relationship thus entails additional costs for the employer such as those related to compliance with employment protection legislation and the payment of (social security) insurance contributions.

Even so, information asymmetries provide incentives for employers to enter into longer-term contractual relationships rather than one-off transactions on an open market (cf. Green 2006; Stanford 2017). If it is costly to monitor workers to determine how much work has been done and to measure the quality of labour output, employers have an incentive to rely on positive incentives to extract effort. A longer-term relationship may foster worker loyalty and a positive attitude to work. Employers may thus prefer to work with workers they already know and trust. Information asymmetries also add to the costs of labour turnover: workers have incentives not to reveal their actual skills and attitudes to work, which adds to the search costs of employers who may, in addition, not even be able to assess workers' ability to perform what is required. Such information may become accessible only through extensive testing, or observation of performance when doing the job. Labour turnover then becomes particularly costly if work involves the use of job-specific and tacit knowledge.

Labour platforms address these costs so that it becomes worthwhile to solicit labour through one-off market transactions. This, again, involves matching and audience building. First, they provide a matching service that cuts search costs and allows trade in tasks. Companies that might essentially be in a client or customer relationship with workers can thereby purchase specific services through separate transactions rather than hiring an employee that performs a bundle of tasks. Second, they build audiences by creating virtual labour exchanges, cutting search costs and making matching more efficient. Third, they reduce search costs by making information on a worker's past performance and its evaluation by clients or customers available to new potential ones. Finally, they provide performance and monitoring systems, making supervision relatively inexpensive. This also reduces the need to rely on positive motivation and well-tried workers. Monitoring systems allow precise measurement of the time spent on a task. Importantly, automated supervision and guidance make it possible to manage high worker turnover by reducing the importance of job-specific knowledge.

This shift from employment to a worthwhile form of market mediation makes labour cheaper by transferring risks to the worker. For instance, it was estimated in 2020 that the employment model would cost Uber $400 million a year if implemented in the United States, a country with a relatively low level of worker protection.[2] Implementing the employment model globally could add another $100 million (Uber's total losses in that year's second quarter, without drivers as employees, reached almost $1.8 billion).[3]

In practice, however, the distinction between complete market contracts and incomplete employment contracts is not clear-cut. In fact, the algorithmic management and monitoring systems that allow the platforms to work with casual labour entail authority structures that put workers into a position of subordination to the platform, and hence in the situation of an incomplete contract akin to an employment relationship (cf. Aloisi 2020; Muehlberger 2005). At the same time, these management and monitoring systems allow flexible working times and piece rates. This complicates the classification of their legal status in many jurisdictions as a result of which the legal status of their workers is contested (see Chapters 8 and 9). This enables platforms to compete through regulatory, labour cost arbitrage (e.g. Rahman and Thelen 2019). Some companies that are considered labour platforms do actually employ workers directly (e.g. Takeaway.com) or through intermediaries (e.g. Uber in some markets and also Deliveroo in the past). Many platform companies appear, however, to be in the business of creating fictitious contractor relationships (e.g. Prassl 2018).

Market making by platforms thus also includes the business of evading employment status by exploiting the gaps within the legal frameworks that determine such status, or by simply ignoring existing regulations altogether. This extends to direct lobbying to influence the determination of employment status (e.g. CEO and AK Europa 2019). This applies in particular to location-based platforms that are directly affected by local labour regulation. Large platforms can additionally take advantage of the potential to mobilize their users to exert political pressure (Rahman and Thelen 2019). A case in point is the 2020 campaign related to Proposition 22 that would exempt platforms from classifying workers as employees under California's AB5 law, possibly setting a model for the United States at large. Ridesharing and delivery platforms contributed some $205 million to the campaign and used their apps successfully to mobilize voters to vote in favour of Proposition 22.[4]

Many platform companies that present themselves as labour platforms seem rather to be service providers that may, or may not, work with self-employed workers. This is most apparent in the case of location-based platforms providing food delivery and ridesharing but, as shown in Chapter 3, it also applies to many business-to-business providers of online services. Rather than matching workers with clients, the latter combine tasks sourced on internal platforms into a service, clearly adding value in the process. In the case of food delivery, the companies serve two-sided markets involving restaurants and their customers but their relationship with workers seems rather similar to that of a postal service and its employees.[5] Rather than matching workers with restaurants and their customers, food delivery platforms are selling restaurants and their customers a delivery service. The worker is then hired to work under the direction of the platform. The latter, rather than any of the other parties involved, determines and monitors how the task is performed. Remuneration may be based on a piece rate system but the price is determined by the platform (see Chapters 16–18). The details of work organization implemented by food delivery platforms vary, often including shift systems and incentives against the refusal of offered tasks. Yet it is difficult to square this with the notion of market transactions between workers, restaurants and customers.

2.5 INFRASTRUCTURAL POWER

As providers of market infrastructure, platforms can use their 'infrastructural power' (Rahman 2016) to generate value beyond the rents gained either from the matching process or from the cost advantages of engaging in regulatory arbitrage. This can be related to three types of capacity: structuring interactions; fostering co-creation; and – crucially – capturing and utilizing data on users. Labour platforms utilize these capacities in their operations too.

First, platforms can use their market power to structure interactions (Rahman and Thelen 2019). The regulatory role of intermediaries is limited to influencing the volume of transactions through the price mechanism. The platform model involves a more active regulatory role, however. Platforms represent regulatory structures that shape the rules and parameters of action (Kenney and Zysman 2016). As critical intermediaries and market makers, they exercise market power by controlling participants on either side of the platform (Srnicek 2016; Rahman and Thelen 2019). They can do so by setting governance rules such as the private codes of conduct that govern interactions on their marketplaces and, in the case of labour platforms, regulate conditions of labour (Aloisi 2020; Silberman and Johnston 2020). Platforms also invest and engage in behaviour design (Choudary 2015). Accordingly, in order to bring new participants to the market, platforms must foster new habits and behavioural patterns so that workers are able and willing to engage directly with their clients or customers. The latter, in turn, must adjust their habits so that they purchase services directly from workers. Ultimately, one can argue that the goal of labour platforms must be to generalize a mentality of network consumer-entrepreneurs (Morozov in Langley and Leyshon 2017). In such a mindset, people see themselves simultaneously as customers and as providers of services traded through the platforms. Consumer-entrepreneurs are disciplined by reputation economies as reputation represents key personal capital on both the consumer and the entrepreneur side of the market (Arvidsson and Peitersen 2013).

Second, and relatedly, platforms foster the co-creation of value through user interaction (Langley and Leyshon 2017). Chase (2015) claims that the growth potential of platforms defies the law of physics because their expansion is based on unlocking and leveraging the co-creation of value by networked peers. Rather than customers, users are in fact producers and creators of value. Platforms thus strive to increase the user base as it is crucial to their capacity to cultivate and capture value. This form of value creation applies most directly to social media – in this context, van Dijck (2013) argues that, rather than just intermediating connections, platforms are in the business of actively 'curating connectivity'. In the case of labour platforms, connectivity is curated into a system of labour control through the active participation of users who produce rankings and feedback that discipline workers.

Third, the mediating role gives platforms access to data about transactions and their customers. The ability to utilize data about the network of their users represents a key feature of digital platforms (Birkinshaw 2018). In fact, it stands out from the other two capacities as it is key to the potential to monetize the platform model.[6] The capability to collect and process big data has indeed allowed tech companies to supercharge the advertiser-supported business model of platform mediation (cf. Evans 2003). In the surveillance business model, platform companies monetize their monopoly on data about users and the predictive and governance power that it gives them (e.g. Pistor 2020).

Data collection and utilization is also key to the operation of labour platforms, some of which characterize platform work as data work (Srnicek 2016; van Doorn and Badger 2020). Labour platforms implement extensive surveillance of their workers. They use the data for their systems of labour control and also to manage labour supply and thus liquidity on the platform. Deliveroo, for instance, uses metrics including cancellation rate, attendance rate and super-peak performance to determine the times at which riders are granted access to its shift scheduling tools (Deliveroo's UK Rider Privacy Policy in van Doorn 2020). As a result, these metrics and the ranking collected by platforms become valuable assets for workers. They can be understood as 'ubercapital' (Fourcade and Healy 2017, p. 14) that reflects one's position and trajectory, as recorded through various scoring, grading and ranking methods. It is a capital that workers accrue but do not control. Crucially they cannot transfer it to another platform which undermines their negotiating position vis-à-vis the platforms.

There is a fundamental difference between platforms that use data as a direct source of revenue, however, and those that use it to improve their services, including for labour management purposes. This distinction relates to two, partially overlapping, revenue models, each associated with different types of rent from circulations (Chase 2015). The first approach to revenue generation is significant for 'unconstrained' platforms, including GAFA and BATX companies, which earn income from targeted advertisements and recommendations.[7] The second approach relies on revenues from the fees and charges that platforms require from users. These 'constrained' or 'closed' platforms provide the infrastructure to facilitate a relatively uniform collaboration or exchange of goods and services. Their added value is in structuring these interactions in a way that facilitates the proliferation of transactions and hence their income from fees.

Labour platforms follow the second model. The data they collect on workers may thus represent a valuable asset for workers who sell their services there, but the information to which they have access is relatively limited. Importantly, neither does it represent a major source of revenue for the platforms. This is reflected in the miniscule market capitalization of labour platforms relative to the big tech platform companies. Among listed platforms, only Uber and Lyft had market capitalizations higher than $10 billion ($104 and $16 billion, respectively, as of January 2021), which dwarfs in comparison with GAFA companies whose capitalization ranges from $744 billion (Facebook) to $2 trillion (Apple). Uber's relatively high market capitalization may be attributed to investors' expectation that the company becomes an 'unconstrained platform'. Management of Uber indeed signalled in the documentation for its initial public offering the ambition to use information about its network of customers as an asset that would generate future revenues. Uber's share price can in fact be accounted for by future income streams which, however, are based purely on arguably optimistic assumptions about the potential for market growth and the possibility to improve margins (see MBI 2020). Its market capitalization can thus make sense if the success of the intermediation business model is assumed.

The boundaries between the models are not set in stone. Labour platforms may never achieve sufficient scale and scope to pursue the surveillance model into the long term but they may be integrated into larger businesses, including the big tech companies. Amazon, for instance, acquired a stake in Deliveroo in 2020. Alternatively, venture capital firms and investment funds can serve as 'meta-platforms' looking to exploit the network effects and synergies that emerge from the integration of platform companies and other data-centred businesses in their portfolios (van Doorn and Badger 2020). Such a vision has been presented by SoftBank's founder and chief executive officer, Masayoshi Son. Accordingly, the strategy of Softbank's Vision Fund, which was believed to control 90 per cent of the ridesharing market in 2019, was centred on building a comprehensive data surveillance capability by integrating information on users collected by a range of companies in its portfolio (Medeiros 2019).

2.6 CONCLUSION

The platform model may indeed be emblematic of twenty-first-century capitalism. GAFA and BATX companies are developing new ways of capturing value through taking advantage of their strategic position as market makers, giving them access to data on large segments of market participants and putting them into the position of determining clients' position on respective markets and, in the case of individuals, influencing their behaviour and preferences. Labour platforms do not seem to be able to compete in the premier league, however: their surveillance power is more limited and the markets they serve are characterized by more limited monopoly potential or the high costs of sustaining the network effects.

The ability to facilitate one-off transactions on the markets for labour can be transformatory in the world of work, but its impact – and the viability of the strategy of regulatory arbitrage – is contingent on the regulatory response. In fact, some have argued that the platform economy – as distinct from the platform model of labour mediation – can 'comfortably coexist' with giving platform workers employment status and rights (Aloisi and De Stefano 2020, p. 56; cf. Drahokoupil and Piasna 2019). This is illustrated by Uber's contingency plan in case Proposition 22 was eventually defeated. In this context, Uber discussed using franchising models with local fleet owners where drivers would be hired locally by third parties who would then use Uber's platform to attract business.[8]

Labour platforms, particularly the location-based ones, are associated with a track record of accumulated losses.[9] These are typically being financed by venture capital funds that take longer-term bets on backing a platform that will achieve either dominance, providing gains from monopoly rents, or, if acquired by an incumbent defending its position, income from cashing in the potential (Langley and Leyshon 2017; Rahman and Thelen 2019; van Doorn and Badger 2020). In any case, the business of matching and that of regulatory arbitrage may prove a viable one. Location-based platforms face much stronger headwinds than online platforms, however, since they face the high costs of sustaining the network effects while also being more vulnerable to worker mobilizing. Finally, their opportunities for regulatory arbitrage can be more easily restricted as both their customers and workers are found in a single location.

NOTES

1. www.ft.com/content/a6418a9e-2dba-4bed-8cd6-6737a26d6ef8.
2. www.ft.com/content/1bf57b48-0af5-4386-af4c-c2005a4dc3e6.
3. More specifically, per driver costs would increase by about 30 per cent (in California). But the higher costs per employee would be offset by having fewer drivers as the higher costs would mean lower demand. Moreover, the impact would be limited by lower training and turnover costs if the ridesharing platform imposed shifts and improved checks on drivers before taking them on board (www.spglobal.com/marketintelligence/en/news-insights/trending/QPH1ad9bdJKlLah3yTDOzA2).
4. https://techcrunch.com/2020/11/05/after-prop-22s-passage-uber-is-taking-its-lobbying-effort-global/.
5. The parcel delivery part of the postal service indeed often works with 'independent contractors' and, in this sense, shows similarities with the 'gig economy'.
6. This has also been emphasized by Uber in the S-1 form within its Initial Public Offering documentation, see www.sec.gov/Archives/edgar/data/1543151/000119312519103850/d647752ds1.htm.
7. Facebook and Google, for instance, earn most of their revenue from advertisements but the revenue model is more diverse and dynamic than in the classical advertiser-supported media model since it allows them to take advantage of the ability to harvest and analyse aggregated real-time data on the activities and movements of platform users.
8. In fact, such a strategy was adopted by FedEx in 2014 when it lost a similar employee–contractor court battle. See www.ft.com/content/1bf57b48-0af5-4386-af4c-c2005a4dc3e6.
9. www.ft.com/content/d75f4dd3-6f48-4fea-93e8-4f60172aa9cf.

REFERENCES

Aloisi, A. (2020), 'Hierarchies without firms? Vertical disintegration, outsourcing and the nature of the platform', *Quaderni 8th Giorgio Rota Best Paper Award*, 11–32.

Aloisi, A. and V. De Stefano (2020), 'Regulation and the future of work: the employment relationship as an innovation facilitator', *International Labour Review*, **159** (1), 47–69.

Arthur, W. B. (1989), 'Competing technologies, increasing returns, and lock-in by historical events', *The Economic Journal*, **99** (394), 116–131.

Arvidsson, A. and N. Peitersen (2013), *The ethical economy: rebuilding value after the crisis*, New York: Columbia University Press.

Berg, J. (2016), 'Income security in the on-demand economy: findings and policy lessons from a survey of crowdworkers', Conditions of Work and Employment Series 74, Geneva: ILO.

Birkinshaw, J. (2018), 'How is technological change affecting the nature of the corporation?', *Journal of the British Academy*, **6** (S1), 185–214.

Bowles, S., R. Edwards and F. Roosevelt (2005), *Understanding capitalism: competition, command, and change*, New York: Oxford University Press.

Cant, C. (2018), 'The warehouse without walls: a workers' inquiry at Deliveroo', *Ephemera*, accessed 3 February 2021 at www.ephemerajournal.org/contribution/warehouse-without-walls-workers%E2%80%99-inquiry-deliveroo.

CEO and AK Europa (2019), *Über-influential? How the gig economy's lobbyists undermine social and workers rights*, Brussels: Corporate Europe Observatory and the Austrian Federal Chamber of Labour, Brussels Office, accessed 3 February 2021 at https://corporateeurope.org/sites/default/files/2019-09/%C3%9Cber-influential%20web.pdf.

Chase, R. (2015), *Peers Inc*, London: Headline.

Choudary, S. P. (2015), *Platform scale: how an emerging business model helps startups build large empires with minimum investment*, Boston, MA: Platform Thinking Labs.

Coase, R. H. (1937), 'The nature of the firm', *Economica*, **4** (16), 386–405.

Drahokoupil, J. and A. Piasna (2017), 'Work in the platform economy: beyond lower transaction costs', *Intereconomics: Review of European Economic Policy*, **52** (6), 335–340.

Drahokoupil, J. and A. Piasna (2019), 'Work in the platform economy: Deliveroo riders in Belgium and the SMart arrangement', Working Paper 2019.01, Brussels: ETUI.

Evans, D. (2003), 'Some empirical aspects of multi-sided platform industries', *Review of Network Economics*, **2** (3), 191–209.

Evans, D. (2009), 'How catalysts ignite: the economics of platform-based start-ups', in A. Gawer (ed.), *Platforms, markets and innovation*, Cheltenham, UK and Northampton, MA, USA: Edward Elgar Publishing.

Evans, D. and R. Schmalensee (2007), 'The industrial organization of markets with two-sided platforms', *Competition Policy International*, **3** (1).

Fourcade, M. and K. Healy (2017), 'Seeing like a market', *Socio-Economic Review*, **15** (1), 9–29.

Green, F. (2006), *Demanding work: the paradox of job quality in the affluent economy*, Princeton, NJ: Princeton University Press.

Katz, M. L. and C. Shapiro (1994), 'Systems competition and network effects', *Journal of Economic Perspectives*, **8** (2), 93–115.

Kenney, M. and J. Zysman (2016), 'The rise of the platform economy', *Issues in Science and Technology*, 32 (3), accessed 3 February 2021 at https://issues.org/the-rise-of-the-platform-economy/.

Langley, P. and A. Leyshon (2017), 'Platform capitalism: the intermediation and capitalization of digital economic circulation', *Finance and Society*, **3** (1), 11–31.

MBI (2020), 'Deep dive on Uber', accessed 3 February 2021 at https://mbi-deepdives.com/deep-dive-on-uber/.

Medeiros, J. (2019), 'How SoftBank ate the world', *Wired UK*, 7 February, accessed 3 February 2021 at www.wired.co.uk/article/softbank-vision-fund.

Mishel, L. (2018), *Uber and the labor market: Uber drivers' compensation, wages, and the scale of Uber and the gig economy*, Washington, DC: Economic Policy Institute.

Muehlberger, U. (2005), 'Hierarchies, relational contracts and new forms of outsourcing', ICER Working Paper 22/2005, International Centre for Economic Research, accessed 3 February 2021 at https://ideas.repec.org/p/icr/wpicer/22-2005.html.

Pistor, K. (2020), 'Rule by data: the end of markets?', *Law and Contemporary Problems*, **83** (2), 101–124.

Prassl, J. (2018), *Humans as a service: the promise and perils of work in the gig economy*, Oxford: Oxford University Press.

Rahman, K. S. (2016), 'The shape of things to come: the on-demand economy and the normative stakes of regulating 21st-century capitalism', *European Journal of Risk Regulation*, **7** (4), 652–663.

Rahman, K. S. and K. Thelen (2019), 'The rise of the platform business model and the transformation of twenty-first-century capitalism', *Politics and Society*, **47** (2), 177–204.

Rochet, J.-C. and J. Tirole (2003), 'Platform competition in two-sided markets', *Journal of the European Economic Association*, **1** (4), 990–1029.

Rochet, J.-C. and J. Tirole (2006), 'Two-sided markets: a progress report', *RAND Journal of Economics*, **37** (3), 645–667.

Shapiro, C. and H. R. Varian (1998), *Information rules: a strategic guide to the network economy*, Boston, MA: Harvard Business Press.

Silberman, M. 'Six' and H. Johnston (2020), 'Using GDPR to improve legal clarity and working conditions on digital labour platforms', Working Paper 2020.05, Brussels: ETUI.

Simon, H. A. (1951), 'A formal theory of the employment relationship', *Econometrica*, **19** (3), 293–305.

Srnicek, N. (2016), *Platform capitalism*, Cambridge: Polity Press.

Stanford, J. (2017), 'The resurgence of gig work: Historical and theoretical perspectives', *Economic and Labour Relations Review*, **28** (3), 382–401.

van Dijck, J. (2013), *The culture of connectivity: a critical history of social media*, Oxford: Oxford University Press.

van Doorn, N. (2020), 'On data assets and meta-platforms', *Platform Labour*, 7 September, accessed 3 February 2021 at https://platformlabor.net/blog/on-data-assets-and-meta-platforms.

van Doorn, N. and A. Badger (2020), 'Platform capitalism's hidden abode: producing data assets in the gig economy', *Antipode*, **52** (5), 1475–1495.

Vandaele, K. (2021), 'Collective resistance and organizational creativity amongst Europe's platform workers: a new power in the labour movement', in J. Haidar and M. Keune (eds), *Work and labour relations in global platform capitalism*, Cheltenham, UK and Northampton, MA, USA: Edward Elgar Publishing.

Williamson, O. E. (1981), 'The economics of organization: the transaction cost approach', *American Journal of Sociology*, **87** (3), 548–577.
Wood, A. J., M. Graham, V. Lehdonvirta and I. Hjorth (2019), 'Networked but commodified: the (dis)embeddedness of digital labour in the gig economy', *Sociology*, **53** (5), 931–950.

3. Moving on, out or up: The externalization of work to B2B platforms

Pamela Meil and Mehtap Akgüç

3.1 INTRODUCTION

Research on the platform economy tends to concentrate on two actors: the platforms themselves and the workers. This leaves out a key third actor, however: the customer. Customers can be individuals making transactions, booking rides, ordering food or buying a service such as cleaning or babysitting. In this chapter the focus is on business to business (B2B) constellations in which customers are companies. Little has been researched on how or why companies use platforms to perform work or tasks. A major difficulty in researching this issue lies in the opacity of work distribution and the content of work on so many platforms. Although platforms proudly advertise their business customers on their sites, for those carrying out a job or task it is often unclear who their contractor actually is. It is difficult to collect data on company use of platforms, not only in terms of numbers but especially in the precise kinds of tasks and activities being performed on platforms. Behind the question of the extent of how much work is being carried out and the types of activities involved, the question looms as to why companies are externalizing work to platforms and how this affects what they outsource.

Turning to the platform economy and its role in externalization, this chapter addresses the following question: is the outsourcing of work to platforms simply a logical extension and continuation of the long-term trend to externalize work or does it contribute to new forms of outsourcing and transformation of value chains? In order to address this question, it is necessary to look at what kinds of platforms companies are using to externalize work and how this process has evolved and become more complex over time; as well as what kind of work is being externalized, particularly which industries, occupations and skill sets, and the implications for skills use and development as well as quality of work.

The chapter maps the activities outsourced to platforms and, as a result, makes the argument that there are three main platform models which can be discerned emerging from the current activity. These models represent different ways of dealing with the question of the reintegration of work into company value chains to present the ultimate customer with an integrated solution. The first is anonymous microtasking, with standardization and codification of tasks and easy reintegration back into company processes; the second is isolation and modularization of tasks that often involve creative or design work or information technology (IT) software development and require little reintegration; and the third is that of intermediary coordinator, similar to that of tier one suppliers in conventional outsourcing models, whereby the platform takes on a managing role in coordinating the complex tasks outsourced by companies as well as some reintegration tasks.

A first step in unravelling the questions surrounding company participation in the platform economy is a close look at the evolving and diverse platform landscape. A recent report by World Economic Forum (2020) provides a useful typology. Accordingly, location-based platforms oriented towards (personal) services are mainly categorized as business-to-end-consumer and are relatively less relevant for companies as customers, although this does not exclude companies making use of services such as cleaning or food delivery. Location-based platforms oriented toward professional services require local presence and are a focus of corporate activity. Online platform work involving professional services of varying complexity (e.g. microtasking, freelancing) done remotely is, however, the type mainly targeted by companies as customer.

3.2 FROM INTERNALIZATION TO THE EXTERNALIZATION OF WORK VIA PLATFORMS

Before we address the role of the platform economy in today's employment landscape and work externalization, we take a look back at the development of work organization and the employment relationship in the conventional economy to put current trends in a broader historical context.

In post-war western economies, the organization of work occurred increasingly in bureaucratic, hierarchically integrated firms (Scott and Davis 2014). For workforces with stable employment contracts in mostly company-bound employment, a certain amount of two-sided loyalty or at least a type of symbiotic relationship emerged: tenure of employment, which could also include skill development and wage security on the one side; and the stability (or stickiness) of labour commitment to the company on the other (Smith 2010; Meil 2012). Obviously, labour market segmentation existed and there were always parts of the workforce left out of this contract – particularly women, minorities and migrant workers (Gordon et al. 1982). The employment contract also manifested itself differently across institutional contexts (Hall and Soskice 2001; Streeck 1998). Nonetheless, this depiction of bureaucratic, firm-based stability reflected the overwhelming societal consensus of the employment relationship being centred around company-bound employment in mainly industrial sectors. It was, additionally, the model for creating systems of labour protection and regulation by institutions of industrial relations.

Forms of control over work and the work process were inherent parts of vertically integrated and hierarchical organizations of work. They also changed over time – from Taylorist divisions of labour and time regimes to more bureaucratically regulated forms of control (Edwards 1979). Thus, within an internalized employment relationship control takes on indirect forms, aiming at eliciting workers' effort through target setting, paying bonuses, promising employment stability and advancement, addressing normative orientations, influencing workers' beliefs, norms and values or creating corporate identities. By the 1980s, this culminated in human resource approaches in which identification with company cultures and empowerment were buzzwords in the management literature. Especially for skilled workforces in dominant industries, the ties to place-bound companies were stronger than ever (Peters and Waterman 1982).

Also beginning in the 1980s but taking off considerably in the 1990s and ongoing since then, there has been a continuing trend in the externalization of work in the form of outsourcing and offshoring and this has characterized the organization of work in several industries of the economy (Porter 1985; Gereffi and Korzeniewicz 1994; Froebel et al. 1981). Outsourcing began with so-called non-core tasks being removed from company headquarters or central production units and relocated at (usually) lower wage destinations with more precarious working conditions. Over time, more and more functions in the organization (production, sales, purchasing, personnel, IT) were externalized across growing value chains (Huws 2006; Huws et al. 2009).

Nonetheless, the tasks, activities and functions in these outsourcing processes were externalized to other place-bound units: intermediaries in the value creation process of the lead firms. Working conditions and job security in these outsourced units often displayed lower wages and precarity, especially in emerging economies such as India, China and eastern Europe. Outsourced companies could also add to the pressure put on workforces across the entire chain, including at company headquarters, in the desire to force wage and other concessions (Doellgast and Greer 2007). The formation and elongation of value chains did not necessarily result in solely negative outcomes such as downgrading across the chain, however. Value chain creation tends to be dynamic and, over time, companies lower in the value chain can potentially upgrade to higher-level tasks and activities and become more integrated into the company network and the overall organization of work (Gereffi 2019).

The focus here is the ways – and models vary widely – that companies externalize paid work to platforms which, acting as an intermediary and facilitated by rapid digitalization, then extend the externalization process to individuals or groups of workers (Bergvall-Kåreborn and Howcroft 2014; Bourdeau and Lakhani 2013; Florisson and Mandl 2018; European Commission 2019). There are certainly some similarities between outsourcing to intermediaries and outsourcing to platforms, including reducing fixed costs in lead firms, lowering wages, and standardizing and breaking down tasks and functions so that they can be more readily externalized. However, there are also various areas of crucial difference: B2B platforms are not always place-bound and are therefore even more difficult to regulate than low-cost intermediaries; moreover, these are mainly market-based transactions driven by price on which bidding for the work is done by, with the work then distributed to, individuals (Williamson 1985).

Organizationally, platforms seem to take us full circle in models of organizing work, away from hierarchical organization and bureaucratic forms of control and back to pre-industrial forms of piecework based on putting-out systems of remuneration (Acquier 2018). In the literature, there continues to be some debate on the meaning of platforms for the organization of work and the resulting effects on governance of the work process and outcomes for individual workers (Meil 2018; Meil and Kirov 2017). For instance, work carried out on platforms – outside of in-house labour and work processes – is sometimes considered a marginal phenomenon that affects only a small portion of the work process. Since outsourcing to platforms diffuses and obscures the critical questions of who is employed and the content and organization of work, it can be difficult to sort out the exact extent of employment. Particularly with regard to microwork and cloud applications, it can seem that work is divorced from the actual persons who carry it out, although there is always real activity behind virtual or digital work, even in clickwork where workers see only a small sliver of the total picture and therefore often do not know what they are working on or to what they are contributing. This is also true of algorithms which may run automatically but which are designed and programmed by real people with inclinations, priorities and agendas and whose products have an impact on jobs, work, processes and organizations.

3.3 ACTIVITIES AND OCCUPATIONS OUTSOURCED THROUGH B2B PLATFORMS AND THE PARTICIPATING INDUSTRIES

This section provides a descriptive overview of the activities and occupations carried out on B2B platforms and then analyses the function of platforms in the global value chain.

The platform landscape incorporates a wide range of work externalized by companies to B2B platforms including: content monitoring; product recognition; coding; IT software development; translation and editing; web design; logo design; human resources; product development; algorithm writing; rental of cloud space for data management; and distribution planning for logistics. Looking at the types of occupations performed on platforms and the way companies are using these gives an indication of what levels of skill and scales of tasks are affected by this process, as well as the occupations and industries involved. Platforms tend to specialize around a set of tasks, the platform workers they recruit and the markets they target. Some platforms have grown very large, like CloudFactory and Upwork, and may have a limited range of applications for a diverse range of industries or a large workforce pool that ranges widely in skill level. Other sites try to survive the volatile platform economy by concentrating on regions and languages (German-speaking or Russian-speaking customer base for instance).

There has been a growing interest in mapping the diverse occupations covered by jobs performed on platforms. One such attempt is the Oxford Internet Institute's Online Labour Index[1] (see Chapter 12) which seeks to 'measure the supply and demand of online freelance labour across countries and occupations by tracking the number of projects and tasks across platforms in real time'. The Index has been steadily growing over time, reflecting the increase in online platform work since data collection started in May 2016 (see Figure 3.1).

Examples of the types of work being carried out through platforms include software development; building/feeding web content; the building/cleaning/management of databases; classifying web pages; the translation and editing of texts; transcribing audio files; classifying or tagging images; and the design of logos (Florisson and Mandl 2018). Figure 3.1 shows that software development and technology-related tasks make up the largest number posted to online platforms, similar to the historical offshoring patterns of service sectors, followed by creative and multimedia design, writing and translation and clerical and data entry tasks.

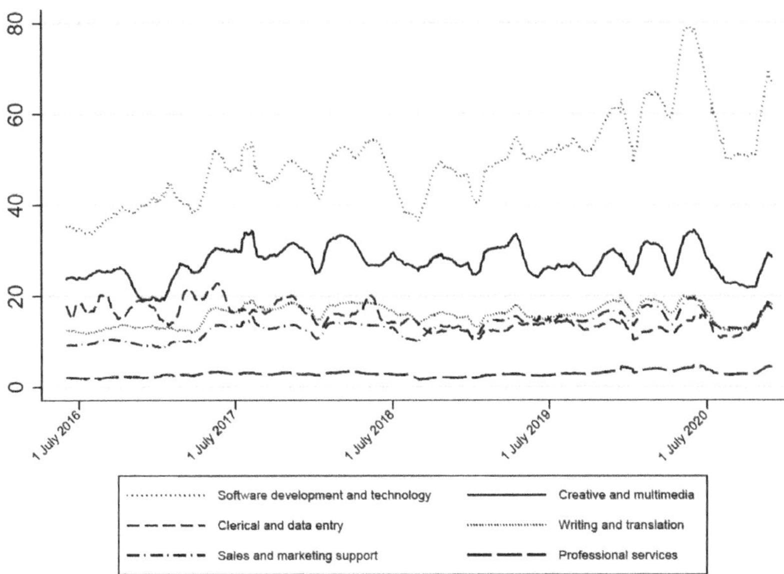

Note: Authors' elaboration using daily data points of the Oxford Internet Institute's Online Labour Index smoothed using 28-day moving average (accessed on 1 December 2020).
Source: Kässi and Lehdonvirta (2016).

Figure 3.1 Trend in Online Labour Index by occupation

Aside from the more traditional types of work externalized to platforms involving microtasks and modular/isolated tasks outside of company operating processes, there is also evidence that platforms have taken over aspects of managing quality assurance and orchestrating the integration of fragmented processes (Schmidt 2019), indicating a move to more complex sets of activities in company value chains. One platform – 10EQS – advertises its role in this way: '10EQS orchestrates crowd-based solutions. We take your issue, break it apart, find the right specialists for each piece, and put everything back together. You get a complete and customized solution – whatever your industry, geography or business challenge.' Topcoder, the computer programming platform, relies on breaking down the traditional steps of a software development project, such as conceptualization, requirements specification, architecture design, component production, assembly, certification and deployment, and then structuring these as a 'game plan'. Multiple crowdworkers take part and the winning output of each round (as determined by the more experienced members of the community) becomes an input to the subsequent stage.

Platforms have also become involved in the manipulation of huge amounts of data in simple production processes (such as textiles or electronics) and logistic distribution and coordination. Data are stored in the clouds of large platforms (Alibaba, Amazon, Pactera), some of which previously specialized in e-commerce or e-marketplaces but are now adapting their business models to become actors in global production networks (Butollo and Lüthje 2017).

There is also a diverse range of industries outsourcing to B2B platforms. Rather than being confined to the digital economy, media and social media or IT, there is also strong involvement from the 'traditional economy': for example, a pharmaceutical company has launched InnoCentive, an innovation crowd-sourcing platform; while the automotive industry is one of the largest clients of digital-related microworking services (Casilli and Posada 2019). Developments in automobiles have been taking place as a result of demand for algorithms used in autonomous vehicle design as well as for voice-activated virtual assistant systems (Schmidt 2019). The enormous amount of data necessary for these artificial intelligence (AI) applications derives from human labour right across the globe to carry out the menial, but necessary, tasks of measurement and photography. This is the explanation for the use of crowd-working sites by carmakers since as early as 2016 (Meil and Schörpf 2017). A range of large, but not publicly well-known, platforms such as Mighty AI, understand.ai, playment, Hive and Pactera are involved in this business, catering almost exclusively to large industrial customers, as well as a number of better-known sites (Clickworker, Streetspotr, AMT and Crowd Guru).

Aside from the development of AI solutions in automobiles, the creative industries – journalism, film, media, design – also outsource to platforms; IT industries and the IT divisions of companies within other industries outsource software development and coding tasks; while marketing companies and divisions outsource product reviews and digital media for content management and editing.

It is apparent that the increasing use of platforms is linked in large part to a broad range of new applications and tasks involving AI and data storage and (re)distribution. Although some of the platforms are not specifically labour platforms, they do shift the relationship between internal and external processes in the organization of work while also facilitating their increasing automation. For many of these tasks, therefore, it is difficult to determine the exact extent of the outsourcing or how (re)integration in company processes might occur.

3.4 NEW FORMS OF OUTSOURCING THROUGH PLATFORMS

Value chains are created when companies outsource to platforms which then distribute this work to one or a group of workers. Just as in the beginning phases of outsourcing in the traditional economy, the first forms of the externalization of work to platforms were simple linear processes: a company posted a task or set of tasks to the platform; the platform found the worker or group of workers to do the work; and then delivered the result. However, this did not fundamentally change the internal processes or business models of the companies involved.

In traditional value chains, research shows that the configuration changes over time in what is a dynamic development of expansion and upgrading or possible downgrading (Gereffi et al. 2011; Fernandez-Stark et al. 2011). Interestingly, evidence is now emerging that value chains on platforms are also becoming more complex and undergoing their own expansion and extension (Tubaro and Casilli 2019).[2] Furthermore, the use of various forms of subcontracting on platforms to create platform-based networks resembles the extension and growth of value chains in the traditional place-bound economy. Platform subcontracting takes place either directly through a customer or via the platforms themselves. For instance, 99designs, a contest site for graphic designers,[3] outsources the organizational tasks behind their contests to freelancers on Upwork. Large automotive companies also outsource the work implied by the development of autonomous vehicle applications to technology companies which then further outsource the work to other platforms or to workers directly.

Some platforms (e.g. 10EQS, CloudFactory and Topcoder), however, are actively promoting a more comprehensive service offer including project management as well as the coordination of dispersed activities, implying the taking up of an enhanced role as integrators on a range of tasks or outsourced processes. It is this 'upgrading' activity which makes such platforms begin to look more and more like online variants of tier one engineering service suppliers. As Kaganer et al. (2013, p. 27) explain it: 'Increasingly, we are seeing entire business models that are fundamentally based on tapping contributions from crowds, where a primary source of value creation is from the crowd. In other cases, platforms are creating value by helping their customers use crowds well.' This new form of outsourcing seeks to 'deliver the work of an in-house team without the burdens of managing one' (quote from the CloudFactory website).

The outsourcing to platforms being undertaken by companies differs by type of work or tasks as well as by the workers who are targeted, as we have demonstrated. It is the degree to which work processes are linked directly to ongoing company operations which determines how easy reintegration into the value chain might be compared to the outsourcing of isolated tasks that are easily separated from ongoing company processes. However, what kinds of value upgrading, downgrading or integration are taking place as a result of platform work is largely non-transparent and, equally, little is known about the extent to which outsourced work can, in practice, be successfully reintegrated into overall business processes (Mrass et al. 2019).

The platforms themselves differ to the extent that they publish tasks to unidentified crowds who may either be individuals or groups or even other platforms; and whether they can offer dedicated pools of labour or particular individual experts. Such platforms often overlap with others providing a more traditional range of work or tasks but are evidently trying to move up the value chain by engaging in a broader range of services based on facilitating integration into company operational and work processes.

Against this broad and changing background, one can discern at least three emerging models of outsourcing in the value chain of companies making use of platforms, with some subcategories. Each model represents a different way of distributing work to platforms and/or dealing with reintegration into the value chain. One should also bear in mind that, as the platform scene is constantly evolving, some types of outsourced work through platforms might also crosscut several models.

First, there is anonymous microtasking by human microworkers, involving small units of work which cannot be carried out by algorithm and which require human decision-making such as identifying objects in photos, writing short descriptions and identifying or tagging content in texts. New applications involving autonomous vehicle development and virtual assistants have also expanded the range and use of microtasks. Such work can be externalized by engaging in an extreme break-down of tasks in which the standardization and simplification required facilitates reintegration into the company.

The second model is the isolation and modularization of tasks which require little reintegration in ongoing processes because they stand on their own. This model is often linked to 'creative' work such as design tasks, website creation and translation, but is also used with technical tasks such as data analysis or software development. Often the use of platform workers is highly personalized in this model. Rather than utilizing automated or anonymous algorithmic matching, profiles of particular workers are advertised with portfolios, evaluations and ratings. Here, reintegration would require greater coordination, but the tasks are usually divorced from the main processes of the company's production or service.

The third model is that of the intermediary coordinator who takes on a management role in the coordination of complex tasks similar to that of tier one suppliers or engineering service contractors in traditional outsourcing models. The work may be broken down, thus making the tasks for the platform worker fragmented and based more on a particular function or speciality. However, in this model it is the platform which takes the responsibility of putting the pieces back together and offering an integrated solution to the customer, allowing them to rise up the value chain. Such platforms are located around the globe, in the Global South but not only so: from Nepal (CloudFactory) to China (Xometry and Pactera) as well as South America, Europe and the United States.

Outsourcing project management to intermediaries is not new but moving coordination and monitoring activities to platforms is a fresh development which is imbued with a number of challenges, including for labour in terms of wage pressures, dispersed management and control systems, and, of course, the lack of employment regulation. Another phenomenon is the creation of informal value chains in which work being carried out does not get remunerated or officially registered. For example, a film studio can outsource animation work to a freelancer via a platform. This task is primarily creative work but can involve the use of complex software. The creative worker potentially reaches out to the collaborative tech community for help and receives assistance due to the open-source mentality of this community or because a film project is perceived as 'fun'. In this way, the film studio benefits from uncompensated labour, having already outsourced activities to lower-cost freelancers outside of their organizations (Forsler and Velkova 2018).

Table 3.1 provides an overview of the types of task and their distribution through platforms, as well as examples of the broad range of companies involved in externalizing work in this way. Each type of outsourcing often uses labour differently and has different implications for how work plugs into company value chains.

Outsourcing to microwork or crowdwork platforms mainly targets anonymous workers (model 1); while outsourcing to freelancing or contest sites often targets particular types of individual or groups of experts but often for modular tasks which remain outside the main value chains (model 2). The innovations that arise may nevertheless be used by the companies and reintegrated into their technology, work or labour processes.

Other forms of the outsourcing taking place on the B2B platforms listed in Table 3.1 deal with cloud applications, AI and algorithm development and appear to be growing in importance. These are complex platform types which use microworkers for some tasks and applications but which may also have their own virtual labour pools. Although they engage in large amounts of outsourcing for a wide array of companies, they are not household names. Some of them, such as CloudFactory, are clearly labour platforms to the extent that they act as intermediaries for the provision of labour for company-defined tasks (model 3). Algorithms and automation play a large role but, although the effects on the labour process are one step removed, they do still exist: new platform services dealing with logistical distribution and management and production capacity planning have an evident impact on company networks, divisions of labour and control of information and, consequently, affect the overall organization of work and the configuration of value chains.

Table 3.1 *Types of task and their distribution through platforms*

Types of task and their distribution	Platform examples	Customer examples
Microwork: content editing and tracing; testing; product review; measurement; data entry; etc.	Clickworker; Streetspotr; Amazon Mechanical Turk (AMT); Microworkers; Crowd Guru; Topcoder	Groupon; Deutsche Telekom; Honda; PayPal; Daimler; Sharewise; Epoq; Kiveda; CastingWords; Channel Intelligence; Facebook; BMW; Daimler
Freelancing: long-term and one-off tasks (creative work; journalism)	Upwork; Jovato; Freelancer; Crowdspring	Opentable; Pinterest; NBC; Panasonic, Unilever; Audi; LG
Contests	InnoCentive; 99designs; fiverr	Thomson Reuters; Proctor and Gamble; SAP; Dupont; Airbus
Integrated, project-based tasks	InnoCentive; Machdudas; CloudFactory; Local Motors	Microsoft; Luminar; Hummingbird Technologies
Cloud applications	Pactera; Amazon; CloudFactory; Xometry	BMW; GE; NASA; Dell Technologies; SNCF Réseau; Bosch
App development; algorithm development	Appen; CloudFactory	pilot.ai; Gospotcheck; Amazon; Adobe; Zefr

Source: Authors' compilation from platform websites (2020).

3.5 CHALLENGES AND RISKS OF OUTSOURCING TO PLATFORMS

While providing innovative business opportunities and creating global value chains, there are also challenges when companies outsource work to platforms. These include (but are not limited to) the skills requirements, the effects on skills use and development (both deskilling and upskilling) and the uncertainties revolving around the quality of work on platforms.

First of all, the skills requirements of online work are highly contingent on the type of tasks intermediated through platforms. While evidence points to platform workers having higher than average levels of educational attainment, this mainly holds true for workers undertaking online work – rather than on-site and local platform work (e.g. Berg 2016; Ilsøe and Madsen 2017). Moreover, platform work involving professional or creative tasks (e.g. software programming, creative design) requires more specialized skills (Florisson and Mandl 2018).

Second, when a more complex project is outsourced via a platform it may be that the worker not only needs to possess the specialized skills to provide the complex service but also needs to have project management skills, be flexible and available to answer the ad hoc demands made by customers. This might encourage platform workers to acquire further skills or discover new ones, allowing the possibility of skills development. However, this happens in a context where workers are not regularly employed by platforms and often have precarious employment arrangements with few opportunities for training or other benefits (exceptions exist).

The literature on the processes of fragmentation and modularization, often (although not always) involving standardization and codification as a preparatory step for outsourcing, is extensive (Sturgeon 2002; Sturgeon et al. 2017; Huws et al. 2009; Flecker and Meil 2010; Gereffi 2019). Where companies outsourcing (part or most of) a complicated job – requiring a wide range of skills – break it down into smaller, codifiable tasks, this leads to less skill requirement per fragmented task. This means that the outsourcing process is more convenient for companies, since work can be externalized more easily, but it also generates tasks requiring few or no skills. This whole process of the simplification and taskification of online work is somewhat in line with the idea of Smith (1997, 1998), who argues that non-standard forms of work are often associated with the deskilling of jobs.

Externalizing part of their activities through online platforms also means that the responsibilities of the human resources department can shift away from training and investment in skills acquisition for internal employees towards looking out for those skills externally and buying them from platforms. This process of buying skills on demand is increasingly feasible as a result of the availability of online platforms and is often cost-effective because of the impact of platforms in encouraging a race to the bottom on pay. This process might also imply a gradual decline in firm-specific skills development in companies (George and Chattopadhyay 2015; Lepak and Snell 2002).

Finally, as regards work quality control, while transparency between platform, customer and worker is not always a given with the presence of information asymmetries in the triangular relationship, platforms nonetheless allow the possibility of control or monitoring over task completion. In some cases, the platform takes the lead in closely monitoring the work and making the surveillance results available. For instance, Upwork provides this option through its work diary app, in which the platform closely monitors platform workers by taking frequent screenshots and making these available for customers to assess workers' productivity (Florisson and Mandl 2018). In other platforms involving microtasks or other types of clickwork, the simplification and standardization of subtasks allow a checking of the quality of each completed task step by step. For example, Clickworker advertises that most completed jobs go through routine quality control in which tasks are double-checked by other clickworkers (peer review).

Another method of controlling work quality is customer ratings. These can cover wide ranges of work across all types of skill requirements. Research does suggest that ratings could lead to controlling the behaviour of platform workers who feel the need continuously to self-optimize to please the customer (Schörpf et al. 2017). In this way, platforms use ratings to transfer responsibility for quality control and the monitoring of work to the customer. For instance, on platforms such as AMT and many contest platforms, very little oversight is provided by the platform (Florisson and Mandl 2018) and quality control is mostly left to customers.

In the case of more complex tasks, quality control sometimes takes place through result-oriented remuneration. For example, workers on contest or bidding platforms propose their bids to win a service contract which, in some cases, requires full-fledged project preparations involving risks with no guarantee of winning – while the preparation phase is not remunerated. In other cases, remuneration of platform work is conditional on completing subtasks. In either case, such quality control can contribute to the precariousness of platform work as there is an increasing expectation among customers of task performance but not always a parallel improvement in working conditions, such as guaranteed payment in complex projects or assurance against risks for platform workers.

3.6 CONCLUSION

This chapter has pointed to a rise in the externalization of work by companies via platforms. Taking stock of the evidence on developments in the B2B platform economy, a number of trends are emerging that indicate the ways in which companies are externalizing work to platforms and the impacts this is having on the organization of work and working conditions.

The companies involved represent a diverse range of industries while the array of tasks is also broad, although an increasing number appear to be linked to algorithm development and apps. Overall, it is clear that we are seeing an increase in the use of platforms by place-bound companies rather than a stagnation or reduction.

Three main platform models seem to be emerging that constitute different ways of dealing with reintegration into new forms of value chains. The first involves anonymous microtasking by human microworkers. In this case, most activities seem to be discrete tasks outside of or divorced from the main business or processes. It is conceivable that such activities could be integrated into normal company organizational processes without much effort because they are the types of task (e.g. clickwork, content editing and translation) that are easy to separate out and do not require complex reintegration.

The second model is the isolation and modularization of tasks requiring little reintegration as they stand on their own. This model is often linked to creative work involving, for instance, IT and software development, design tasks or mobility concepts that might require greater efforts at and difficulties with reintegration but which companies are nevertheless externalizing to platforms.

The third model is the one of intermediary coordinator taking on a task management and coordination role similar to that of tier one suppliers in traditional outsourcing models. In this case, the intermediary takes on a more managing role for coordinating complex tasks. The work may be broken down into component parts, but it is the responsibility of the platform to put the pieces back together and ultimately offer an integrated solution.

Such trends in the increasing externalization of work to platforms are not to the benefit of workers. There are few standards or rules governing platforms and those that do exist are generally voluntary while platforms are difficult to regulate. Moving work out of company processes necessarily involves a certain amount of standardization and codification of tasks, making them less skilled and usually less well paid. There are signs that some upskilling is occurring on higher-end platform sites, but this is rare, and the employment relationship is still precarious. Collective representation efforts have seen some success mainly on location-based platforms but have proven impossible to implement on B2B platforms other than in respect of the voluntary acceptance of codes of conduct in isolated cases.

As in the traditional place-bound economy there is not one B2B platform worker profile. Anonymous crowdworkers being paid by hits or microtasks have different challenges and needs – not to mention earnings potential – than workers who are expressly marketing themselves, their skills and their ratings. Nonetheless, remote-global 'virtual' workers all lack the benefits of shared work experience, of being deeply integrated in company processes and organizations and usually of having access to social benefits such as health insurance or pensions. Since it appears that the externalization of work to B2B platforms by companies is not disappearing, and is in fact not a marginal phenomenon but a growing one, the consequences for labour and the quality of work life are all too real.

NOTES

1. For more details, see https://ilabour.oii.ox.ac.uk/online-labour-index/.
2. On contest-based crowd-sourcing platforms, customers post a briefing with a fixed 'prize' attached and crowdworkers have the opportunity to submit finished or near-finished entries in the hope of winning the contest. This version of awarding jobs is more common for more complex jobs with higher payments (at least, as regards the successful entry). Some platforms offer 'shop features' – often to complement a bid- or contest-based mediation of work – where online workers can offer ready-made products at fixed prices. These may include goods and services such as handmade craft products, graphic designs and video or audio clips. Some platforms, for instance fiverr, focus exclusively on ready-made work packages. Contest-based sites often target creative tasks and innovative solutions in a range of industries including aerospace, automobiles and pharmaceuticals.
3. In employment terms, this is making the identification of the employer and those who carry out the work and under what conditions even more difficult.

REFERENCES

Acquier, A. (2018), 'Uberization meets organizational theory: platform capitalism and the rebirth of the putting-out system', in N. Davison, M. Finck and J. Infranca (eds), *Cambridge Handbook on Law and Regulation of the Sharing Economy*, Cambridge: Cambridge University Press, pp. 13–26.

Berg, J. (2016), 'Income security in the on-demand economy: findings and policy lessons from a survey of crowdworkers', Conditions of Work and Employment Series No. 74, Geneva: ILO.

Bergvall-Kåreborn, B. and D. Howcroft (2014), 'Amazon Mechanical Turk and the commodification of labour', *New Technology, Work and Employment*, **29** (3), 213–223.

Bourdeau, K. and K. Lakhani (2013), 'Using the crowd as an innovation partner', *Harvard Business Review*, **91** (4), 60–69.

Butollo, F. and B. Lüthje (2017), 'Made in China 2025: intelligent manufacturing and work', in K. Briken, S. Chillas, M. Krzywdzinski and A. Marks (eds), *The new digital workplace: how new technologies revolutionise work*, Basingstoke: Palgrave Macmillan, pp. 42–61.

Casilli, A. and J. Posada (2019), 'The platformization of labor and society', in M. Graham and W. H. Dutton (eds), *Society and the internet: how networks of information and communication are changing our lives*, 2nd edition, Oxford: Oxford University Press, pp. 293–306.

Doellgast, V. and I. Greer (2007), 'Vertical disintegration and the disorganisation of German industrial relations', *British Journal of Industrial Relations*, **45** (1), 55–76.

Edwards, R. (1979), *Contested terrain: the transformation of the workplace in the 20th century*, New York: Basic Books.

European Commission (2019), *The changing nature of work and skills in the digital age*, Luxembourg: Publications Office of the European Union.

Fernandez-Stark, K., P. Bamber and G. Gereffi (2011), 'The offshore services value chain: upgrading trajectories in developing countries', *International Journal of Technological Learning, Innovation and Development*, **4** (1–3), 206–234.

Flecker, J. and P. Meil (2010), 'Organisational restructuring and emerging service value chains: implications for work and employment', *Work, Employment and Society*, **24** (4), 680–698.

Florisson R. and I. Mandl (2018), *Platform work: types and implications for work and employment – literature review*, Dublin: Eurofound.

Forsler, I. and J. Velkova (2018), 'Efficient worker or reflective practitioner? Competing technical rationalities of media software tools', in P. Bilic, J. Primorac and B. Valtysson (eds), *Technologies of labour and the politics of contradiction*, Cham: Palgrave Macmillan, pp. 99–119.

Froebel, F., J. Heinrichs and O. Krey (1981), *The new international division of labour*, Cambridge: Cambridge University Press.

George, E. and P. Chattopadhyay (2015), *Non-standard work and workers: organizational implications*, Geneva: ILO.

Gereffi, G. (2019), *Global value chains and development: redefining the contours of 21st century capitalism*, Cambridge: Cambridge University Press.

Gereffi, G. and M. Korzeniewicz (eds) (1994), *Commodity chains and global capitalism*, London, CT: Greenwood.

Gereffi, G., K. Fernandez-Stark and P. Psilos (2011), *Skills for upgrading: workforce development and global value chains in developing countries*, Durham, NC: Duke Center on Globalization, Governance and Competitiveness.

Gordon D. M., R. Edwards and M. Reich (1982), *Segmented work, divided workers: the historical transformation of labor in the United States*, Cambridge: Cambridge University Press.

Hall, P. A. and D. Soskice (eds) (2001), *Varieties of capitalism: the institutional foundations of comparative advantage*, Oxford: Oxford University Press.

Huws, U. (2006), 'The restructuring of global value chains and the creation of a cybertariat', in C. May (ed.), *Global corporate power*, Boulder, CO: Lynne Rienner, pp. 65–82.

Huws, U., S. Dahlmann, J. Flecker, U. Holtgrewe, A. Schönauer, M. Ramioul and K. Geurts (2009), *Value chain restructuring in Europe in a global economy*, Leuven: HIVA.

Ilsøe, A. and L. W. Madsen (2017), 'Digitalization of work and digital platforms in Denmark', FAOS Forskningsnotat 157, Copenhagen: University of Copenhagen.

Kaganer E., E. Carmel, R. Hirscheim and T. Olsen (2013), 'Managing the human cloud', *MIT Sloan Management Review*, **54** (2), 23–32.

Kässi, O. and V. Lehdonvirta (2016), 'Online Labour Index: measuring the online gig economy for policy and research', *Technological Forecasting and Social Change*, **137**, 241–248.

Lepak, D. P. and S. A. Snell (2002), 'Examining the human resource architecture: the relationships among human capital, employment, and human resource configurations', *Journal of Management*, **28** (4), 517–543.

Meil, P. (2012), 'Consent and content: effects of value chain restructuring on work and conflict among highly skilled workforces', *Work Organisation, Labour and Globalisation*, **6** (2), 8–23.

Meil, P. (2018), 'Spinning the web: the contradictions of researching and regulating digital work and labour', in P. Bilic, J. Primorac and B. Valtysson (eds), *Technologies of labour and the politics of contradiction*, Basingstoke: Palgrave Macmillan, pp. 271–290.

Meil, P. and V. Kirov (2017), 'Introduction', in P. Meil and V. Kirov (eds), *The policy implications of virtual work*, Basingstoke: Palgrave Macmillan, pp. 3–28.

Meil, P. and P. Schörpf (2017), 'Winner takes all: bidding and contesting for highly skilled work on internet platforms', Paper presented at the International Labour Process Conference, Sheffield, 4–6 April.

Mrass, V., C. Peters, and J. M. Leimeister (2019), 'Crowdworking-Plattformen und die Digitalisierung der Arbeit', in A. Boes and B. Langes (eds), *Die Cloud und der digitale Umbruch in Wirtschaft und Arbeit. Strategien, Best Practices und Gestaltungsimpulse*, Freiburg: Haufe Group, pp. 173–190.

Peters, T. and R. H. Waterman (1982), *In search of excellence*, New York: HarperBusiness Essentials.

Porter, M. (1985), *Competitive advantage: creating and sustaining superior performance*, New York: Free Press.

Schmidt, F. A. (2019), 'Crowdproduction von Trainingsdaten: Zur Rolle von Online-Arbeit beim Trainieren autonomer Fahrzeuge', Study 417, Düsseldorf: Hans-Böckler-Stiftung.

Schörpf, P., J. Flecker, A. Schönauer and H. Eichmann (2017), 'Triangular love-hate: management and control in creative crowdworking', *New Technology, Work and Employment*, **32** (1), 43–58.

Scott, W. R. and G. F. Davis (2014), *Organizations and organizing: rational, natural, and open system perspectives*, Harlow: Pearson Education.

Smith, C. (2010), 'Go with the flow: labour power mobility and labour process theory', in P. Thompson and C. Smith (eds), *Working life: renewing labour process analysis*, Basingstoke: Palgrave MacMillan, pp. 269–296.

Smith, V. (1997), 'New forms of work organization', *Annual Review of Sociology*, **23** (1), 315–339.

Smith, V. (1998), 'The fractured world of the temporary worker: power, participation, and fragmentation in the contemporary workplace', *Social Problems*, **45** (4), 411–430.

Streeck, W. (1998), 'Industrielle Beziehungen in einer internationalisierten Wirtschaft', in U. Beck (ed.), *Politik der Globalisierung*, Frankfurt: Suhrkamp, pp. 169–202.

Sturgeon, T. (2002), 'Turn-key production networks: a new American model of industrial organization', *Industrial and Corporate Change*, **11** (3), 451–496.

Sturgeon, T., T. Fredriksson and D. Korka (2017), 'The "new" digital economy and development', UNCTAD Technical Notes on ICT for Development 8, New York: UNCTAD.

Tubaro, P. and A. A. Casilli (2019), 'Micro-work, artificial intelligence and the automotive industry', *Journal of Industrial and Business Economics*, **46** (3), 333–345.

Williamson, O. E. (1985), *The economic institutions of capitalism*, New York: Free Press.

World Economic Forum (2020), *The promise of platform work: understanding the ecosystem*, Geneva: World Economic Forum.

4. Measuring the platform economy: Different approaches to estimating the size of the online platform workforce

Agnieszka Piasna

4.1 INTRODUCTION

In recent years, the emergence of online labour platforms that use digital technologies to match workers with clients on a per-task basis has sparked an intense debate about their economic and social implications. Research in this area has exploded equally rapidly, primarily in the form of qualitative or case study investigations, on the issues that most capture the imagination such as algorithmic management, extremely flexible work models, the dismantling of long fought-for worker protections, legal cases or worker struggles (e.g. Berg and De Stefano 2017; Drahokoupil and Piasna 2019; Graham et al. 2017; Vandaele et al. 2019; Wood et al. 2019). However, little is still known about the true scale of platform work, which is especially puzzling given that, as opposed to the traditional informal sector, all transactions mediated by online platforms are digitally recorded. Thus, questions on the proportion of workers engaged in platform work, whether such workers differ from the general workforce and the identification of the countries in which they are more common remain largely unanswered (Codagnone and Martens 2016; Healy et al. 2017). Existing official labour market statistics are not well suited to measuring the platform economy as they are generally not sensitive enough to catch sporadic or secondary employment or fail to distinguish it from other economic activities. Ad hoc modules to national employment surveys use different question wordings and are thus difficult to compare, while rare cross-national surveys provide such divergent results that they raise even more questions than they set out to answer (see discussion in Piasna and Drahokoupil 2019).

This chapter provides a critical assessment of different approaches to estimating the scale of the engagement in platform work of the general population. The aim is to examine the main obstacles encountered in previous studies, the reasons for surprising or contradictory results and possible sources of error, but also the lessons learned for future research. This is illustrated with key research in this area. The analysis ranges from the use of secondary data, produced in abundance by simple virtue of the operations of the platforms, to the collection of primary data through dedicated surveys. It is not an exhaustive analysis of all published studies, but rather an analytical review of various approaches illustrated with a selection of examples.

4.2 ABUNDANCE OF (HARD-TO-ACCESS) DATA

The paradox in measuring the platform economy is that, although its operations generate a wealth of data, with all transactions being digitally recorded, one of the biggest unknowns is still the scale of platform work (Codagnone et al. 2016). A good starting point for a review of methods for measuring the platform economy is thus initiatives that have attempted to access such data, either directly from the platforms or by tapping into other sources of big data generated by their operations.

Platforms are generally highly protective of their proprietary databases and thus research that uses such data is scarce. One of the early examples is a study by Hall and Krueger (2018), who used anonymized administrative data from Uber on the work histories, schedules and earnings of its drivers in the United States (US) covering the period 2012–2014. Its strength undoubtedly lies in charting in great detail the extent of work for one of the largest platforms. However, as the study was carried out at Uber's request and one of the authors worked for Uber Technologies at the time, it remains unfeasible for independent researchers to replicate such an analysis over time or in other countries. Another example of the use of administrative data is a study of Deliveroo riders in Belgium carried out by Drahokoupil and Piasna (2019). In this case, a rare opportunity to access comprehensive administrative records containing information on hours worked, earnings, age, gender and student status of workers was based on cooperation with SMart (Société mutuelle pour artistes), an additional intermediary that hired Deliveroo riders and billed the platform on their behalf. However, Deliveroo ended its agreement with SMart soon after the research was carried out, so such data collection cannot now be repeated.

Insofar as access to the administrative records of one platform provides the precise number of workers on that particular platform, and usually allows the separation of registered users from active ones, it can serve as a basis for estimates of the size of the platform economy at national level. Nevertheless, such estimates are extremely rough. A complete picture would require information from all platforms and some indication on the scale of overlap; that is, how many workers are registered on more than one platform. As this is currently unattainable, other sources of data can be used to impute missing information. Kuek et al. (2015) complemented the publicly available data disclosed by online labour platforms with expert interviews; while Harris and Krueger (2015) supplemented data from Uber on the number of workers with the frequency of Google searches for the names of selected labour platforms. Their approach rested on the assumption that the number of workers providing services through a platform is proportionate to the frequency of its Google searches, even though the latter may be driven by a variety of factors, including media interest, litigation or academic research, and are likely to be skewed in favour of the most recognized platforms. Nonetheless, Harris and Krueger's (2015) conclusion that labour platforms accounted for 0.4 per cent of total employment in the US is very close to the results from other studies of that period.

Digitally mediated transactions also leave records outside the platform, such as in financial institutions or, at least in theory, in tax records. A rare example of the use of tax returns data is a study by Collins et al. (2019), tracing independent work mediated by the 50 biggest online labour platforms in the US between 2010 and 2016. It revealed that, by 2016, about 1 per cent of the US workforce registered income from platform work, even though it could not, by design, include informal revenues and those falling below a certain threshold. An interesting illustration of the use of financial records is a report by Farrell and Greig (2016) from JPMorgan Chase Institute. Having access to a full database of the clients of a major bank in the US, they counted how many accounts received any payments from one of 30 online platforms (expanded to include 128 platforms in a follow-up study by Farrell et al. (2018)). Their analysis revealed that, by 2015, 1 per cent of adults earned income from online platforms in that month and 4.2 per cent had done so in the past three years. The clear advantage of such approaches lies in the large number of platforms that can be included in the analysis and the possibility of replicating and repeating measurements over time. However, such studies will miss payments not coming directly from platforms' accounts (i.e. through PayPal or Amazon vouchers) and, in the case of bank records, produce data not strictly at an individual level as families may have joint bank accounts, also raising ethical concerns where data are used without clients' explicit consent.

Another approach to gathering the data produced by platforms, in principle not contingent on access to exclusive sources, is web 'scraping' – automatically accessing and downloading publicly available data from the platform's web user interface. The most comprehensive initiative of this sort to date is probably the Online Labour Index (OLI) produced by the Oxford Internet Institute (Kässi and Lehdonvirta 2018), which tracks new vacancies (i.e. projects or tasks) posted on five major English-speaking online labour platforms. Continuously updated data provide a consistent time series of the number of vacancies, their occupational category and the country of the requester. However, as the OLI and other similar projects (see e.g. Ipeirotis 2010) count posted job offers and not the number of workers completing them, they might confuse an increasing fragmentation of tasks for an increase in the size of the platform workforce. Some tasks might also be completed by multiple workers. Acknowledging that this measure of online labour utilization is incomplete, the authors of the OLI choose to present it as an indexed trend rather than in absolute numbers of vacancies. Consequently, while valuable in mapping trends in online platform work and its occupational heterogeneity, the OLI does not provide answers to the scale of such work.

Therefore, the use of secondary data generated by platforms' operations seems a good way to sketch the contours of the platform economy, although it is not best suited for mapping the prevalence of platform work at an individual (worker) level. To investigate how widespread are experiences with platforms, how often and to what extent individuals engage in platform work and the role of this type of work in supporting their livelihoods, a collection of primary data is necessary. This has usually been done through social surveys.

4.3 COLLECTION OF DATA THROUGH SURVEYS

Data on the involvement of individuals in the labour market are typically derived from official labour market statistics. Ideally collected through frequent large-scale population surveys, with carefully designed methodologies that are consistent over time, they are considered the gold standard of labour market statistics. However, in principle, they risk overlooking a large chunk of platform work because of the way work and employment are defined by national statistical offices. According to the universally applied International Labour Organization guidelines, only those who worked for pay for at least one hour in the previous week (or day) are counted as employed. This definition fails to capture those who engage in platform work only sporadically, or who regularly perform platform work but who did not do so during the reference week. Moreover, the focus of most official employment statistics is the main paid job (Gazier and Babet 2018) while platform work is, in the majority of cases, a supplementary paid activity (Hall and Krueger 2018; Piasna and Drahokoupil 2019).

Nevertheless, there have been attempts to gauge, from existing official labour market statistics, the size of the platform workforce or the impact of digitalization on the labour market more broadly, for instance by looking at the extent of freelance work, solo self-employment or multiple jobholding (see e.g. the use of European Union (EU) Labour Force Survey by Eichhorst et al. 2016; Piasna and Drahokoupil 2017). While these non-standard forms of work might, to some extent, overlap with the platform economy, only dedicated questions on platform work can provide sufficiently accurate estimates.

However, national statistical offices have long been rather hesitant to include direct questions on platform work in conventional labour force surveys. This is mostly motivated by a very small target population, and thus an expected very low rate of response to these questions, as well as a lack of an agreed definition and operationalization of platform work (see e.g. ONS 2017). Another challenge faced by labour force surveys is that employed persons are assigned to sectoral and occupational classes, yet there is a lack of guidelines on where to position platform work in the ISCO and NACE systems in current usage. This would prevent platform work from being fully integrated into the existing statistical frameworks.

Despite these concerns, there have been attempts to measure platform work with dedicated surveys, or ad hoc modules to official data collections. In what follows, the three steps that need to be taken in the design of such studies and which play a major role in their success, or lack thereof, are discussed: the definition of concepts; the formulation of questions; and the selection of respondents.

4.3.1 Conceptual Clarity and Shared Definitions

Conceptual clarity and agreed definitions are the first prerequisite for a successful measurement of any concept in labour market statistics (see discussion in Piasna et al. 2017). The very slow progress in measuring the platform economy through harmonized labour force surveys at the level of the EU is a good illustration of this point: the major hurdle has been a lack of conceptual clarity as to what exactly should be measured. The European Commission (2020) continues to use the term 'online platforms' in a very broad sense, which includes search engines, social media and e-commerce. Online labour platforms, as intended in this volume, are referred to by the loose term 'collaborative economy' which includes both for-profit and not-for-profit activities encouraging meaningful peer-to-peer interactions and trust (European Commission 2016; see also Hawley 2018).

Eurostat, one of the Directorates-General of the European Commission, launched its efforts to devise a measurement of platform work starting from the same terminology and conceptual framework. In order to operationalize the fuzzy concept 'collaborative economy', a special Eurostat task force narrowed it down to digitally mediated transactions that make temporary use of the idle capacity of assets and/or labour without a change of ownership or ongoing employment relationship (Eurostat 2018). However, the use of the idle capacity of assets as a part of the definition created obvious difficulties in establishing a boundary to the collaborative economy. Drawing a distinction between freelance services provided through a platform and those obtained via traditional agencies using a digital marketplace would also be very difficult.

Given that even the experts involved in conceptual work on measuring the collaborative economy have found it very tricky to classify particular activities (Eurostat 2018; see also ONS 2017), such a conceptual framework is simply not suitable for the formulation of survey questions that would prompt respondents to classify their work appropriately.

The inclusion of normative and fairly subjective features in the definition of the collaborative economy, such as the use of idle capacity or peer-to-peer transactions, has turned out to be a significant barrier to devising official and comparative measurements of platform work at EU level. This contrasts with a burgeoning academic literature and independent research on platforms that have proposed reasonably consistent definitions and classifications (e.g. Berg et al. 2018; Bergvall-Kåreborn and Howcroft 2014; Drahokoupil and Piasna 2017; Vallas and Schor 2020; Wood et al. 2019). In the literature, there is generally little disagreement as to which intermediaries can be categorized as online platforms, with the main challenge faced by empirical attempts to measure platform work being one of how to convey these definitions to respondents and mould them into a survey question.

4.3.2 Formulating Questions

It is relatively easy to describe any given online labour platform by listing the features it shares with other platforms, such as digital intermediation, task-based work or digital payment processing. An obvious choice in formulating survey questions about platform work is thus to start from the defining features of platforms and ask respondents whether they have done work of this sort. However, this has proven remarkably difficult. Describing an online labour platform in plain language, in a concise manner and in a way that avoids any confusion with job search websites, professional social networks, online search engines or the use of information and communications technology at work in a standard job has been particularly challenging.

An approach based on formulating a direct question as to whether respondents work on online platforms was chosen, among others, in the Collaborative Economy (COLLEEM) survey, fielded in 16 EU countries (Pesole et al. 2018; Urzí Brancati et al. 2020). Respondents were asked whether they gained income from ‘providing services via online platforms, where you and the client are matched digitally, payment is conducted digitally via the platform’, further distinguishing between work that is ‘location-independent (web-based)’ and work that is ‘performed on location’ (Pesole et al. 2018, p. 14).

The risk with such questions is that they are rather complex and crammed with technical terms and thus not easy to understand for non-specialist audiences. Understanding can be improved with simpler and less specific wording, but at the expense of the precision achieved with the use of jargon. Such simpler wording can be found in the 2017 US Contingent Worker Supplement (CWS) (Current Population Survey 2018), which asked respondents about on-location platform work as follows: 'Some people find short, in-person tasks or jobs through companies that connect them directly with customers using a website or mobile app. These companies also coordinate payment for the service through the app or website. Does this describe any work you did last week?' There might be no jargon used, but the question is nonetheless rather complex and, at the same time, not very precise. A cleaner working through a traditional work agency that assigns tasks electronically could probably feel this describes their job too (for similar issues with classification, see the ad hoc module to the French Labour Force Survey in Gazier and Babet 2018). Another example is a survey commissioned by the Federal Ministry of Labour and Social Affairs in Germany, which used a long (seven sentence) definition of platform work, albeit written in accessible language and providing many examples of tasks that constitute platform work followed by the question 'Do you currently do any paid work assignments that you obtained via the Internet or an app?' (Bonin and Rinne 2017; own translation).

One way to assess whether such questions work well is to pilot test them. An extensive cognitive testing preceded the fieldwork for the 2017 CWS but did not spark major concerns. Nonetheless, the survey resulted in many incorrect 'yes' answers (Current Population Survey 2018). The questions were so complex that respondents could not understand them, often soliciting prompts from interviewers who then gave examples of tasks (driving own car, data entry, etc.). As it turned out, these were often mistaken for tasks performed in the main offline job. After careful data cleaning, only about one-quarter of positive responses were considered valid.

Similarly in the German survey, about three-quarters of responses were false positives. Respondents who acknowledged having ever done platform work, when asked to give the name of the used platform, cited websites and apps such as eBay, WhatsApp, Facebook, LinkedIn or Google, or even job offers received in an email (Bonin and Rinne 2017, pp. 26–27). A misclassification of websites and apps as platforms has surfaced also in other studies using open questions to validate self-assessment as a platform worker (e.g. Newlands et al. 2018).

In view of such difficulties, another approach is to ask respondents whether they work through a given platform citing its name. This would avoid any issues of misclassification. Starting from such a premise, Statistics Finland added to the 2017 Labour Force Survey a simply worded question: 'Have you during the past 12 months worked or otherwise earned income through the following platforms?', with multiple choice responses '1. Airbnb; 2. Uber; 3. Tori.fi/Huuto.net; 4. Solved; 5. Other (which?); 6. None of the above' (Sutela 2018). The fifth option allowed respondents to indicate the name of the company but, as in the studies described above, the most common responses were Facebook and websites advertising cars or homes for sale.

While simple, such questions would need to be adapted over time, with actors in the platform economy in a constant flux, and to particular countries, limiting the comparability of findings across time and space. Moreover, as we are interested in mapping the entire platform economy, not just one or two platforms, a survey question should ideally list as many names of platforms as possible. However, apart from creating an unfeasibly long questionnaire, such a list would hardly ever be complete.

There are alternatives that might avoid overestimation arising from an over-complex question on the one hand and a narrow focus only on a handful of platforms on the other. Katz and Krueger (2019), who conducted their own version of the US CWS in 2015, asked a series of filtering questions starting from a general 'Do you do direct selling to customers?', and then narrowed this down to selling through an intermediary and, finally, through an online intermediary, citing Uber and TaskRabbit as examples. Such a way of formulating questions gave an estimated 0.5 per cent of the US workforce engaged in the online platform economy, almost identical to the estimates of other studies (Farrell and Greig 2016; Harris and Krueger 2015).

Such an approach can be expanded by distinguishing between types of tasks. A special module of the Swiss Labour Force Survey in 2019 differentiated between four types of service with two examples of company names for each: accommodation rental; taxi services; the selling of goods; and the provision of other services (OFS 2020). Huws et al. (2019), in a survey in 13 EU countries, asked about 13 separate online activities, two of which were expected to capture platform work. They asked about 'looking for work' online, citing examples of eight online labour platforms' names (e.g. Upwork, Freelancer, Handy). Those who responded positively were then asked what type of work this was. However, it is still conceivable that 'looking for work online' could be understood as also including job search websites or any other online information gathering.

Recognizing the difficulty of distinguishing platform work from other forms of using the internet to generate income, the Internet and Platform Work Survey of the European Trade Union Institute (ETUI) (Piasna and Drahokoupil 2019) positioned the platform economy in a wider labour market and economic context. The approach adopted here is to look also at other forms of economic activity that are mediated (to a different degree) by digital technologies and which are not based on standard dependent employment contracts (see Vallas and Schor 2020).

The ETUI survey, thus far conducted in five countries in central and eastern Europe and with a second EU-wide wave planned for early 2021, first asked respondents about internet work (a broad range of paid activities that can be found or carried out online, typically on a freelance basis), and then about platform work *sensu stricto*. The question about internet work listed ten types of paid activity such as accommodation rental, taxi services and freelance information technology (IT) work, but also content production/creation and the generation of income as influencers through blogging and social media. This was followed by a question about platform work which contained a detailed description of online labour platforms using fairly simple language: ‘Online platforms are internet websites or apps through which workers can find short jobs or tasks such as IT work, data entry, delivery, driving, personal services, etc. Online platforms both connect workers with customers and arrange payment for the tasks. They usually charge a fee for each transaction’, with examples of the most recognizable platforms adapted to each country.

The relative success of these latter approaches is not straightforward to judge as there is no additional information on testing that would allow an examination of how the questionnaires worked, while their results are remarkably diverse. The survey of the ETUI (Piasna and Drahokoupil 2019) showed that between 0.4 per cent of respondents in Poland and 3.0 per cent in Hungary engaged in platform work at least monthly. Similarly, in the Swiss survey (OFS 2020), 0.4 per cent of the population reported having worked via an internet platform in the past 12 months, while only 0.2 per cent provided regular and consistent services via a platform. In contrast, the Huws et al. (2019) survey showed that, in Switzerland, as much as 12.7 per cent of the population engaged in platform work on a monthly basis. In other countries the results were also generally very high, up to a whopping 33.9 per cent in Czechia (Huws et al. 2019).

While we cannot dismiss the role of variations in the questionnaire, the method of selecting respondents to take part in the survey is a more likely explanation of these disparities in results.

4.3.3 Sampling of Respondents

In studies that focus on platform workers and which set out to examine their working conditions or pay, the aim is to get good representation *within* the group of platform workers. The sample can then consist entirely of those who work on platforms and the challenge, therefore, is to recruit as many of them as possible to the study. An essentially different approach needs to be taken when the aim is to map the true prevalence of platform work in the population as a whole. Then the sample of respondents should be as close a representation of the general population as possible, with everyone having equal (and known) chances of being selected in a process called random probability sampling (Groves 1989).

Random probability sampling offers the most inclusive, robust and representative methodology and has been used in the collection of official labour market statistics, including in the ad hoc modules on the platform economy in the Finnish (Statistics Finland 2018), French (Gazier and Babet 2018) and Swiss (OFS 2020) labour force surveys. It has also been used to select a nationally representative sample for the CWS in the US (Current Population Survey 2018), as well as in two big survey projects in the United Kingdom (UK) (Balaram et al. 2017; Lepanjuuri et al. 2018). Remarkably, the ETUI's Internet and Platform Work Survey (Piasna and Drahokoupil 2019) is the only study thus far that has used random probability sampling in a comparative cross-national analysis of platform work. What all these studies have in common is quite consistent results that place the prevalence of platform work generally around 0.5–5.0 per cent of the adult population. However, such studies are also relatively costly and are thus not likely to be undertaken frequently or on a large cross-national scale.

This has prompted the use of other, essentially cheaper and faster, methods of data collection. Among them, non-probability online samples have been used most frequently (e.g. Huws et al. 2019; Katz and Krueger 2019; Newlands et al. 2018; Pesole et al. 2018). However, these suffer from a number of limitations and inherent biases and require a thorough methodological understanding to tackle likely sources of error (see e.g. Lehdonvirta et al. 2020).

In general, as long as internet access and use are not universal, online surveys will exclude those demographics with limited internet access (or very infrequent internet use) and which, at the same time, are less likely to engage in platform work. This will lead to an overestimation of the prevalence of platform work. There have been attempts to correct for such bias. For instance, the COLLEEM team used quota-stratified sampling and post-stratification weights based on age, gender, education and the proportion of frequent internet users in a country (Urzí Brancati et al. 2020). Indeed, Schneider and Harknett (2019), having compared non-probability web-based surveys carried out on Facebook with standard probability samples, concluded that post-stratification weighting on the basis of demographic characteristics yielded fairly accurate results.

Nevertheless, such adjustments do not address the major methodological weakness of existing online surveys, namely that respondents are usually recruited via commercial polling panels and participate in such surveys in return for compensation on a per-task basis and without any formal employment contract. In such cases, completing the online survey can itself be considered an example of online gig work with the whole sample of respondents consisting of online gig workers (see discussion in Piasna and Drahokoupil 2019). This will result in an over-claim of the extent of platform work.

Moreover, online paid surveys may represent certain types of platform work more than others, especially those performed entirely online and involving microtasking. Workers performing such tasks via platforms are much more likely to come across paid online surveys whereas those whose services are generally performed offline, or highly skilled professionals working on bigger projects, have very low chances of being recruited to such surveys (see also Vallas and Schor 2020).

The reliability issues with online paid surveys can be illustrated with the example of the COLLEEM and Huws et al. studies. Both these surveys collected data through the Cint network that relies on commercial panels of self-selected respondents who typically receive some type of reward, including cash payments, for completing various online surveys. In Czechia, despite almost identical methodologies involving the use of Cint panels and a coincident timing of data collection, Huws et al. found that 28.5 per cent of adults work on platforms on a weekly basis whereas the COLLEEM survey reported that only 5.9 per cent of Czechs have ever tried platform work (Huws et al. 2019; Urzí Brancati et al. 2020). Such discrepancies in the results can be attributed to the methodological problems that come with a reliance on opt-in (self-selected) online samples as well as to different approaches to data cleaning and weighting.

This does not mean that online surveys cannot provide a good representation of all internet users, or even the whole population, in a given country. Huws et al. (2019, pp. 50–51) compared the responses to their online questionnaire with offline surveys administered in the UK (face to face) and in Switzerland (telephone), concluding that the results were broadly consistent. The use of representative sampling frames, instead of the opt-in and uncontrolled recruitment of participants online, can be a step in the right direction, as can complementing online surveys with telephone or face-to-face interviews among non-internet users or others not able or willing to read questions on a screen.

4.4 CONCLUSIONS

The potential transformation of labour markets by the emergence of online labour platforms has triggered an intense academic, media and policy debate, but its true scale remains speculation. Previous studies reviewed in this chapter, mapping the extent of platform work, certainly provide us with many important lessons to guide future research. First, experimenting with new methods of data collection, mostly in the US, has provided interesting and largely consistent results (compare Farrell and Greig 2016; Harris and Krueger 2015; Katz and Krueger 2019) and further efforts in this direction should be encouraged also in other countries. Second, independent and ad hoc surveys provide a vast library of questions, with some indication of their reliability, that can be further used and tested by other researchers and in other countries. Nevertheless, a patchwork of statistics from singular surveys is certainly insufficient to inform and guide policy. A practical way forward would be the development of a harmonized instrument to be implemented in official, regularly repeated labour force surveys with unmatched sample sizes.

Official labour market statistics have been very slow to devise a measurement of platform work and have been additionally hampered by a lack of agreed definitions and clear methodological guidance from international institutions. A fruitful avenue for future labour market statistics would be to measure platform work within the broader framework of precarious and casual work, where the only difference is technological intermediation (Aleksynska et al. 2019; Berg and De Stefano 2017). An organizing framework for statistics classified by status at work, put forward during the last International Conference of Labour Statisticians (ILO 2018), can be seen as a step in this direction (although its proposal for a new and refined classification of various forms of non-standard work still left platform workers grouped within a broader category of dependent contractors).

Adopting such an approach could also address the broader need for an improved measurement of casual work, capturing the complexity of subcontracting, freelance contracting, multiple jobholding and precarious forms of work such as day labour. This will only grow in importance with the labour market crisis created by the Covid-19 pandemic. With technologically enabled remote work, growing demand for services such as food delivery or care, as well as rising unemployment, platform work may resume its rapid growth. Adequate policy responses to this will hinge on a good understanding of these dynamics.

REFERENCES

Aleksynska, M., A. Bastrakova and N. Kharchenko (2019), 'Working conditions on digital labour platforms: evidence from a leading labour supply economy', Discussion Paper 12245, Bonn: IZA.

Balaram, B., J. Warden and F. Wallace-Stephens (2017), *Good gigs: a fairer future for the UK's gig economy*, London: RSA.

Berg, J. and V. De Stefano (2017), 'It's time to regulate the gig economy', *Open Democracy*, 18 April, accessed 5 October 2020 at www.opendemocracy.net/en/beyond-trafficking-and-slavery/it-s-time-to-regulate-gig-economy/.

Berg, J., M. Furrer, E. Harmon, U. Rani and S. Silberman (2018), *Digital labour platforms and the future of work: towards decent work in the online world*, Geneva: ILO.

Bergvall-Kåreborn, B. and D. Howcroft (2014), 'Amazon Mechanical Turk and the commodification of labour', *New Technology, Work and Employment*, **29** (3), 213–223.

Bonin, H. and U. Rinne (2017), 'Omnibusbefragung zur Verbesserung der Datenlage neuer Beschäftigungsformen', Research Report 80, Bonn: IZA.

Codagnone, C. and B. Martens (2016), 'Scoping the sharing economy: origins, definitions, impact and regulatory issues', Institute for Prospective Technological Studies Digital Economy Working Paper 2016/01, JRC100369, European Union.

Codagnone, C., F. Biagi and F. Abadie (2016), *The passions and the interests: unpacking the 'sharing economy'*, JRC Science for Policy Report EUR 27914 EN, Luxembourg: Publications Office of the European Union.
Collins, B., A. Garin, E. Jackson, D. Koustas and M. Paynek (2019), *Is gig work replacing traditional employment? Evidence from two decades of tax returns*, accessed 5 October 2020 at www.irs.gov/pub/irs-soi/19rpgigworkreplacingtraditional employment.pdf.
Current Population Survey (2018), 'Electronically mediated work: new questions in the Contingent Worker Supplement', *Monthly Labor Review*, accessed 6 October 2020 at www.bls.gov/opub/mlr/2018/article/electronically-mediated-work-new-questions-in-the-contingent-worker-supplement.htm.
Drahokoupil, J. and A Piasna (2017), 'Work in the platform economy: beyond lower transaction costs', *Intereconomics: Review of European Economic Policy*, **52** (6), 335–340.
Drahokoupil, J. and A. Piasna (2019), 'Work in the platform economy: Deliveroo riders in Belgium and the SMart arrangement', Working Paper 2019.01, Brussels: ETUI.
Eichhorst, W., H. Hinte, U. Rinne and V. Tobsch (2016), 'How big is the gig? Assessing the preliminary evidence on the effects of digitalization on the labor market', Policy Paper 117, Bonn: IZA.
European Commission (2016), *Communication from the Commission to the European Parliament, the Council, the European Economic and Social Committee and the Committee of the Regions. A European agenda for the collaborative economy, COM (2016) 356 final, 2 June 2016*, Brussels: European Commission.
European Commission (2020), 'EU Observatory on the Online Platform Economy', accessed 6 October 2020 at https://ec.europa.eu/digital-single-market/en/eu-observatory-online-platform-economy.
Eurostat (2018), *Note on Measuring the Digital Collaborative Economy*, Eurostat Workshop Luxembourg, 29–30 May.
Farrell, D. and F. Greig (2016), *Paychecks, paydays, and the online platform economy: big data on income volatility*, JPMorgan Chase & Co. Institute, accessed 6 October 2020 at https://ssrn.com/abstract=2911293.
Farrell, D., F. Greig and A. Hamoudi (2018), *The online platform economy in 2018: drivers, workers, sellers, and lessors*, JPMorgan Chase & Co. Institute, accessed 6 October 2020 at https://ssrn.com/abstract=3252994.
Gazier, B. and D. Babet (2018), 'Nouvelles formes d'emploi liées au numérique et mesure de l'emploi, Colloque du CNIS: L'économie numérique: enjeux pour la statistique publique', Paris, 7 March.
Graham, M., I. Hjorthand V. Lehdonvirta (2017), 'Digital labour and development: impacts of global digital labour platforms and the gig economy on worker livelihoods', *Transfer*, **23** (2), 135–162.
Groves, R. M. (1989), *Survey errors and survey costs,* Hoboken, NJ: John Wiley & Sons.
Hall, J. V. and A. B. Krueger (2018), 'An analysis of the labor market for Uber's driver-partners in the United States', *ILR Review*, **71** (3), 705–732.
Harris, S. D. and A. B. Krueger (2015), 'A proposal for modernizing labor laws for twenty-first-century work: the "independent worker"', Hamilton Project Discussion Paper 2015-10, Washington, DC: Brookings Institute.
Hawley, A. J. (2018), 'Regulating labour platforms, the data deficit', *European Journal of Government and Economics*, **7** (1), 5–23.

Healy, J., D. Nicholson and A. Pekarek (2017), 'Should we take the gig economy seriously?', *Labour and Industry: A Journal of the Social and Economic Relations of Work*, **27** (3), 232–248.

Huws, U., N. H. Spencer and M. Coates (2019), *The platformisation of work in Europe: results from research in 13 European countries*, Brussels: FEPS.

ILO (2018), *Statistics on work relationships, ICLS/20/2018/2*, Geneva: ILO.

Ipeirotis, P. G. (2010), 'Analyzing the Amazon Mechanical Turk marketplace', *Crossroads*, **17** (2), 16–21.

Kässi, O. and V. Lehdonvirta (2018), 'Online labour index: measuring the online gig economy for policy and research', *Technological Forecasting and Social Change*, **137**, 241–248.

Katz, L. F. and A. B. Krueger (2019), 'The rise and nature of alternative work arrangements in the United States, 1995–2015', *ILR Review*, **72** (2), 382–416.

Kuek, S. C., C. M. Paradi-Guilford, T. Fayomi, S. Imaizumi, P. Ipeirotis, P. Pina and M. Singh (2015), *The global opportunity in online outsourcing*, Washington, DC: World Bank.

Lehdonvirta, V., A. Oksanen, P. Räsänen and G. Blank (2020), 'Social media, web, and panel surveys: using non-probability samples in social and policy research', *Policy and Internet*, accessed 6 October 2020 at doi:10.1002/poi3.238.

Lepanjuuri, K., R. Wishart and P. Cornick (2018), *The characteristics of those in the gig economy*, London: Department for Business, Energy and Industrial Strategy.

Newlands, G., C. Lutz and C. Fieseler (2018), 'Collective action and provider classification in the sharing economy', *New Technology, Work and Employment*, **33** (3), 250–267.

OFS (2020), *Travailleurs des plateformes numériques en 2019*, Neuchâtel: Office Fédéral de la Statistique.

ONS (2017), *The feasibility of measuring the sharing economy: November 2017 progress update*, London: Office for National Statistics.

Pesole, A., M. C. Urzí Brancati, E. Fernández-Macías, F. Biagi and I. González Vázquez (2018), *Platform workers in Europe: evidence from the COLLEEM Survey. EUR29275 EN*, Luxembourg: Publications Office of the European Union.

Piasna, A. and J. Drahokoupil (2017), 'Gender inequalities in the new world of work', *Transfer*, **23** (3), 313–332.

Piasna, A. and J. Drahokoupil (2019), 'Digital labour in central and eastern Europe: evidence from the ETUI Internet and Platform Work Survey', Working Paper 2019.12, Brussels: ETUI.

Piasna, A., B. Burchell, K. Sehnbruch and N. Agloni (2017), 'Job quality: conceptual and methodological challenges for comparative analysis', in D. Grimshaw, C. Fagan, G. Hebson and I. Tavora (eds), *Making work more equal: a new labour market segmentation approach*, Manchester: Manchester University Press, pp. 168–187.

Schneider, D. and K. Harknett (2019), 'What's to like? Facebook as a tool for survey data collection', *Sociological Methods and Research*, accessed 6 October 2020 at https://doi.org/10.1177/0049124119882477.

Statistics Finland (2018), *Labour Force Survey 2017: platform jobs*, Helsinki, accessed 6 October 2020 at www.stat.fi/til/tyti/2017/14/tyti_2017_14_2018-04-17_tie_001_en.html.

Sutela, H. (2018), 'Platform jobs are here to stay–how to measure them?', Statistics Finland, 17 April, accessed 6 October 2020 at www.stat.fi/tietotrendit/blogit/2018/platform-jobs-are-here-to-stay-how-to-measure-them/.

Urzí Brancati, M. C., A. Pesole and E. Férnandéz-Macías (2020), *New evidence on platform workers in Europe: results from the second COLLEEM survey*, Luxembourg: Publications Office of the European Union.
Vallas, S. and J. B. Schor (2020), 'What do platforms do? Understanding the gig economy', *Annual Review of Sociology*, **46**, accessed 6 October 2020 at https://doi.org/10.1146/annurev-soc-121919-054857.
Vandaele, K., A. Piasna and J. Drahokoupil (2019), '"Algorithm breakers" are not a different "species": attitudes towards trade unions of Deliveroo riders in Belgium', Working Paper 2019.06, Brussels: ETUI.
Wood, A. J., M. Graham, V. Lehdonvirta and I. Hjorth (2019), 'Good gig, bad gig: autonomy and algorithmic control in the global gig economy', *Work, Employment and Society*, **33** (1), 56–75.

5. A historical perspective on the drivers of digital labour platforms

Gérard Valenduc

5.1 INTRODUCTION

To what extent are the organizational work practices of digital labour platforms really new and to what extent do they differ from continuous changes in work organization enabled by information and communications technologies (ICT)? The development of flexible work practices and atypical work contracts linked to the diffusion of ICT were already on the research agenda more than 30 years ago. Digital platforms can be analysed both as a continuing trend in the expansion of internet-based online services, starting in the 1990s, and as the disruptive emergence of new business models at the beginning of the 2010s. Elements of continuity and disruption are intertwined.

This chapter considers two successive periods. The first can be characterized as the expansion of online services; it covers the 'long nineties' from the preparation of the European Single Market at the very end of the 1980s to the bursting of the dot-com bubble in 2000–2001. The second period covers the shift from online services to digital platforms, from 2008 onwards, following the accelerated spread of smartphones and the increasing power of the major current players in the digital economy. In each period, the key drivers of innovation, both institutional and technological, are analysed followed by the new features of work organization. While the first period was shaped by deregulation and liberalization, the second period highlights the drawbacks of these policies and calls for new models of regulation and public policy. New work forms which emerged in the first period have been continuously expanding during the second period. However, the economic models of former online services and current platforms are quite different.

5.2 ONLINE SERVICES AND FLEXIBLE WORK FORMS DURING THE 'LONG NINETIES'

The expression 'long nineties' covers a slightly longer period than the 1990s decade; that is, from the end of the 1980s to the beginning of the 2000s. This period was characterized by intensive institutional and technological changes which shaped the development of the service economy and gave a decisive impetus to online services. In western countries, the decade following the end of the Fordist 'golden age' of 1945–1975 was marked by a strong decline in employment in manufacturing and a rapid growth in services. The development of the latter was enabled by the widespread use of personal computers and data communications systems from the beginning of the 1980s.

5.2.1 Institutional and Political Drivers of Innovation in Online Services

The European case highlights the role of policies and institutions in the development of the 'information society'. Among the aims of the Single European Act, which came into force in July 1987 as a follow-up to Jacques Delors's White Paper, a high priority was given to the removal of the technical and regulatory barriers to an open market for telecommunications services. Telecommunications were formerly the subject of state monopolies and administrative regulation. The completion of the European Single Market in 1992 required a series of structural changes in this sector from both technical and regulatory aspects: the end of monopolies; opening up to international competition; common standards for the digitalization of networks and data transmission; and the interoperability of value-added services. Between 1987 and 1992, the European telecommunications landscape radically changed. Former monopolies were either transformed into state-owned companies or privatized. Significant European Union (EU) investments were made in ICT research, also covering the socioeconomic dimensions of the information society (Guzzetti 1995).

Two key policy initiatives also appeared in the 1990s. In the United States (US), Al Gore's report 'The national information infrastructure: an agenda for action' (1993) paved the way for public and private investments in information highways and the development of online digital services. In the EU, the Bangemann report 'Europe and the global information society' (1994) was a decisive step in the liberalization of telecommunications markets. The implementation of a European-wide market for all kinds of online services – business to business and business to consumer – became a top priority on the political agenda. The European directive organizing this open market was adopted in 1996 and in force from early 1998 (Musso 2008). The implementation of the Global System for Mobile Communications (GSM) network (1994) and the opening of internet access to commercial companies and the general public (1993–1995) happened in this context of accelerated liberalization. Eastern and central European countries experienced a direct transition from bureaucratic and state-controlled telecommunications systems to liberalized and privatized systems.

Beyond several well-known consequences – like the end of monopolies and administrative regulations, partial or full privatization, independent regulatory bodies, increasing international competition and changes in workers' status and job stability in the telecoms sector – the liberalization of telecommunications also enabled the international development of devices and services which were previously limited by administrative barriers, such as automated call distribution systems, computer-assisted telephony and call centres. Although technically available, call centre systems could not have become widespread without a new regulation of telecommunications, as call diversion was considered illegal under monopolistic regulation. Within a few years, call centres had become a cornerstone in the development of online services.

The development of online services raises new questions related to free trade in services as far as online services can be produced, managed, provided and consumed in different countries subject to different national regulations. The project of the liberalization of service provision, covering a wider scope than online services, was launched in 2000 together with the Lisbon strategy. The first Commission proposal, known as the 'Bolkestein directive', raised huge political controversies and strong opposition from trade unions and non-governmental organizations. After long negotiations and compromises, the revised version of the directive on service provision was adopted in 2006 and came into force in 2009. A set of public or social services, including education and labour market intermediation, were made subject to limitations on free trade.

5.2.2 Technological Drivers for Innovation in Online Services

Several technological innovations converged towards the development of online services during the long nineties. The full digitalization of telecommunications networks started in the mid-1980s and required a period of about 10–15 years before completion depending on the size of the country, the quality of the existing analogue networks and the investment capacities of state operators. This was a pre-condition for the implementation of end-to-end digital services. In the area of mobile telecommunications, a decisive step was passed in 1991 with the adoption by the EU of the GSM standard, developed in the labs of the International Telecommunications Union in Geneva and experimented on in the Nordic countries. The digital GSM standard immediately replaced former national systems, which were not interoperable, and spread around the world. Still in the area of mobile communications, the US government allowed in 1996 the 'dual use' (that is, for both civilian and military purposes) of the US Army's Global Positioning System, opening the way for geolocation in business, transport and leisure (Musso 2008).

In parallel, the opening of the internet in the public domain and to commercial activities, after 1993, inaugurated a new period in the development of online services. In the first place internet services were mainly information services with a low level of interactivity. Using the principles of the World Wide Web, designed in Geneva in 1993, the first browsers were developed in 1996 and became increasingly popular. At the same time, new infrastructures for digital data communications (asymmetric digital subscriber line) were progressively implemented in European networks, allowing for higher data transfer rates. The potential of web services started to be harnessed.

5.2.3 Work Organization Patterns Shaped by Information Communications Technology in Online Services

The shift towards online services entailed many changes in work organization in services industries. During the long nineties several sectors were extended from face-to-face towards online service provision. Banking and insurance, real estate, travel and accommodation, retail trade, education and ticketing for cultural events are typical examples of the services which were concerned with this shift. Either such services became directly accessible for consumers through websites or call centres, or face-to-face service provision became 'telemediated' through computers and networks.

There are three main changes which will have further impacts on the development of the platform economy: the extension of the market sphere in services; the rise of a '24/7' or 'round the clock' society; and the increasing enrolment of the customer in service provision.

1. ICT makes services increasingly tradable as they can be produced in one location and consumed elsewhere. Production and consumption may happen at different times and different places, such as for material goods, leading to an industrialization of services. Not only are market services concerned, but also non-profit ones such as education and training, public administration and social assistance. The online model is progressively shaping all domains of service provision and consumption.
2. Through the dissociation of the production and consumption of services, and their increasing tradability, continuous availability has become a key factor in competition and non-profit services have been forced to follow

this trend. Call centres play an important part in the development of this extended availability, 24 hours a day and seven days a week. Service provision has been externalized from core companies to call centre companies located in different countries. Jobs have consequently been transferred from core companies to subcontracting call centres, covered by lower-level collective agreements and offering lower wages.

3. Online services rely on an increasing involvement of customers in the service process. In banking, insurance and travel, as well as in public administration, customers are pushed to encode their data themselves, check validity, ensure data security, manage outcomes and archive their files (Dujarier 2014). Several tasks have been shifting from workers to consumers.

Workers have been deeply affected by this shift towards online service provision. They are required to be flexible, i.e. to carry the adjustment to market strategies and technological hype themselves (Gadrey 2003). Flexibility has become a key word in human resource management – and a matter for social conflict and bargaining in many service enterprises – whereas flexibility issues were formerly concentrated in the manufacturing sector.

A study carried out for the European Parliament at the end of the 1990s (Gillespie et al. 1999) on 'technology-induced atypical work forms' highlighted the role of ICT as an enabling factor in flexible work patterns. In his influential book *The Rise of the Network Society*, Castells (1996) analysed this enabling role of ICT. Although organizational changes are implemented independently of technology, in order to cope with a constantly changing operational and economic environment, Castells suggests that, once these changes start to take place, the feasibility of organizational change is 'extraordinarily enhanced' by ICT. During this period, an abundant literature started to question the technological determinism in the relation between ICT and work organization (Orlikowski 1992; Kling and Lamb 1999) and to highlight that, although designers included an 'implicit organizational design' in technology, it is shaped by social relations (Alsène 1990).

Indeed, this period was characterized by an increasing diversification of flexible work forms in which ICT had played an enabling role. More than just information and communications, ICT is, first of all, a set of technologies for the management of time and space. Therefore, basic concepts of working time and work location are deeply challenged by ICT.

Table 5.1 summarizes the key trends in the development of flexible work forms in four areas: working time; work location; work contracts; and subordination links. During the 1990s particular emphasis was put on spatial flexibility and distance working, with a strong focus on telework or 'e-work'. Working time and work contracts also became subject to new trends, for instance the development of on-call work, zero-hours contracts and online standby. Among other dimensions of ICT-enabled flexible work practices, particular attention was also paid to new forms of subordination links: secondment; detachment; freelancing; subcontracting to the self-employed; and a work status existing in-between employment and self-employment (Serrano Pascual and Crespo 2001).

It appears clearly that some features of work organization in the current platform economy are rooted in the organizational changes summarized in Table 5.1. On-call working, remote co-working, task-based contracts, piecework, agency work, freelance work and the blurring of subordination links were emerging trends at that time while some of them are even older (Stanford 2017).

Table 5.1 Information communication technology-enabled flexible work forms

Working time	**Work location**
Flexible work forms	*Flexible work forms*
Variable part-time working	Mobile working, itinerant work
Flexi-time working	Remote office working, remote call centres
Shift work (twilight or night)	Homeworking, telecommuting
Variable overtime working	Working in telecentres or co-working spaces
On-call working	Remote computer-supported teamwork
Online stand-by	
Enabling roles of ICT	*Enabling roles of ICT*
Extending services accessibility, lengthening the working day	Extension and diversification of distance working and itinerant working
Tuned management of task flows and quantitative manpower needs	Remote organization and planning of project work and management by objectives
Just-in-time production in services	Ubiquitous work
Work contracts	**Subordination links**
Flexible work forms	*Flexible work forms*
Fixed-term, temporary or seasonal contracts	Working for an agency
Contracts with annualized working hours	Self-employed subcontractor, freelance
Zero-hours contracts (without fixed volume)	Employed by third-party supplier
Piecework	Work contract transferred to third-party supplier
Performance-related pay	Working for several employers
Task-based contracts	Franchising
Enabling roles of ICT	*Enabling roles of ICT*
'Just-in-case' manpower management	Support for subcontracting and externalization
Modelling and planning of atypical work contract management	Coordination of remote independent subcontractors
Electronic performance monitoring coupled to performance-related pay	Co-working platforms

Source: Adapted from Valenduc and Vendramin (2002), pp. 186–187.

5.2.4 Call Centres: Precursors of Platforms?

Call centres were mentioned above as a cornerstone in the development of online services. There are different categories of call centre. Inbound call centres manage incoming calls and workers have to provide responses to customers: help desk; after-sales; assistance; and e-commerce. Outbound call centres initiate phone calls: surveys; marketing campaigns; and advertisements. Several studies highlight a clear trend of the externalization and subcontracting of the business functions supported by call centres (Flecker 2009; Doellgast and Pannini 2015). During the 1990s, many inbound call centres evolved towards contact centres combining phone, email and chat, harnessing a more efficient integration of computer-assisted telephony, internet and databases.

Work in call centres is given particular attention in socioeconomic research for at least three reasons. Firstly, it is considered as illustrative of ICT-enabled flexible work forms. Although case studies of call centres observe some variations in work organization patterns, most of them underline that call centres are a laboratory for flexible work practices (Pichault and Zune 2001; Greenan et al. 2009). Concerning work location, the externalization of call centres entails a relocation of service provision and often offshoring. Wage differentials are a key driver: not only lower wage levels in destination countries but also wage differentials in the same country; and between the provisions of collective agreements in core and outsourced firms. On the issue of working time, most call centres operate with extended working hours, including evenings, nights and weekends. Regarding work contracts, a high proportion of call centre employees have temporary or agency contracts, some of them being zero hours (Doellgast and Pannini 2015). In connection with subordination links, externalized call centres are typical examples of a triangular employment relationship in which workers are confronted with the requirements both of their employer and of client companies. Secondly, work organization and work content are strongly prescribed and formatted by software tools: there is a standardization and fragmentation of tasks; tasks are prescribed by scripts; and there is online performance monitoring and continuous control over work methods and procedures. Workers have to follow the procedures imposed by algorithms and feed databases of frequently asked questions (Greenan et al. 2009). Thirdly, the expansion of externalized, subcontracted and offshored call centres reflects the growing globalization of service industries and the impacts that this has had on workers (Flecker 2009).

Many services (for example, travel and accommodation, delivery and labour market intermediation) which are nowadays provided by platforms were provided through call centres and contact centres during the long nineties. Work organization in call centres prefigured a series of features of work organization in platform-based services: remote work; the adjustment of working time to customer expectations; fragmented employment; blurred subordination links; the prescription of tasks by software; intrusive supervision through computers and networks; the enrolment of the customer in the service provision process; and the management of local services on a global scale.

5.3 FROM ONLINE SERVICES TO DIGITAL PLATFORMS

The second period considered in this chapter starts from the worldwide financial crisis in 2008 and the eruption of the smartphone into the market of end-user technologies in 2007–2008. How and why are these two distinct events related and with what consequences in the area of work organization?

5.3.1 Institutional and Economic Drivers Leading to the Platform Economy

Scholars in evolutionary economics highlight that the financial crisis of 2008 cannot be attributed solely to the collapse of the sovereign debt and mortgage lending markets; it was also the outcome of the unfettered digitalization of the financial system (Perez 2013, 2016). The first warning shots were fired when the speculative dot-com bubble burst in 2000–2001. However, the 2008 crisis marked a turning point in the development and implementation of the techno-economic paradigm, or 'great surge', based on digital technologies: between the period of installation, from the emergence of ICT to the frenzy for digital networks in the long nineties; and the period of deployment, starting after the crisis and requiring a deep institutional reconfiguration (Perez 2013; Valenduc 2018). Other authors refer to this turning point as the transition from the 'information age' to the 'internet age' (Huws 2013; Veltz 2017), occurring between 2000 and 2010 and heralding an explosion in virtual work which, in this context, refers to online and networked tasks which are easy to outsource at global level.

This turning point was marked by tensions and conflicts between the logic of liberalization and deregulation which characterized the deployment period of the paradigm; and the need for regulation and new roles for public institutions required by the development period. The advance of the platform economy is a typical case of such tensions and conflicts. In the European context, platform owners systematically enter into conflict with the regulatory system in the areas of labour law, professional regulation, privacy protection, commercial regulations and so on.

Another significant economic driver was the development of two-sided markets, according to a recent theoretical model that appears particularly relevant to the digital economy (Wauthy 2008; Tirole 2016; Veltz 2017, pp. 45–48; see also Chapter 2 in this volume). In this model, the platform owner delivers products or services to two distinct user groups at the same time – the two sides of the market – through an online system. One side of the market is made up of consumers who benefit from access to low-cost or free services and positive network externalities, since services become more attractive as user numbers grow; by accessing these services, however, and whether they realize it or not, they are supplying the platform with sets of data on their personal profile, location and consumer habits. The other side of the market comprises economic players which are involved in the provision of platform-based services and which also benefit from positive network externalities in proportion to the size of the consumer base. The value of a service for actors on one side of the market is correlated to the number and the quality of actors on the other. Economists refer to such phenomena as 'cross network externalities' and regard them as a typical feature of two-sided markets. Platforms of this kind are funded by levies on the transactions between the two sides of the market, but the information which is collected is valuable to actors on both sides, representing not only a source of data but also a body of knowledge. The platform itself is, therefore, the primary location of value creation for both sides.

Examples of platforms which correspond to this description include Google, Booking.com, Uber, Amazon, Facebook and many others (Casilli 2019, pp. 63–91). However, online platforms have also developed in the labour market as intermediaries in the demand and supply of fragmented or freelance work.

Finally, another decisive change since 2008 has been the increasing power of the new major players in the global digital economy; that is, the so-called GAFA (Google, Amazon, Facebook, Apple). This raises increasing challenges for the European economy and for EU policies (Finck 2017).

5.3.2 Technological Drivers Leading to the Platform Economy

It is worthwhile questioning the effective novelty of a series of innovations that arrived on the market in the second part of the 2000s. Several authors have pointed to a threshold effect for digital technologies, in connection with their exponential leaps in performance: 'little by little, then all at once' (Brynjolfsson and McAfee 2014); whereas others wonder if leaps in performance are likely to trigger effective changes in the workplace (Holtgrewe 2014).

Six key innovation areas can be identified regarding their potential impact on workplaces and jobs: cloud storage and cloud computing; big data and algorithms; smartphones and mobile apps; geolocation of people and objects; the Internet of Things and connected objects; deep learning and artificial intelligence systems (Valenduc and Vendramin 2016). The cloud is both a driver for the growth of all forms of remote and virtual work as well as a tool for implementing outsourcing and offshoring strategies. Big data collection and analysis expand the possibilities of surveillance, monitoring and tracking. The use of consumer-generated big data is also transforming work practices in the fields of commerce, marketing and financial services, and indeed all client-facing activities. Geolocation has a major impact in terms of the planning, monitoring and tracking of mobile workers in delivery, maintenance and repair, as well as in inspection operations, transport and logistics. Artificial intelligence systems raise issues of competition or complementarity between human capabilities and machine performance.

When taken separately, each of these areas of innovation can be analysed as a mix of elements of continuity (incremental innovations) and factors of change (radical innovations), given the different pace of the diffusion of innovations into economy and society. The effective novelty comes from the extended possibilities provided by the combination and recombination of these six areas. The development of the platform economy intensively harnesses the potential of this new set of innovations.

5.3.3 New Forms of Work in the Platform Economy

Key aspects of platform work such as workers' status, working conditions, the risks of precariousness, workers' rights, challenges for collective bargaining and the need for renewed labour regulation are extensively developed here in this volume. They are inherited from ICT-enabled flexible work forms during the long nineties but have been enhanced in the platform economy (Stanford 2017; Valenduc and Vendramin 2017) and sometimes 'relabelled'. For example, the word 'gig' was rarely used previously to describe fragmented and precarious work, but it is often used nowadays. Crowdworking and platform-based on-demand work can be characterized as virtual work, but the latter also includes any kind of ICT-mediated mobile work. A broader concept, virtual work thus points to global outsourcing and the fragmentation processes of dematerialized tasks that can be carried out anytime, anywhere (Huws 2013). It is linked to the processes of globalization and the restructuring of value chains.

The focus will be put here on three new concepts which are typical of the platform economy: digital labour; 'prosumer' work; and open work.

The concept of digital labour covers a continuous spectrum between non-remunerated online work, underpaid or precarious activities such as for platforms, casual work and zero-hours contracts, and flexible jobs in outsourced tasks, at local or global level (Casilli 2019). Casilli distinguishes three types of digital labour. The first is on-demand work, organized through digital platforms. Uber, Deliveroo or Task Rabbit belong to this category. On-demand digital labour amplifies several flexible work practices which emerged during the long nineties: contractual precariousness, unpredictable working time and tight and permanent supervision, often by means of algorithms and geolocation. In the platform model, workers are not only delivering a service but also feeding databases and algorithms.

The second category consists of micro-work platforms. Such platforms, like Amazon Mechanical Turk, establish a global marketplace for micro-tasks in areas including web development, design, software, image recognition, data replication, translation, audio transcription and data mining. They result from a process of virtual Taylorism, separating design from execution and fragmenting the work process into elementary units that can be outsourced to the global crowd. Tasks can be outsourced to anybody, amateur or professional, without any control over skills; only the rating is relevant. Work is turned into drudge. Casilli emphasizes the role of 'click workers', frequently located in low-pay countries and carrying out all the tasks that bots, apps or artificial intelligence systems are not able to do accurately; in this case, humans replacing robots. The development of an offshored, low-paid and precarious manpower reserve was already underlined some years ago by Huws (2003) when she introduced the word 'cybertariat' as a portmanteau drawn from cyber and proletariat.

These two categories, on-demand work and micro-tasks, fall under crowdworking or crowdsourcing, but the latter also encompasses global platforms for freelance work, addressing freelance professionals who must be accredited as such by the platforms. Some platforms are marketplaces for freelance tasks on which prices and deadlines can be negotiated; while others play a role of intermediary in the labour market for self-employed professionals, like an agency. The border between freelance crowdworking and micro-work is, however, thin as the outsourcing process often leads to an increasing fragmentation of work.

The third category refers to the participation of users in producing and sharing information in social media or similar settings. Any kind of user-generated content contributes to the process of value creation on the platform. As such tasks are often compared to leisure, sociability, creativity or fun, they partially overlap the concepts of 'prosumer' or open work.[1]

The concept of prosumer work is linked to the business model of digital platforms, refreshing the role of individuals who both produce and consume digitized information: prosumers. Although rarely paid, prosumers carry out work by supplying data and services for which employees were previously at least partly responsible, such as reviews of services or products, the rating of services and data entry. In a less explicit way, a wide range of our daily digital activities as users of social networks, mobile apps, search engines and connected objects contributes to the production of economic value which is captured by the platform owners. Such activities could be considered 'work' if they comply with three conditions: they produce value captured by economic players; they are subject to a minimal level of contractual agreement, usually accepted by ticking a box; and they provide performance indicators as regards rating, reputation, consumer satisfaction, quality assessment and popularity measurements (Cardon and Casilli 2015, pp. 12–14). Indeed, referring to the Marxist approach, an activity which generates value for a capital owner, which is subject to a contract and which produces marketable outputs has to be considered as labour (Scholtz 2012).

Prosumer work can be regarded as an extension of the enrolment of the consumer in service provision which started with the development of online services and call centres (Dujarier 2014). Another trend in prosumer work is the development of evaluations and ratings by lay people instead of professionals (Pasquier 2014) and its consequences for platform workers. Several problems are raised by prosumers' evaluations: biases in the selection of evaluators, and in the scope and methods of rating; control of the authenticity, validity and relevance of ratings; and the use of ratings by market players and employers. Platform workers often report the pernicious influence of consumers' rating on their working conditions, reputation or employability (Eurofound 2018; Drahokoupil and Piasna 2019).

The concept of open work raises questions about the boundaries between paid and unpaid work, between formal and informal work and between creative and productive activities carried out at work and those carried out elsewhere – boundaries which are blurred by digitalization. Digitalization can lead to a blossoming of 'making' activities or, in other words, a broad spectrum of pursuits which allows people to get active or creative outside the formal world of work – ranging from do-it-yourself to volunteering in the non-profit sector and skill-swapping networks – which many individuals find more conducive to personal fulfilment than jobs which are increasingly standardized and dehumanized (Flichy 2017). Flichy names 'open work' among these new forms of 'making' activities facilitated by platforms outside the spheres of wage employment or formal self-employment. This perspective addresses the meaning of work. While a number of jobs are increasingly under pressure, open work gives more space to self-accomplishment and gratification. The likely outcome of this process is a gradual decline in the centrality of formal work in people's lives. Work loses its presumed and actual importance as it becomes just one of many different ways of 'making'.

The concept of open work raises, however, the question of the actual autonomy of workers. Empirical evidence does not allow the conclusion that increasing autonomy could be a general trend in the platform economy (Abdelnour and Méda 2019). The downgrading of working conditions, increasing precariousness and threats to solidarity appear to be widespread whereas positive improvements seem limited to high-skilled workers or creative activities.

5.4 CONCLUSIONS

What can we learn from this historical overview? Evolution, disruption or both? Work organization on platforms amplifies a series of features that were already emerging with the development of online services during the long nineties, in the areas of working time flexibility, flexible work location, atypical work contracts and blurred subordination links. This amplification was not unexpected, as it was in line with the evolution of managerial strategies and the weakening of the bargaining power of trade unions, already visible at that time. Twenty years ago, however, the expansion of flexible work practices was regarded in the framework of continuing trends in economic development: perspectives of European enlargement; globalization of value chains; and the new international division of labour in services. The economic theory of two-sided markets was not even published yet and the technological infrastructure was not ready. Looking back at various foresight publications at the turn of the millennium, the future of work was regarded in the context of the so-called e-economy, without having the potential for significant disruption. The first studies of crowdsourcing only appeared at the beginning of the 2000s and they were regarded with scepticism. In other words, the platform economy has not so much created flexible work forms, but these existing forms have been extraordinarily enhanced by the development of platforms.

Both evolution and disruption are at stake. At first glance, most of the flexible features of platform work had already been implemented in online services – but not all. Platforms rely on a radically new economic model, new predominant global companies, new power relations and a changing approach to the meaning of work.

Now, what about the future prospects of platform work in the framework of digital transition? Instead of a technological revolution, the current digital transition can rather be considered a turning point between the development and the deployment phases of the techno-economic paradigm of digital technologies. The transition towards the deployment phase should require significant institutional reconfiguration and adapted regulation, as well as a revival of the dominant role of public institutions (Valenduc 2018). If platform work is only conceptualized as an amplified continuation of past trends, there is a risk that policymakers and social actors might develop reactive answers, mainly on the basis of the adaptation of existing regulatory frameworks. Proactive regulatory answers would be required, however, if the disruptive aspects of platform work are also taken into account. Such proactive answers must consider the changing meaning of work, the place of work in life trajectories, changes in our working and living environment and new forms of solidarity and engagement, as well as the new roles expected of public authorities and social protection institutions.

NOTE

1. Contributions to Wikipedia represent an example of user-generated content that does not fall under prosumer work.

REFERENCES

Abdelnour, S. and D. Méda (2019), *Les nouveaux travailleurs des applis*, Paris: PUF.
Alsène, E. (1990), 'Les impacts de la technologie sur l'organisation', *Sociologie du travail*, **32** (3), 321–337.
Brynjolfsson, E. and A. McAfee (2014), *The second machine age: work, progress and prosperity in a time of brilliant technologies*, New York: W.W. Norton & Company.
Cardon, D. and A. Casilli (2015), *Qu'est-ce que le digital labor?*, Bry-sur-Marne: INA.
Casilli, A. (2019), *En attendant les robots: enquête sur le travail du clic*, Paris: Seuil.
Castells, M. (1996), *The rise of the network society*, Cambridge, MA: Blackwell.
Doellgast, V. and E. Pannini (2015), 'The impact of outsourcing on job quality for call centre workers in the telecommunications and call centre subcontractor industries', in J. Drahokoupil (ed.), *The outsourcing challenge*, Brussels: European Trade Union Institute, pp. 117–136.
Drahokoupil, J. and A. Piasna (2019), 'Work in the platform economy: Deliveroo riders in Belgium and the SMART arrangement', Working Paper ETUI 2019.01, Brussels: European Trade Union Institute.
Dujarier, M.-A. (2014), *Le travail du consommateur. de McDo à eBay: comment nous coproduisons ce que nous achetons*, Paris: La Découverte.
Eurofound (2018), 'Platform work: types and implications for work and employment: literature review', Working Paper WPEF18004, Dublin: Eurofound.
Finck, M. (2017), 'Digital co-regulation: designing a supranational legal framework for the platform economy', LSE Legal StudiesWorking Papers 15/2017, accessed 25 November 2020 at http://dx.doi.org/10.2139/ssrn.2990043.
Flecker, J. (2009), 'Outsourcing, spatial relocation and the fragmentation of employment', *Competition and Change*, **13** (3), 251–266.
Flichy, P. (2017), *Les nouvelles frontières du travail à l'ère numérique*, Paris: Seuil.
Gadrey, J. (2003), *Socio-économie des services*, Paris: La Découverte.
Gillespie, A., R. Richardson, G. Valenduc and P. Vendramin (1999), 'Technology-induced atypical work forms [PE 167.794]', STOA Report, Luxembourg: European Parliament.
Greenan, N., Y. Kocuglu, E. Walkowiak, P. Csizmadia and C. Makó (2009), 'The role of technology in value chain restructuring', WORKS report for the European Commission, Leuven: HIVA.
Guzzetti, L. (1995), *A brief history of European Union research policy*, Luxembourg: Office of Publications of the European Communities.
Holtgrewe, U. (2014), 'New "new technologies": the future and the present of work in information and communication technology', *New Technology, Work and Employment*, **29** (1), 9–24.
Huws, U. (2003), *The making of a cybertariat: virtual work in a real world*, New York: New York University Press.
Huws, U. (2013), 'Working online, living offline: labour in the Internet age', *Work Organisation, Labour and Globalisation*, **7** (1), 1–11.

Kling, R. and R. Lamb (1999), 'IT and organizational change in digital economies: a socio-technical approach', *SIGCAS Computers and Society*, **29** (3), 17–25.
Musso, P. (2008), *Les télécommunications*, Paris: La Découverte.
Orlikowski, W. (1992), 'The duality of technology: rethinking the concept of technology in organizations', *Organization Science*, **3** (3), 398–427.
Pasquier, D. (2014), 'Les jugements profanes en ligne sous le regard des sciences sociales', *Réseaux*, **32** (183), 9–25.
Pichault, F. and M. Zune (2001), 'Une figure de la déréglementation du marché du travail: le cas des centres d'appels', *Les Enjeux de l'information et de la communication*, **1**, 83–96.
Perez, C. (2013), 'Unleashing a golden age after the financial collapse: drawing lessons from history', *Environmental Innovations and Societal Transitions*, **6**, 9–23.
Perez, C. (2016), 'Capitalism, technology and a green global golden age: the role of history in helping to shape the future', in M. Jacobs and M. Mazzucato (eds), *Rethinking capitalism: economics and policy for sustainable and inclusive growth*, London: Wiley-Blackwell.
Scholtz, T. (ed.) (2012), *Digital labor: the internet as a playground and factory*, New York: Routledge.
Serrano Pascual, A. and E. Crespo (2001), 'Emerging experiences of work in a changing economy', *Transfer*, **7** (2), 183–196.
Stanford, J. (2017), 'The resurgence of gig work: historical and theoretical perspectives', *Economic and Labour Relations Review*, **28** (3), 382–401.
Tirole, J. (2016), *L'économie du bien commun*, Paris: PUF.
Valenduc, G. (2018), 'Technological revolutions and societal transitions', *ETUI Foresight Brief* 4, Brussels: European Trade Union Institute.
Valenduc, G. and P. Vendramin (2002), 'ICT, flexible working and quality of life', in K. Verlaeckt and V. Vitorino (eds), *Unity and diversity: the contribution of social sciences and humanities to the European research* area, EUR 20484, Luxembourg: Office of Publications of the European Communities, pp. 186–190.
Valenduc, G. and P. Vendramin (2016), 'Work in the digital economy: sorting the old from the new', Working Paper 2016.3, Brussels: European Trade Union Institute.
Valenduc, G. and P. Vendramin (2017), 'Digitalisation, between disruption and evolution', *Transfer*, **23** (2), 121–134.
Veltz, P. (2017), *La société hyper-industrielle: le nouveau capitalisme productif*, Paris: Seuil.
Wauthy, X. (2008), 'Concurrence et régulation sur les marchés de plateforme: une introduction', *Reflets et perspectives de la vie économique*, **47** (1), 39–54.

6. The platform economy at the forefront of a changing world of work: Implications for occupational health and safety

Pierre Bérastégui and Sacha Garben

6.1 INTRODUCTION

The types of work offered through digital platforms are ever increasing, as are the challenges for existing regulatory frameworks. Since the initial launch of Amazon Mechanical Turk in 2005, the number and diversity of digital labour platforms has dramatically increased and they now cover a wide range of activities including transportation, food delivery, accounting, illustration and data processing. In addition to the specific hazards entailed by these different types of activities there are also risks related to the way platform work is organized, designed and managed. Most platform companies have deliberately and strategically self-positioned themselves as neutral intermediaries in order to evade labour protection laws and social welfare. Yet a growing corpus of studies demonstrates that platforms have the power to organize work, to control its execution and to reprimand workers not complying with its rules. In that sense, platform work should be regarded as a new paradigm for labour relations and responsibilities which, as we shall go on to examine, gives rise to a range of physical and psychosocial risks.

The types of activities mediated through digital platforms may be grouped into three primary categories: location-based platform work; and online platform work with a distinction being drawn between micro- and macro-tasks. The first involves location-based applications fulfilling consumer demands placed online via immediate and convenient access to services performed offline. Micro-task platforms facilitate the division of virtual services into tiny and low-skilled assignments distributed through a large pool of candidates. Platforms mediating macro-tasks enable organizations or individual requesters to access a network of online freelancers with high and specialized skills, delivering assignments directly through the platform. Although more detailed classifications have been proposed, these main categories succeed in covering the wide scope of the platform economy.

At first glance, working conditions in the platform economy may be regarded as 'atypical' due to the innovative technology involved. When subjected to closer scrutiny, however, it appears that the structural characteristics of platform-mediated work are not entirely new but rather constitute radical extensions of pre-existing trends (see Chapter 5). Specifically, platform work combines and extends three well-established developments in order to reinvent labour relations and responsibilities: greater control and surveillance; greater job uncertainty and volatility; and greater worker isolation and workplace fragmentation.

The consequences of platform work for occupational health and safety are not yet well documented. Research is at an early stage and much remains to be done to grasp fully how the platform economy is disrupting the work environment. The current literature is mostly comprised of descriptive studies or opinion papers while the influx of new data sources remains limited (Bérastégui 2021). Established models and theories can, however, inform us of the physical and psychosocial implications of platform work as a result of the well-established roots of the concept. This approach may only provide indirect evidence, but it does succeed in laying the foundations for future research by setting out the priority areas and relevant variables of interest. In this chapter, we analyse how these three trends translate to the platform economy and the potential risks they entail for platform workers' health and safety.

6.2 DIGITAL SURVEILLANCE AND ALGORITHMIC CONTROL

The platform economy has essentially been made possible by the concurrent advances in digitalization and telecommunications (see Chapter 5). Digital platforms not only allow the remote connection of customers and contractors anywhere in the world but also the highest possible degree of standardization in the organization and delivery of work. This is especially salient in micro-work where the entire process of contracting, executing and delivering assignments is mediated through automated management techniques (Cabrelli and Graveling 2019). By endorsing human relations-related duties, algorithms are given the responsibility for making decisions that affect work, thereby limiting human involvement in the labour process (Duggan et al. 2019).

Digital surveillance is an essential component of algorithmic control. Indeed, automated or semi-automated decision-making requires a substantial amount of accurate data which can only be achieved by intensively tracking workers' activities and whereabouts. Constant monitoring allows predictions about workers' future behaviours which are then turned into operational decisions, such as work scheduling or fitness for employment (Mateescu and Nguyen 2019). This aspect of supervision is often illustrated by the 'panopticon' metaphor – a prison system allowing a single observer to watch each prisoner simultaneously from a central point. Such an architecture is intended to 'internalize' the supervisory function as the prisoner cannot know when the observer is watching and so assumes it could happen at any point (Woodcock 2020). Similarly, the constant monitoring facilitated by platform apps practically means that any missteps could have serious consequences for the worker. While most platform workers are unclear about what data are being collected and how they are used by the platform (Anderson 2016), the internalization of the supervisory function is potent enough to create an overall climate of discipline and control (Chan and Humphreys 2018).

The foundations of digital surveillance and algorithmic control may be found in the principles of scientific management introduced by F. W. Taylor in the 1880s. It was probably the first iteration of the 'panopticon' principle as the omnipresence of managers was intended to internalize the supervisory function. In micro-work, the analogy goes beyond simple authoritarian control to encompass the idea of breaking down jobs into tiny, simple and repetitive tasks (Degryse 2016). Overall job completion is thus turned into a machine-like process with each worker completing micro-tasks and following a strict *modus operandi*. Taylorism was then followed by a profusion of spiritual successors of which the common denominator is the articulation of formulae-driven work processes and close monitoring. Algorithmic control and digital surveillance can thus be regarded as an evolution of established trends toward greater control of labour. Preliminary evidence suggests that constant monitoring and automated management are contributing to an increasingly hectic pace of work, a lack of trust in the platform and pronounced power asymmetries which limit workers' opportunities to resist or develop effective forms of internal voice.

Micro-taskers may be especially at risk of cognitive overload, a situation where the volume of information to be treated exceeds the processing capacity of the individual. The financial insecurity experienced by such workers requires them to monitor several platforms, complete simultaneous tasks and control various sources of information at the same time (Poutanen et al. 2019). Micro-taskers are thus continuously challenged to discriminate and filter information for importance as well as to reconcile different strategies for maximizing proficiency. This heightened vulnerability is reflected in the way micro-taskers strive to enhance productivity. For instance, veteran workers rely on 'catchers' – third-party screening tools that automatically find lucrative assignments across multiple platforms (Kaplan et al. 2018). While catchers are recognized as being vital to identify the path of greatest reward, they are also reported as particularly stressful because they imply additional work to manage the queue of assignments (Williams 2020). Managing the queue is a particularly demanding task as it requires workers to systematically compare queued assignments for task demands, time constraints and rewards.

On location-based and macro-task platforms, workload may be compounded by the widespread use of pervasive surveillance technologies. For instance, Upwork allows clients to track the time spent on tasks and captures time-stamped screenshots of the worker's computer (Anwar and Graham 2019). Preliminary evidence indicates that digital surveillance prompts platform workers to work long hours out of fear of not getting paid for the work accomplished. Similarly, location-based platforms such as Deliveroo and Uber monitor the amount of time spent on every stage of the delivery process (Rosenblat and Stark 2016). Another example comes from Honor, a home-care platform, which monitors caregivers to check the time of arrival as well as various 'suspicious' activities such as sitting down for some time, making phone calls or checking social media (Choudary 2018).

Although data are still lacking on the consequences of these practices, it is arguable that constant monitoring, coupled with monetary sanctions, encourages platform workers to maintain a hectic work pace. In addition, encouraging rapid pace without breaks is likely to induce accidents (Huws 2015). For instance, one survey highlighted that many platform-based couriers experience pressure to break speed limits, with 42 per cent even admitting being involved in a collision with 10 per cent saying that someone had been injured – usually themselves – as a result (Christie and Ward 2019).

On all three types of platform, authority is embodied by algorithms and mediated through the platform's interface. Danaher (2016) refers to this organizational model as the 'algocracy', whereby decisions related to work allocation, monitoring and assessment are endorsed by algorithms. Algocracy eliminates the need and therefore the possibility to develop trust-based relationships, as the psychological contract is lifted from the social sphere and reimagined as a machine process (Zuboff 2015). A large body of literature highlights that algorithmic control and digital surveillance contribute to workers' distrust of digital labour platforms (Bérastégui, 2021). Furthermore, the psychological distance created between workers and decision-makers contributes toward a dehumanization of the work relationship (McInnis et al. 2016). For workers, this distance is perceived as demotivating and isolating (Marlow and Dabbish 2014); while for platform and clients alike it forms an excuse to forget that workers are real human beings deserving fair labour practices (McInnis et al. 2016).

On location-based platforms, common sources of mistrust include inconsistencies in the calculation of workers' commissions and platform fees; platforms' reluctance to communicate on algorithmic rules; and an ecosystem heavily biased toward clients – even unethical ones (Malhotra 2020).

Procedural opacity is equally common on micro-task platforms. For instance, Amazon Mechanical Turk does not even require requesters to share the reasons for their rejection of assignments. Opacity also applies to decisions such as account suspension or banning (Martin et al. 2016). Similarly, the criteria to be met in order to be part of the 'Masters' – an elite group of workers demonstrating superior performance – are unclear and seemingly at the firm's whim (Deng et al. 2016).

One distinct characteristic of micro-task platforms resides in the extent to which workers are exposed to privacy issues (Xia et al. 2017). Although it is typically assumed that requesters and micro-taskers are anonymous to each other, workers' identity can be revealed as a consequence of breaches of security protocols. For instance, data triangulation can be used to associate a worker's ID with a shopping profile on Amazon (Lease et al. 2013). Alternatively, requesters can deliberately de-anonymize workers using a sequence of surveys posted on the platform (Kandappu et al. 2015). All these issues act as major sources of frustration and contribute to micro-taskers' perceptions that the platform does not consider them to be meaningful actors in the labour process (Martin et al. 2016; Deng et al. 2016).

When it comes to macro-task platforms, learning the behaviour of algorithms may be a source of competitive advantage. Online freelancers engage in experimentation to make sense of their functions or by exchanging ideas with other more experienced workers (Sutherland et al. 2019).

There is growing consensus that location-based platforms are able to foster power asymmetries using a diversified portfolio of techniques. For instance, Uber drivers are not shown destination or fare information before they accept a ride which prevents them from turning down trips that they would consider insufficiently lucrative (Dunn 2018), while workers cancelling unprofitable rides bear the risk of being temporarily or permanently removed from the platform. There is also evidence that Uber has at least experimented with deceptive market manipulation practices (Calo and Rosenblat 2017). For instance, Uber's app shows a passenger an upfront price by estimating the time the ride will take but calculates the driver's fare according to the distance and time the ride actually takes. Multiple cases have been documented where the Uber app has shown passengers and drivers drastically different fares for the same ride. Setting aside Uber's commission, drivers earned substantially less than the customer paid, even when the ride took approximately the time and distance estimated (Muller 2020). Finally, ratings are used to create a climate of discipline and to ensure that workers' behaviour is aligned (Florisson and Mandl 2018). Requesters' feedback is processed by algorithms to identify and dismiss unproductive or 'borderline' workers with no right to recourse (Aloisi 2016). Arbitrary ratings and low correctability take the 'customer is always right' saying to a whole new level while exposing workers to enormous risks of unfair, arbitrary or inaccurate evaluations.

On both location-based and macro-work platforms, mechanisms used to influence a worker's behaviour also include elements of gamification and nudging intended to foster productivity (Gandini 2019; see also Chapter 18 in this volume). Lehdonvirta et al. (2019) show that online freelancers on Upwork are induced to score 'personal bests' by their metrics on the platform. Similarly, Scheiber (2017) describes how Uber uses game-based 'psychological tricks' to ensure drivers work more hours, thus discouraging disconnection from the app and incentivizing the achievement of personal bests for daily earnings or distances.

When coming from an employer, such nudges have a stronger managerial element of control. What Woodcock and Johnson (2018) characterize as 'gamification-from-above' is particularly relevant for securing cooperation where arm's length relationships restrict platforms' ability to supervise workers closely. Such practices limit workers' informed decision-making and hinder forms of resistance that could potentially challenge existing work arrangements (Scheiber 2017).

In sum, much of the research points towards a power structure dominated by platform owners and their customers. These persistent attempts to influence workers' behaviours further complicate platforms' claims that they operate as neutral intermediaries (Rosenblat and Stark 2016).

6.3 JOB UNCERTAINTY AND VOLATILITY

These last few decades have been marked by radical changes in the economic environment, leading to growing concern about decreasing job stability and the disappearance of the 'job for life' (Eurofound 2015a). Although the 2007–2008 financial and economic crisis is assumed to be partly responsible for this mounting instability, there is also a long-run trend towards shorter job tenure and, overall, it appears that the incidence of precarity at work is increasing (Eichhorst et al. 2018).

Workers in the conventional economy can still expect forms of continuity in employment as well as a certain degree of clarity regarding expected career paths along which they can anticipate moving. In the platform economy, however, these boundaries are becoming increasingly fluid and platforms are using this flexibility as a strategic asset. Platform-mediated work is mostly comprised of short-term assignments that guarantee work for a limited period of time and leave future work relationships uncertain. Moreover, the self-employed nature of platform work implies that 'users' are solely responsible for their own economic upkeep and career planning. What is often presented as an opportunity for variety actually burdens workers with managing the growing complexity of their working lives while trying to generate a steady income flow. Browsing multiple platforms, combining multiple sources of income, installing third-party tools, exchanging tips on virtual communities and exploring new types of tasks or requesters are typical examples of the strategies platform workers carry out to ensure a decent living (Bérastégui 2021). From an occupational health and safety perspective this uncertainty, combined with the requirement to cope with the sheer complexity of this new world of work, gives rise to two psychosocial risk factors: job insecurity and emotional demands.

On all three types of platform, job insecurity represents the norm. A work relationship can be ended without notice or any form of dismissal protection and, even when the relationship is active, there is no guaranteed minimum wage since it is dependent on performing assignments (Garben 2019). Similarly, there are no guaranteed hours, sick pay, pensions, parental leave, overtime or redundancy entitlements. Platform workers assume the cost of periods of inactivity, any lack of demand, delays and any malfunctioning of the app or software (Fabrellas 2019). Other risk determinants include the short duration of tasks, volatility in demand (Smith and Leberstein 2015), the high degree of competition between platform workers (Florisson and Mandl 2018) and the endemic lack of training (Garben 2017). There is evidence that platforms do little to mitigate these risks, with some of the largest highlighting that platform workers are on-demand and can be fired at any time (Graham et al. 2017).

Persistent feelings of insecurity are common among particularly low-skilled platform workers. One recent survey identifies that 43 per cent of micro-workers feel easily replaceable (Graham et al. 2017); a figure that is significantly higher than the overall level of perceived job insecurity among European workers, estimated in 2015 at 16 per cent (Eurofound 2016). Another survey established that only 30 per cent of micro-workers report satisfactory levels of job security while only 20 per cent are satisfied with their future professional prospects (Forde et al. 2017). Similarly, Shokoohyar (2018) has highlighted that job security and advancement are the most problematic work aspects for Lyft and Uber drivers.

Regarding macro-work platforms, findings from a qualitative study underline the erratic professional trajectories of online freelancers, and how they navigate in such context with little security about whether and when they will get paid (Risi et al. 2019). Beside the fear of job loss, platform workers also lose some aspects of the work relationship (Graham and Shaw 2017). Regardless of type, most platforms reserve the right to change the terms of service at any time, rendering the work relationship highly insecure (Rosenblat et al. 2017; Garben 2017; Van Doorn 2017). Furthermore, terms and conditions agreements typically give platforms the right to terminate workers' accounts when they consider that the agreement has been breached. All this not only adds to insecurity but also makes platform workers reluctant to speak out because they are fearful of being disconnected (Florisson and Mandl 2018). Furthermore, job insecurity creates higher stress levels and work intensity as individuals tend to work harder in the hope of greater security if they prove to be 'good soldiers' (Eurofound 2015b).

Workers in an environment characterized by low job security favour 'surface acting' as a coping strategy (Vereycken and Lamberts 2019), displaying the required emotions without changing how they actually feel (Hochschild 1983). This applies to casual or temporary workers having no guarantee of future employment and thriving to keep their employer satisfied in the hope of a more stable contract (Vereycken and Lamberts 2019). Current evidence suggests that platform workers are also required to engage in surface acting to preserve employability. In ride-hailing, drivers feel pressured to be exceptionally affable, tolerate inappropriate behaviour and leave no wish unanswered, which can be emotionally exhausting and stressful (Bajwa et al. 2018). Similarly, online freelancers put extra effort and thought into maintaining good virtual connections with their clients. They feel the need to develop emotional skills to strengthen the typically transient relationships they have with clients and consider emotional labour as essential for ensuring the durability of employment (Sutherland et al. 2019). In that sense, job insecurity makes these jobs far more emotionally demanding than their counterparts outside the platform economy (Raval and Dourish 2016). This constant pressure is further compounded by constant monitoring, public evaluation and fierce competition between workers.

Platform workers share many similarities with both temporary workers and agency workers and are probably exposed to similar levels of physical risk. Studies consistently show higher injury rates among workers on non-standard arrangements (Howard 2017). The lack of appropriate training associated with contingent work further increases the risk of accidents, while several location-based platforms offer work in occupations that are notoriously dangerous, such as transport (Huws 2015). For online freelancers and micro-taskers, the endemic lack of training and of adequate equipment is likely to give rise to fatigue and musculoskeletal disorders (Brancati et al. 2020). Moreover, platform workers being denied the right to paid sick leave further increases the risk of injury at work (Howard 2017): workers with paid sick leave benefits are 28 per cent less likely to sustain a work-related injury than workers without access to paid sick leave (Asfaw et al. 2012).

It has been suggested that health problems may actually form a significant reason to engage in online platform work (Berg et al. 2018). On the one hand, this means that online platform work may provide an alternative way for people with health impairments to carry on in work and therefore contribute to their social and labour market inclusion. On the other hand, however, it also means that many online platform workers are already in a vulnerable position from an occupational health and safety perspective and this may be aggravated by platform work.

6.4 WORKPLACE FRAGMENTATION AND WORKER ISOLATION

The implementation of automated and remote management practices eliminates the need for shared premises. Physical interactions with supervisors or co-workers are considered obsolete and even counterproductive as they introduce undesirable variability in the process of the matching of demand and supply. While remote working may be presented as advantageous, many platform workers feel that they are physically and mentally detached from other human beings (Graham et al. 2017). Even if they were proactively to seek to counter social isolation, the piece rate model and the precarious aspects of platform work favour a high productivity approach, discouraging any attempt to engage in light conversation. In this regard, platform workers are likely to experience a lack of various forms of social support but also encounter difficulties in establishing a consistent professional identity on top of a blurring of the boundaries between work and personal life.

Most platform workers carry out their tasks individually, separated from and often in competition with fellow workers. The lack of spontaneous and mutual exchange is detrimental to workplace social support as it prevents individuals from sharing work-related concerns (Vendramin and Valenduc 2018). Rather, multiple studies highlight a logic of 'everyone for themselves' leading to disputes between platform workers. For instance, it has been shown that food delivery platforms use various incentives to push workers to complete as many deliveries as possible within an hour, ultimately leading to a decreased sense of solidarity among colleagues (De Stefano and Aloisi 2018). Similarly, Uber drivers are facing growing competition from other drivers and are unable to earn the same pay in the same amount of time compared to the situation three or four years previously (Polkowska 2019). Such competitive pressure is not favourable to camaraderie or any other manifestations of peer support found in traditional jobs.

Platform workers also lack organizational forms of support such as coaching or career mentoring. Interactions with supervisors are scarce and, in most cases, embodied by impersonal algorithms and data-driven procedures. Such a work environment lacks the warmth of the face-to-face interactions which are crucial for developing a sense of oneness and of belonging (Vayre and Pignault 2014). In fact, app-based management practices are mainly focused on task support and leave little room for any other forms of workplace support (Tran and Sokas 2017; Vendramin and Valenduc 2018).

The way individuals perceive themselves within the occupational context is mainly determined by processes of socialization (Joynes 2018). Working in isolation is, therefore, detrimental to professional identity: an important cognitive mechanism that not only influences workers' attitudes and behaviours both on the job and beyond, but which is also a way for individuals to assign meaning to themselves and their lives. Professional identity also has a known impact on workers' psychological wellbeing. Without the protective role of professional identity, workers are more likely to experience occupational stress and to suffer from anxiety, burnout and depression. As Supiot (2019, p. 30) writes, 'a bleak despair threatens all those individuals whose work has no other reasons than financial ones.' In this regard, preliminary evidence suggests that micro-workers represent an especially vulnerable population with a professional identity made fragile by lack of meaning and of role models (Bérastégui 2021).

Several empirical studies have attempted to compare depression rates between micro-workers and the general population but have yielded contradictory results which can mainly be attributed to insufficient data quality assurance procedures. A recent study controlled for such biases and showed that 19 per cent of micro-workers suffer from major depression, which is 1.6 to 3.6 times higher than estimates for the general population (Ophir et al. 2019). Sampling differences in sociodemographics, health and lifestyle account for only around one-half of this discrepancy. Similarly, two other studies have reported that some 50 per cent of micro-workers operating on Amazon Mechanical Turk suffer from clinical levels of social anxiety (Arditte et al. 2016; Shapiro et al. 2013), which is significantly higher than the prevalence estimates of 7–8 per cent for the general population (American Psychiatric Association 2013). While the direction of causality is not clear, such evidence further hints that the structural characteristics of micro-work may be detrimental to occupational health and safety.

Platforms' remote technologies are also redefining the boundaries between private and public space. Indeed, the way platforms are designed encourages workers to be always on stand-by to opt for potential upcoming assignments. 'Fear of missing out' on lucrative assignments leads to an obsessional relationship with professional communications tools and therefore encourages an 'always and everywhere' mindset (Degryse 2016). Floridi (2015) refers to this model as the 'onlife paradigm', 'a fluid reality … that exposes our everyday experience and even our personal asset to financialization or value extractive strategies'. Having more permeable boundaries allows work to interrupt, increasing overtime and the risk of work–family conflicts (Tremblay and Thomsin 2012). For instance, one study showed that 55 per cent of micro-workers report overwork and long hours (Graham et al. 2017). Similarly, another study showed that platform delivery workers seek to maximize deliveries in the face of the constant unpredictability of the labour process, working up to 12 hours straight (Griesbach 2018).

Finally, any physical risk could be anticipated to be worse because of the loss of the protective effect of working in a public workplace as work is transacted either on public thoroughfares or in private homes (Tran and Sokas 2017; Rockefeller Foundation 2013). This also exposes the worker to the risk of equipment not meeting ergonomic criteria and other environmental factors not being optimized for the work situation.

6.5 CONCLUSION

Platform work combines and extends three long-standing trends in the reinvention of labour relations and responsibilities: increased control and surveillance; increased job uncertainty and volatility; and increased worker isolation and workplace fragmentation. Although each of these key characteristics should be regarded as a continuum, with some degree of variation across the different types of platform, they are nevertheless part and parcel of the same reality.

Research on platform work is still at an early stage, but it has long been acknowledged that these trends have the potential to give rise to multiple psychosocial risk factors, ultimately leading potentially to several negative worker outcomes. Nevertheless, the extent to which platforms have built their entire business model around these dimensions, and with the aid of technologies which compound the resulting precarity, has lifted the overall set of problems to a higher level.

The guiding thread behind the specific risks exposed in this chapter is a greater imbalance between the job demands placed upon workers and the resources available to deal with them. On the resources side, platforms provide low levels of organizational support and job security, no channels to voice concerns or exercise agency and little means of contesting unfair decisions or unethical behaviour. On the demand side, platforms nevertheless have high standards of performance and require workers to be highly flexible, affable and productive. The interaction between high job demands and low organizational resources is known to generate strain, and, in this regard, preliminary evidence can be found for each of the risk factors described in this chapter. For all these reasons, we should be careful not to overestimate the novelty of the working arrangements emerging from platform work while neither should we underestimate the scale and depth of the issue.

These risks make it all the more important that occupational health and safety regulations apply to platform work. Applying occupational health and safety rules and employment law to platform work is not an easy task. The involvement of platforms in the organization and provision of labour tends to complicate the classification and regulation of the responsibilities applying to the work in question. Meanwhile the triangular nature of the arrangements, their transient nature, the seemingly high autonomy of the worker in terms of working time and the absence of a common workplace all challenge the application of the concept of workplace health and safety regulation developed in the context of standard, permanent, binary employment relationships.

Moreover, the precarious position of platform workers is further aggravated by these features tending to hamper the collective organization of workers, and thus the defence of their rights and interests, as well as the development of social dialogue. Overcoming these challenges is key to improving regulatory and legal environments in a way that is conducive to the welfare of platform workers. That said, there is some evidence that cooperative models are increasing in all types of platform work, although mostly they are at an embryonic stage (Vandaele, forthcoming). Owing to physical distance, workers' resistance is expected to develop the most in location-based platform work in public settings, but less so in other types of platform work. Cooperation in such areas between unions, activist groups and other types of organization could, as a result, point towards a broader transition of trade unionism and its possible revitalization.

REFERENCES

Aloisi, A. (2016), 'Commoditized workers: case study research on labour law issues arising from a set of "on-demand/gig economy" platforms', *Comparative Labor Law and Policy Journal*, **37** (3), 653–690.

American Psychiatric Association (ed.) (2013), *Diagnostic and statistical manual of mental disorders (DSM-5)*, 5th edition, Arlington, VA: American Psychiatric Association.

Anderson, D. N. (2016), 'Wheels in the head: Ridesharing as monitored performance', *Surveillance and Society*, **14** (2), 240–258.

Anwar, M. A. and M. Graham (2019), 'Hidden transcripts of the gig economy: labour agency and the new art of resistance among African gig workers', *Environment and Planning A: Economy and Space*, **52** (7), 1269–1291.

Arditte, K. A., D. Çek, A. M. Shaw and K. R. Timpano (2016), 'The importance of assessing clinical phenomena in Mechanical Turk research', *Psychological Assessment*, **28** (6), 684–691.

Asfaw, A., R. Pana-Cryan and R. Sosa (2012), 'Paid sick leave and nonfatal occupational injuries', American Journal of Public Health, **102** (9), e59–e64.

Bajwa, U., L. Knorr, E. Di Ruggiero, D. Gastaldo and A. Zendel (2018), *Towards an understanding of workers' experiences in the global gig economy*, Toronto: University of Toronto.

Bérastégui, P. (2021), 'Exposure to psychosocial risk factors in the gig economy: a systematic review', Report 144, Brussels: ETUI.

Berg, J., M. Furrer, E. Harmon, U. Rani and S. Silberman (2018), *Digital labour platforms and the future of work: towards decent work in the online world*, Geneva: ILO.
Brancati, C. U., A. Pesole and E. Fernández-Macías (2020), *New evidence on platform workers in Europe: results from the second COLLEEM survey*, Luxembourg: Publications Office of the European Union.
Cabrelli, D. and R. Graveling (2019), 'Health and safety in the workplace of the future', Briefing, accessed 15 December 2020 at www.europarl.europa.eu/RegData/etudes/BRIE/2019/638434/IPOL_BRI(2019)638434_EN.pdf.
Calo, R. and A. Rosenblat (2017), 'The taking economy: Uber, information, and power', *Columbina Law Review*, **117** (6), 1623–1690.
Chan, N. K. and L. Humphreys (2018), 'Mediatization of social space and the case of Uber drivers', *Media and Communication*, **6** (2), 29–38.
Choudary, S. P. (2018), 'The architecture of digital labour platforms: policy recommendations on platform design for worker well-being', ILO Future of Work Research Paper Series 3, Geneva: ILO.
Christie, N. and H. Ward (2019), 'The health and safety risks for people who drive for work in the gig economy', *Journal of Transport and Health*, **13**, 115–127.
Danaher, J. (2016), 'The threat of algocracy: reality, resistance and accommodation', *Philosophy and Technology*, **29** (3), 245–268.
De Stefano, V. and A. Aloisi (2018), *European legal framework for digital labour platforms*, Luxembourg: Publications Office of the European Union.
Degryse, C. (2016), 'Digitalisation of the economy and its impact on labour markets', Working Paper 2016.02, Brussels: ETUI.
Deng, X. N., K. D. Joshi and R. D. Galliers (2016), 'The duality of empowerment and marginalization in microtask crowdsourcing: giving voice to the less powerful through value sensitive design', *MIS Quarterly*, **40** (2), 279–302.
Duggan J., U. Sherman, R. Carbery and A. McDonnell (2019), 'Algorithmic management and app-work in the gig economy: a research agenda for employment relations and HRM', *Human Resource Management Journal*, **30** (1), 114–132.
Dunn, M. (2018), *Making gigs work: career strategies, job quality and migration in the gig economy*, Chapel Hill, NC: University of North Carolina.
Eichhorst, W., P. Marx, A. Broughton, P. De Beer, C. Linckh and G. Bassani (2018), *Mitigating labour market dualism: single open-ended contracts and other instruments*, Brussels: European Parliament.
Eurofound (2015a), *Job tenure in turbulent times*, Luxembourg: Publications Office of the European Union.
Eurofound (2015b), *New forms of employment*, Luxembourg: Publications Office of the European Union.
Eurofound (2016), *Sixth European Working Conditions Survey*, Luxembourg: Publications Office of the European Union.
Fabrellas, A. G. (2019), 'The zero-hour contract in platform work. Should we ban it or embrace it?', *Revista de Internet, Derecho y Politica*, **28**, 1–15.
Floridi, L. (ed.) (2015), *The onlife manifesto*, London: Springer International.
Florisson, R. and I. Mandl (2018), *Platform work: types and implications for work and employment: literature review*, Dublin: Eurofound.
Forde, C., M. Stuart, S. Joyce, L. Oliver, D. Valizade, G. Alberti, K. Hardy, V. Trappmann, C. Umney and C. Carson (2017), *The social protection of workers in the platform economy*, Luxembourg: Publications Office of the European Union.
Gandini, A. (2019), 'Labour process theory and the gig economy', *Human Relations*, **72** (6), 1039–1056.

Garben, S. (2017), *Protecting workers in the online platform economy: an overview of regulatory and policy developments in the EU*, Luxembourg: Publications Office of the European Union.

Garben, S. (2019), 'The regulatory challenge of occupational safety and health in the online platform economy', *International Social Security Review*, **72** (3), 95–112.

Graham, M. and J. Shaw (2017), *Towards a fairer gig economy*, London: Meatspace Press.

Graham, M., V. Lehdonvirta, A. J. Wood, H. Barnard, I. Hjorth and D. P. Simon (2017), *The risks and rewards of online gig work at the global margins*, Oxford: Oxford Internet Institute.

Griesbach, K. (2018), 'Just trying to keep my customers satisfied? Time struggle, managerial control, and the social and spatial dimensions of adjunct and platform work', Paper presented at the 36th International Labour Process Conference, Buenos Aires, 21 March, accessed 15 December 2020 at www.ilpc.org.uk/Portals/7/2018/Documents/PaperUpload/ILPC2018paper-TimeStruggleAdjunctsandPlatformWorkers_Griesbach_ILPCDraft_Feb272018_20180228_015237.pdf.

Hochschild, A. R. (1983), *The managed heart: commercialization of human feeling*, Berkeley, CA: University of California Press.

Howard, J. (2017), 'Nonstandard work arrangements and worker health and safety', *American Journal of Industrial Medicine*, **60** (1), 1–10.

Huws, U. (2015), *A review on the future of work: online labour exchanges, or 'crowdsourcing': implications for occupational health and safety*, Bilboa: EU OSHA.

Joynes, V. C. T. (2018), 'Defining and understanding the relationship between professional identity and interprofessional responsibility: implications for educating health and social care students', *Advances in Health Sciences Education: Theory and Practice*, **23** (1), 133–149.

Kandappu, T., A. Friedman, V. Sivaraman and R. Boreli (2015), 'Privacy in crowdsourced platforms', in S. Zeadally and M. Badra (eds), *Privacy in a digital, networked world*, Berlin: Springer.

Kaplan, T., S. Saito, K. Hara and J. P. Bigham (2018), 'Striving to earn more: a survey of work strategies and tool use among crowd workers', Paper presented at the 6th AAAI Conference on Human Computation and Crowdsourcing, Zurich, 5 July, Palo Alto, CA: AAAI Press.

Lease, M., J. Hullman, J. Bigham, M. Bernstein, J. Kim, W. Lasecki, S. Bakhshi, T. Mitra and R. Miller (2013), *Mechanical Turk is not anonymous*, accessed 18 December 2020 at https://ssrn.com/abstract=2228728 and http://dx.doi.org/10.2139/ssrn.2228728.

Lehdonvirta, V., O. Kässi, I. Hjorth, H. Barnard and M. Graham (2019), 'The global platform economy: a new offshoring institution enabling emerging-economy microproviders', *Journal of Management*, **45** (2), 567–599.

Malhotra, A. (2020), 'Making the one-sided gig economy really two-sided: implications for future of work', in S. Nambisan, K. Lyytinen and Y. Yoo (eds), *Handbook of digital innovation*, Cheltenham, UK and Northampton, MA, USA: Edward Elgar Publishing.

Marlow, J. and L. Dabbish (2014), 'Who's the boss? Requester transparency and motivation in a microtask marketplace', in *CHI EA '14: extended abstracts on human factors in computing systems*, New York: Association for Computing Machinery, pp. 2533–2538.

Martin, D., J. O'Neill, N. Gupta and B. V. Hanrahan (2016), 'Turking in a global labour market', *Computer Supported Cooperative Work*, **25** (1), 39–77.

Mateescu, A. and A. Nguyen (2019), *Explainer: algorithmic management in the workplace*, New York: Data and Society, accessed 15 December 2020 at https://datasociety.net/wp-content/uploads/2019/02/DS_Algorithmic_Management_Explainer.pdf.

McInnis, B., D. Cosley, C. Nam and G. Leshed (2016), 'Taking a HIT: designing around rejection, mistrust, risk, and workers' experiences in Amazon Mechanical Turk', in *CHI '16: Proceeding of the 2016 CHI conference on human factors in computing systems*, New York: Association for Computing Machinery, pp. 2271–2282.

Muller, Z. (2020), 'Algorithmic harms to workers in the platform economy: the case of Uber', *Columbia Journal of Law and Social Problems*, **53** (2), 167–210.

Ophir, Y., I. Sisso, C. S. C. Asterhan, R. Tikochinski and R. Reichart (2019), 'The Turker blues: hidden factors behind increased depression rates among Amazon's Mechanical Turkers', *Clinical Psychological Science*, **8** (1), 65–83.

Polkowska, D. (2019), 'Does the app contribute to the precarization of work? The case of Uber drivers in Poland', *Partecipazione e Conflitto*, **12** (3), 717–741.

Poutanen, S., A. Kovalainen and P. Rouvinen (2019), *Digital work and the platform economy: understanding tasks, skills and capabilities in the new era*, London: Routledge.

Raval, N. and P. Dourish (2016), 'Standing out from the crowd: emotional labor, body labor, and temporal labor in ridesharing', in *CSCW '16: Proceedings of the 19th ACM Conference on Computer-Supported Cooperative Work and Social Computing*, New York: Association for Computing Machinery, pp. 97–107.

Risi, E., M. Briziarelli and E. Armano (2019), 'Crowdsourcing platforms as devices to activate subjectivities: narratives on digital precarity and freelance knowledge workers', *Open Journal of Sociopolitical Studies*, **12** (3), 767–793.

Rockefeller Foundation (2013), *Health vulnerabilities of informal workers*, Washington, DC: Rockefeller Foundation.

Rosenblat, A. and L. Stark (2016), 'Algorithmic labor and information asymmetries: a case study of Uber's drivers', *International Journal of Communication*, **10**, 3758–3784.

Rosenblat, A., K. E. C. Levy, S. Barocas and T. Hwang (2017), 'Discriminating tastes: Uber's customer ratings as vehicles for workplace discrimination', *Policy and Internet*, **9** (3), 256–279.

Scheiber, N. (2017), 'How Uber uses psychological tricks to push its drivers' buttons', *New York Times*, 2 April, accessed 15 December 2020 at www.nytimes.com/interactive/2017/04/02/technology/uber-drivers-psychological-tricks.html.

Shapiro, D. N., J. Chandler and P. A. Mueller (2013), 'Using Mechanical Turk to study clinical populations', *Clinical Psychological Science*, **1** (2), 213–220.

Shokoohyar, S. (2018), 'Ride-sharing platforms from drivers' perspective: evidence from Uber and Lyft drivers', *International Journal of Data and Network Science*, **2** (4), 89–98.

Smith, R. and S. Leberstein (2015), *Rights on demand: ensuring workplace standards and worker security in the on-demand economy*, New York: NELP.

Supiot, A. (2019), *Le travail n'est pas une marchandise. Contenu et sens du travail au XXIe siècle*, Paris: Collège de France.

Sutherland, W., M. H. Jarrahi, M. Dunn and S. B. Nelson (2019), 'Work precarity and gig literacies in online freelancing', *Work, Employment and Society*, **34** (3), 457–475.

Tran, M. and R. K. Sokas (2017), 'The gig economy and contingent work: An occupational health assessment', *Journal of Occupational and Environmental Medicine*, **59** (4), 63–66.

Tremblay, D. and L. Thomsin (2012), 'Telework and mobile working: analysis of its benefits and drawbacks', *International Journal of Work Innovation*, **1** (1), 100–113.

Van Doorn, N. (2017), 'Platform labor: on the gendered and racialized exploitation of low-income service work in the "on-demand" economy', *Information, Communication and Society*, **20** (6), 898–914.

Vandaele, K. (forthcoming), 'Collective resistance and organizational creativity amongst Europe's platform workers: a new power in the labour movement?' in J. Haidar and M. Keune (eds), *Work and labour relations in global platform capitalism*, Cheltenham, UK and Northampton, MA, USA: Edward Elgar Publishing and Geneva: ILO.

Vayre, E. and A. Pignault (2014), 'A systemic approach to interpersonal relationships and activities among French teleworkers', *New Technology, Work and Employment*, **29** (2), 177–192.

Vendramin, P. and G. Valenduc (2018), 'Giga-bits et micro-jobs: l'expansion des petits boulots dans l'économie digitale', in M. Somers (ed.), *Vorm geven aan digitale tijden*, Brussels: Minerva, pp. 78–95.

Vereycken, Y. and M. Lamberts (2019), *Borders between dependent and atypical employment in Europe*, Leuven: HIVA.

Williams A. C. (2020), 'Systems for managing work-related transitions', accessed 18 December 2020 at https://acw.io/pubs/awilliams-dissertation-2020.pdf.

Woodcock, J. (2020), 'The algorithmic panopticon at Deliveroo: measurement, precarity, and the illusion of control', *Ephemera*, **20** (3), 67–95.

Woodcock, J. and M. R. Johnson (2018), 'Gamification: what it is, and how to fight it', *The Sociological Review*, **66** (3), 542–558.

Xia, H., Y. Wang, Y. Huang and A. Shah (2017), '"Our privacy needs to be protected at all costs": crowd workers' privacy experiences on Amazon Mechanical Turk', in *Proceedings of the ACM on Human–Computer Interaction*, New York: Association for Computing Machinery.

Zuboff, S. (2015), 'Big other: surveillance capitalism and the prospects of an information civilization', *Journal of Information Technology*, **30** (1), 75–89.

7. How place and space matter to union organizing in the platform economy

Benjamin Herr, Philip Schörpf and Jörg Flecker

7.1 INTRODUCTION

New forms of work organization associated with platform labour challenge the notion of the traditional workplace, the employment relationship and the given institutional setting of a nation state. Such conventional spatial and institutional 'containers', where work traditionally takes place, are clearly losing importance in the platform economy. In particular, online platform labour seems to be 'placeless', which is not actually the case because the work takes place somewhere, its spatial distribution is not arbitrary and employment is influenced by local and national conditions. Location-based platform work, on the other hand, is not fully limited to the local scale not least because many platforms are multinational corporations. While place and space still matter the platform economy clearly has an impact on labour relations by, among other things, altering spatial structures.

In this chapter we develop the argument that workers are active spatial agents who are able to overcome restrictions at various scales. Following this line of thinking this chapter asks about the challenges and opportunities for trade union organizing in the platform economy. To do so we apply a framework for interpreting union strategies in the platform economy from a spatial perspective by taking account of the social relationships between platform operators, clients, workers and unions at different geographical scales: local, national and transnational. Workers pursue 'politics of scale', for instance by scaling up organizing activities from the local to the transnational. This chapter tackles worker agency and thereby overcomes a narrow analysis of capital–labour relations that restricts the focus to movements of capital (Herod 1997). The main interest of the chapter is how workers and organizations overcome their spatial dispersion and isolation; therefore, this chapter looks at organizing through trade unions and through other forms of collective organization.

7.2 PLACE AND SPACE INTERLINK IN THE PLATFORM ECONOMY

There is a fundamental spatial difference between work on online platforms and location-based platform work. Online labour platforms potentially address a global pool of labour and thereby seem to be 'placeless'. However, path dependencies relating to language areas or former colonial ties spatially structure the triangular relationships between platform, clients and workers (Eurofound 2015; Will-Zocholl et al. 2019). Despite the evidence of global competition and processes of social and institutional detachment there are strong indications that place remains important even for online platform work. As shown in Graham et al. (2017) and the Online Labour Index (Kässi and Lehdonvirta 2018), by far the largest numbers of digital workers are located in India, Bangladesh, Pakistan and China while the 'employer countries' are dominated by the United States (US) and European countries.[1] Furthermore, such work – as with any other type – takes place somewhere even if online platform workers work from home (Graham and Anwar 2019). Additionally, even if they are isolated from other workers in the country, when engaging in online platform work they might check the payment offered against local or national income levels or costs of living. They are, thus, to some extent, always 'territorially embedded' (Hess 2004). Clients, in turn, may show a preference for particular nationalities and places of residence of workers (see Chapter 14). Thus, place and space do not become irrelevant in the context of online platform labour.

In contrast, location-based platform work seems to operate only at a local scale between platform clients and workers. At the same time platform companies such as Uber or Deliveroo are multinational corporations which again has an impact on workers and unions. The case of Lieferando in Vienna, discussed below, shows that they can engage in politics of scale to undermine the power of workers' representatives. Moreover, some local platforms centralize certain functions across their operations. Deliveroo, for instance, provides customer service and also certain services for riders remotely from a low-cost location. However, local platforms are not in a position to influence local regulations by playing states against each other as is common with multinationals producing tradeable goods or services (Dicken 1998). If faced with regulations they do not like, they rather tend to ignore them and, if the local regulator insists on applying them, may opt to leave the market. There have also been cases of local platforms leaving markets in reaction to workers' organizing efforts, as seen with Delivery Hero in Canada (see Chapter 17).

As platform labour is neither 'placeless' nor 'only local', social relationships between platform operators, clients, workers and unions may span various spatial scales from the local to the global. In the literature, the interlinkages between place and space have been subject to considerable debate for quite some time. Soja (1988) claims that place and space are interlinked because, even though social relations are place-bound – as they are carried out in a specific geographically defined entity – they cannot be separated from wider processes. This idea has been taken up by Robertson (1994) who, in a similar vein, points to the local and the global being mutually constitutive. With this in mind, the global has been introduced (Tarrow 2001) in an attempt to overcome the nation state as the primary angle of analysis. To think outside the 'national container' (Massey 1995) has therefore become an important step towards a relational view on place and space highlighting the interconnection of spatial scales (Brenner 2001; Massey 1999; Swyngedouw 1997).

What is more, place and space are always politically contested and the politics of scale is a relevant concept to union organizing. Scale is a particular political terrain (Jonas 1994) where social action takes place (Swyngedouw 2004). In this line of thought, scales – the interrelationships between various geographical entities – are produced and reproduced through political struggle (Brenner 2001) in which labour is an active geographical agent (Herod 1997). As in other industries, union representation and organizing may thus imply activities at multiple scales combining, for example, local support with transnational negotiations but also tapping into virtual spaces where workers overcome the lack of a common workplace. Platforms' spatial flexibility and their interlinkages of different scales calls for a relational view on place and space. Hence, what is needed is a space-sensitive analysis which takes potential multi-scalar activities into account.

As online platform labour is not simply placeless and global and location-based platform work is not simply local, it is necessary to look at the spatial aspects of this type of labour when analysing the challenges and opportunities for union organizing. The development of workers' solidarity and their organizing in unions has always been facilitated, at least in the case of industrial unions, by the spatial co-location of large numbers of workers in one factory or office building. Conversely this was hampered by the scattering of workers over many small branch establishments or by work being mobile, for example among lorry drivers. In addition, economic transnationalization and the increase in outsourcing and subcontracting turn the workplace into a necessary but insufficient base for union organizing (Anderson et al. 2010). The spread of transnational companies has resulted in a loss of power for unions who, in turn, have resorted to politics of scale and tried to raise negotiations at the transnational scale.

7.3 SCALES AT WORK

Unions have experience in multi-scalar activities 'ranging from the workplace … through national structures … to regional organisations and international structures' (Berglund and Waddington 2020, p. 241). Typically, they link different localities and bundle them to a higher degree of organization in order both to enact solidarity and to bridge up to the national scale as a means of creating centralized confederations (Morgan and Pulignano 2020). While some sets of social relationships and institutions are scale specific, others stretch over various scales (Castree et al. 2004). In this chapter, we aggregate the scales to three: workplace/local; national; and transnational.

The most concrete scale is the workplace which is located in a particular community or city; that is, a discrete piece of a larger geographical mosaic (Crang 1999). Rather than being the least influential scale, workers there are active spatial agents and potentially shape work relations through their activities (Herod 1997). However, the spatially fragmenting practice of platform work causes difficulties for unions.

The national scale embeds workplaces in terms of regulation (Johnston 2020) and thereby provides particular paths for unions. Joyce et al. (2020), for instance, indicate that the particular set of industrial relations shapes the ways in which workers gain collective voice. In contrast, if people work remotely across borders, i.e. on a transnational scale, difficulties arise in applying national regulation (Gerasimova et al. 2017). De Stefano and Wouters in Chapter 8 in this volume, for example, highlight the difficulties in applying national labour law to online platform work.

By transnational we understand relations that connect workplaces to larger geographical scales (Castree et al. 2004) ranging from the regional to the European or larger global level. Due to the geographic expansion of companies and production networks and, in particular, because of economic transnationalization and globalization, unions have increasingly targeted several spatial scales. For example, Anderson et al. argue that in light of the 'wider spatial architecture of capital … unions now need to develop deeper and qualitatively different relationships across national borders' (2010, p. 383). In their understanding, multi-scalar union practices focus scales beyond the workplace through 'the development of extra-workplace networks' (2010, p. 384). Overcoming the 'national container' (Massey 1995) includes, for instance, transnational campaigns (Bronfenbrenner 2007) or focusing on key political-economic nodes to reach zonal agreements (Anderson et al. 2010). Other multi-scalar activities are transnational workers' organizations such as the European Trade Union Confederation, the European trade union federations, European works councils, the International Trade Union Confederation and the global union federations. Within these structures, unions and workers' representatives may pursue transnational strategies and cooperate in solidarity (Pernicka 2015).

The following space-sensitive analysis of labour agency and union mobilizing and organizing in the realm of platform labour distinguishes the local, national and transnational scales and looks at the location and the spatial characteristics of work in location-based and online platforms. These spatial characteristics include the setting of the workspace, the co-location of workers, the geographic distances between workers and between clients and workers, and the location of platform operators. In doing so, it also describes the complaints and issues relating to place and space, including discrimination on the grounds of a particular nationality, the availability of workspace or the absence of rooms in which to meet. It considers the territorial embeddedness (Hess 2004) of both employment relations and union activities such as workers' local ties and relationships, unions' local and regional resources and the characteristics of interest associations. In addition, it analyses the politics of scale (Swyngedouw 1997); that is, how social actors constrain, create and shift spatial scales in order to consolidate or fight power relations, for instance by escalating local conflicts to a national or supranational scale or through a dispute over the applicability of national legislation.

7.4 LOCATION-BASED PLATFORM WORK: OVERCOMING INVISIBILITY AT THE WORKPLACE AND SCALING UP TRANSNATIONALLY

Local platform work encompasses social relations, institutions and work practices, all of which are place-bound. This means the service is bound to the local scale and, in the main, concentrated in cities. Thus far, research on local platform labour has largely focused on delivery services and transportation (Rosenblat 2018; Vandaele 2018; Veen et al. 2020; Shapiro 2018; Griesbach et al. 2019; Chai and Scully 2019) while less visible services located in the private sphere, such as care work, have been largely neglected (Ticona and Mateescu 2018). Space interlinks with challenges and opportunities for organized labour at several scales: the workplace; the regional or national; and the transnational.

At a workplace scale, the absence of co-location with other workers contributes to workers' isolation and a weakening of workers' power. Spatial proximity between workers promotes the gaining of a collective voice whereas a geographically dispersed workforce consisting of isolated workers, together with huge turnover, hampers that (Johnston and Land-Kazlauskas 2018). As union initiatives aim to strengthen workers' power, overcoming spatial fragmentation is key. Hatton (2017) argues that several intersecting mechanisms make work 'invisible'; notably, spatial dispersion and a lack of co-location increase workers' 'invisibility'. Platforms further this isolation through geographical dispersion. For instance, the taxi platform company Uber does not maintain central dispatch locations where workers could meet (Choudary 2018). Where platform-provided recreation rooms for workers were in place, platforms have eventually shut them down because of the alleged high costs (Herr 2017).

Platform workers in the domestic sphere do not have opportunities to get together with others, in contrast to platform workers in the public one. One such opportunity is provided by the waiting locations in the platform food delivery sector, where workers wait for new deliveries, which offer spatial proximity to workers and which have contributed to workers' self-organization over the past few years (Vandaele 2018). The waiting zones at Deliveroo in London, for example, were the breeding ground for the London protests in the summer of 2016 that started a series of protests at food delivery platforms in Europe (Waters and Woodcock 2017). In consequence, it is easier for the publicly visible segment of local platform workers to aggregate, mobilize and organize than it is for the segment of local platform workers working in the private sphere (Tassinari and Maccarrone 2020).

Although platforms shape the labour process in a way that isolates workers from each other (Choudary 2018), workers could engage in digital communities to connect with fellow workers. These include, for instance, workers using online forums and social media to share work-related information (Howcroft and Bergvall-Kåreborn 2019, p. 32). While local platforms apply digital technologies in a particular way that serves their profit interest and thereby accordingly directs work practices, workers have adopted these technologies to overcome spatial restrictions (Howcroft and Bergvall-Kåreborn 2019). Social media (Instagram, Facebook and Twitter) and messenger services (WhatsApp, Telegram and Signal) potentially bridge the missing element of co-location in the labour process. Maffie (2020) argues that, through these interactions, workers not only learn more about their industry but also develop links to other workers which could eventually translate into a collective identity and collective action, echoing earlier findings on how digital networks have an impact on collective action in the private service sector (Wood 2015). For local platform workers on private sites these digitally mediated networks currently serve practical purposes, such as getting advice or information (Ticona et al. 2018; for location-based platform work in the public sphere see Griesbach et al. 2019), but local platform workers in the public sphere have extended these into collective action (Joyce et al. 2020). Providing such a digital place also facilitates interaction between workers which becomes apparent by the various forums that platform workers maintain to cope with their jobs (Gerber 2020).

These are potential networks that unions could tap into in an attempt to overcome platform workers' spatial fragmentation on jobs at private sites and to reach a greater number of workers through virtual means.

Face-to-face contact between workers and its resulting collective visibility facilitates the building of solidarities and therefore establishes union organizing in the workplace (Morgan and Pulignano 2020). For some segments of local platform work it is easier to gain collective visibility than for others. In food delivery, workers are visible on motor scooters and bicycles, often wearing the branding of the respective firm. In contrast, workers in US food delivery platforms and generally in transportation are less visible because they spend most of their active working time in private cars where the connection to the platform is less evident (Griesbach et al. 2019). Workers on private sites have especially low visibility, an area which is traditionally prone to irregular forms of recruitment. Platforms somewhat increase those workers' individual visibility by providing a central dispatch (Ticona and Mateescu 2018).

At the regional or national scale, it has proven to be challenging to apply existing regulations to location-based platform work (Howcroft and Bergvall-Kåreborn 2019). While online platform workers do not share a common national jurisdiction, location-based platform workers are subject to a common regulatory framework (Johnston 2020), providing an appropriate scale for regulation. For instance, Danish service union 3F and cleaning platform Hilfr signed the first collective agreement in the platform economy, followed by several others not only in Denmark but also in Sweden and Norway (Jesnes et al. 2019; see also Chapter 16 in this volume). Also, in Austria, the social partners signed a collective agreement for bicycle riders covering platform couriers in permanent employment (Vida 2019). This indicates that regionally developed institutional and legal resources shape the available and prospective ways for organized labour to go (Joyce et al. 2020). However, the United Kingdom (UK) Fast Food Shutdown of 15 October 2018 shows that institutionalized measures are not the only possible means for union action in the local platform economy. The Fast Food Shutdown linked the demands of platform delivery workers at Uber Eats with the demands of workers in hospitality. Partly organized by Trades Union Congress-affiliated unions and partly by grassroots unions, this protest did not only connect several UK regions but also put strong reliance on workers' self-organization rather than representation (Cant and Woodcock 2020).

At a transnational scale, we see both challenges and opportunities for labour. The transnational nature of labour platforms poses challenges to workers' organizing initiatives. The case of Lieferando in Vienna, for instance, points out how the interplay of different scales challenges the applicability of national legislation. In this case, the Dutch group Takeaway.com, owner of the delivery platform Lieferando, sued the works council in its Viennese branch arguing that the respective branch was not an entity of itself and thus that the founding of the works council was against the law. Shifting the politics of scale, the company denied the existence of a separate local organizational level, thereby depriving workers' representatives of their institutional power.

Multi-scalar union activities have successfully connected workplaces at different local scales with each other – at least for the segment of the location-based platform economy where the workplace is in the public sphere. One such example of the politics of scale is provided by the courier network Transnational Couriers Federation. Founded in October 2018 by 100 couriers from 34 self-organized courier collectives or unions, the network aims to build couriers' organizing potential and enforce key demands such as job security, data transparency and a guaranteed minimum income per hour.

While the Transnational Couriers Federation connects several national scales in Europe, the annual Riders' Days of the German platform delivery workers collective *Liefern am Limit* primarily overcomes the absence of co-location by connecting platform delivery workers in the German-speaking region. *Liefern am Limit* started as a self-organized campaign by platform delivery workers to improve working conditions in this segment of the location-based platform economy and is now a project of NGG, a German service union.

Completely independent of official union structures were the Rideshare Drivers United protests of 8 May 2019. Initially planned to take place in Los Angeles, an international call for drivers to shut off the app on 8 May in support of the Los Angeles protest reached other worker interest groups in large US cities such as Boston, Atlanta and New York City, as well as in the UK which eventually took part in the protest. From a spatial perspective this shows how local forms of labour action, linked through virtual space, potentially scales to transnational forms of protest within an industry. Such mobilization may also translate into transnational workers' organizations, such as the International Alliance of App-Based Transport Workers.[2]

7.5 ONLINE PLATFORM WORK: HOPES FOR SUPRANATIONAL SCALE AND VIRTUAL SPACES

Digital work mediated through online platforms is archetypal of the debate about the delocalization of work. In principle, information and communications technologies allow and facilitate fast and easy communications and coordination across large distances. From a spatial perspective the intermediation between supply and demand over the internet allows for workers and potential co-workers and employers or customers not to be at the same location, or in the same area, region or state, and also shows in the frequently transnational or global operation of platforms (Pongratz 2018).

At the scale of the workplace and in spite of the global reach of online platform work, the work itself is happening somewhere, i.e. in (shared) offices or – most likely – in workers' own homes (Huws et al. 2017). Arguably, compared to more traditional office work in a company, online platform labour is characterized to a lesser extent by regional or national social and institutional contexts. The lack of embeddedness in a firm and the absence of colleagues and temporal structures is subsumed under 'organizational detachment' (Lehdonvirta 2016, p. 56).[3] However, Wood et al. (2019) show the local embeddedness of online platform workers who redistribute work to family members, friends and local co-workers in order to keep up with tight deadlines set by platform clients.

Being part of a global labour market does not mean that the national or regional scale and its social and institutional contexts no longer play a role. When labour is traded like a commodity over the internet (Bergvall-Kåreborn and Howcroft 2014), we see a detachment of labour from the territorial legal framework. For instance, it is often unclear which jurisdiction should be applied to the categorization of online workers as employees or as self-employed (De Stefano 2017). Nonetheless, research suggests that regional or national contexts do have an impact on online platform work (see Chapters 12 and 14). For instance, median wages on the online platform studied by Graham et al. are 'low in low-income countries and … significantly higher in medium- and high-income countries' (2017, p. 142). One explanation for this is online platform workers' actual embeddedness in local institutional structures, laws and juridical decisions and the different costs of living; another explanation relates to what is termed 'liabilities of origin' (Lehdonvirta et al. 2015), meaning that digital workers from developing countries face discrimination based on the country of residence displayed on their online profile. The pattern of the geographical imbalance, with low-income countries delivering services and higher-income countries buying them, may also be interpreted from a workers' perspective. This means that online platform labour allows for 'skill arbitrage' where workers are able to sell their work where they receive the highest wage which, in turn, leaves digital workers less reliant on the local labour market (Graham et al. 2017).

For online platform work the transnational scale is the most obvious: the availability of an online workforce and the possibility to perform jobs more or less anywhere has led to the notion that digital online platforms contribute to the creation of a 'planetary labour market' (Graham et al. 2017). Work is globally connected and online platforms facilitate 'temporary states of co-presence between workers and employers' (Graham and Anwar 2019). In the case of a high substitutability of workers, it makes sense to speak about a 'global reserve army' for virtual work (Huws 2003). To create a global labour market, work needs to be easily tradable over large distances. This often means that work needs to be reorganized, mostly through standardization and codification, into tasks that can easily be outsourced over the web and performed with little or no communication, complex instruction or the need for extensive control.

When it comes to organizing or forming a collective voice, much of the debate on remote platform labour gives the impression of a decisive shift of spatial scales, as the transnational scale is becoming dominant at the expense of local and national scales. In the environment of online platforms, workers share no common shopfloor and have little way of meeting physically while communication between workers over the platforms' own communications channels proves difficult (Wood et al. 2018). Even though there are accounts of digital workers using their own local networks to recruit friends and family members to carry out digital work for them (Wood et al. 2019), worker-to-worker communication happens predominantly through virtual channels. Taking account of the new spatial configuration digital workers seem to 'reunite in "virtual places"' (Lehdonvirta 2016, p. 69), in forums, on blogs or watching tutorials; furthermore, networking may become a multi-scalar activity and transnational connections to other digital workers may exist alongside ties in local communities. Wood et al. (2018) found that online freelancers tend to organize in social media groups while micro-workers prefer to use forums. Workers on Amazon Mechanical Turk seem to be part of online communities using forums or Facebook (Kuek et al. 2015), while workers on other platforms are engaged in virtual places, including the built-in chat and messaging systems provided by the platforms themselves (Lehdonvirta et al. 2015). Such online communities are primarily used for exchanging information about digital labour, about experiences with platforms and employers and for forming networks among workers. However, virtual communication between digital workers is not always possible on the platforms' messaging systems and in some cases is even actively prevented (Choudary 2018). Where the creation of communications spaces on the platforms' own forums or chats is supported, such spaces have rather been used to exert indirect management control through 'a form of decentralized self-regulation' and do not seem to improve workers' power (Gerber 2020, p. 20).

There are many examples which show how unions and other initiatives transcend national borders and go beyond their territorial area of responsibility. In spite of the increased importance unions have recently attached to online platform labour, unions obviously face difficulties in addressing and organizing digital workers locally as the mostly anonymous workforce is hard to identify, lacks organizational attachment and cannot easily be approached locally via traditional channels.

7.6 CONCLUSIONS

Platform labour represents an escalation of recent developments towards less protected and precarious work as it combines the fragmentation of employment, predominant self-employment or bogus self-employment and frequent low wages. In addition, the spatial scattering of workers and their separation from each other makes it difficult to organize or be addressed by existing unions. At first glance, space and place seem to be of little importance to platform labour as digital work over online platforms seems to be 'placeless' whereas platform-mediated service work appears only as local. In both cases, however, the location and spatial characteristics of work and employment relations are important for working conditions, opportunities for workers' solidarity and union strategies. In location-based platform labour, work takes place in public spaces or private households implying particular challenges for workers. Online platform work may be carried out everywhere the internet can be accessed. The place of work has consequences for workers' possibilities for face-to-face contact with each other and for meeting regularly and developing solidarity. Analysis shows that different spatial scales – the local, the national and the transnational – need to be considered when it comes to assessing working conditions, labour agency and union organizing.

In location-based platform labour, space and place are often contentious issues and, as such, lead to workers' grievances. In transport and delivery work, this relates to the distances that couriers need to cover, as well as meeting places, waiting rooms and bicycle repair workshops. Experiences show that meeting places provide opportunities for workers to communicate and create a collective voice. This may result in protests, as in the case of platform food delivery workers in many European cities. There are indications that platform companies rarely provide such places and have abolished them to cut costs. Occasionally, unions do step in and provide locations for bottom-up organizing. Workers also try to replace physical places with virtual ones when using communications tools to share information. Through the medium of the internet, local initiatives are being connected to form transnational networks. Such multi-scalar activities provide mutual support and help to address transnational employers in the sector.

In contrast, online platforms for labour do not address a local workforce but provide opportunities to employers to widen their geographical reach and often to tap into a global labour market. Workers thus rely on virtual means if they want to get in touch with each other. However, platform companies usually do not provide opportunities for workers' communication, often strictly limiting their channels to the exchange between clients and workers or otherwise using them to enhance managerial control. To overcome these barriers, bottom-up initiatives have developed websites, forums and social media groups to share information between online platform workers. The focus of union activities is on providing information to workers over the internet while they seem unable to overcome the spatial dispersion of workers at local level.

Although both local and online platform labour appear to be disembedded, the institutional context of a particular city, region or nation state does play a role. Local service platforms were created to disrupt service markets and to circumvent employment relations. Yet, these business models remain contested. Legislation in California in 2019 clearly showed that disembedding from territorial employment relations may be reversed (see Chapter 8). Online platform labour is characterized by a larger degree of detachment from cultural and legal norms. Nevertheless, workers assess the conditions of platform work against local and national practices. Similarly, employers may not only have preferences regarding workers' country of residence but also discriminate between countries when it comes to cost and performance expectations. As a result, the global labour market is not homogenous. Rather, what interests employers are the very possibilities of conducting labour arbitrage.

Spatial fragmentation in platform work limits the opportunities for union organizing. In this chapter we have shown that workers are active spatial agents able to overcome restrictions at various scales. Digital infrastructures have become central tools to workers' (self-)organizing efforts. Research into platform labour shows that space and place do play a prominent role when it comes to working conditions, labour agency and union organizing. Anwar and Graham (2019) show how even unorganized micro-struggles among online platform workers could inform political contestation at various scales. Using crowdsourcing tools to enhance social dialogue in the spatially fragmented work settings of platform work is another possible tool for enhancing union organizing.

What is more, the relationship between different spatial scales and the capacities to bridge those scales turns out to be highly important. Location-based platform workers have faced transnational employers and gained countervailing power by turning local protests into transnational ones. Unions have joined forces to create initiatives evaluating platforms in order to provide information to workers and, as a consequence, to influence conditions on the platforms. Both employers and labour activists are thus engaged in a politics of scale, i.e. trying to shift scales in order to challenge power relations. While activists and unions try to enforce national legislation or attempt to escalate conflicts to a transnational scale, employers may question the applicability of national law to their transnational operations. Spatial scales thus not only provide important framework conditions for employment relations, they are also continuously contested.

NOTES

1. https://ilabour.oii.ox.ac.uk/online-labour-index/.
2. http://iaatw.org.
3. When it comes to organizational detachment, this does not seem to be alleviated by identification with the platform as the new employer or with the customer.

REFERENCES

Anderson, J., P. Hamilton and J. Wills (2010), 'The multi-scalarity of trade union practice', in S. McGrath-Champ, A. Herod and A. Rainnie (eds), *Handbook of employment and society: working space*, Cheltenham, UK and Northampton, MA, USA: Edward Elgar Publishing, pp. 383–397.

Anwar, M. A. and M. Graham (2019), 'Hidden transcripts of the gig economy: labour agency and the new art of resistance among African gig workers', *Environment and Planning A: Economy and Space*, **52** (7), 1269–1291.

Berglund, T. and J. Waddington (2020), 'Editorial', *Transfer*, **26** (3), 241–243.

Bergvall-Kåreborn, B. and D. Howcroft (2014), 'Amazon Mechanical Turk and the commodification of labour', *New Technology, Work and Employment*, **29** (3), 213–223.

Brenner, N. (2001), 'The limits to scale? Methodological reflections on scalar structuration', *Progress in Human Geography*, **25** (4), 591–614.

Bronfenbrenner, K. (ed.) (2007), *Global unions: challenging transnational capital through cross-border campaigns*, Ithaca, NY: Cornell University Press.

Cant, C. and J. Woodcock (2020), 'Fast food shutdown: from disorganisation to action in the service sector', *Capital and Class*, **44** (4), 513–521.

Castree, N., N. M. Coe, K. Ward and M. Samers (2004), *Spaces of work: global capitalism and the geographies of labour*, London: Sage.

Chai, S. and M. A. Scully (2019), 'It's about distributing rather than sharing: using labor process theory to probe the "sharing" economy', *Journal of Business Ethics*, **159** (4), 943–960.

Choudary, S. P. (2018), 'The architecture of digital labour platforms: policy recommendations on platform design for worker well-being', ILO Future of Work Research Paper Series 3, Geneva: ILO.

Crang, P. (1999), 'Local-global', in P. Cloke, P. Crang and M. Goodwin (eds), *Introducing human geographies*, London: Routledge, pp. 24–34.

De Stefano, V. (2017), 'Labour is not a technology: reasserting the declaration of Philadelphia in times of platform-work and gig-economy', *IUSLabor*, **2**, 1–16.

Dicken, P. (1998), *Global shift: transforming the world economy*, 3rd edition, New York: Guilford Press.

Eurofound (2015), *New forms of employment*, Luxembourg: Publications Office of the European Union.

Gerasimova, E., T. Korshunova and D. Chernyaeva (2017), 'New Russian legislation on employment of teleworkers: comparative assessment and implications for future development', *Pravo. Zhurnal Vyssheyshkoly Ekonomiki*, **2**, 116–129.

Gerber, C. (2020), 'Community building on crowdwork platforms: autonomy and control of online workers?', *Competition and Change*, **25** (2), 190–211.

Graham, M. and M. A. Anwar (2019), 'The global gig economy: towards a planetary labour market?', *First Monday*, **24** (4), https://doi.org/10.5210/fm.v24i4.9913.

Graham, M., I. Hjorth and V. Lehdonvirta (2017), 'Digital labour and development: impacts of global digital labour platforms and the gig economy on worker livelihoods', *Transfer*, **23** (2), 135–162.

Griesbach, K., A. Reich, L. Elliott-Negri and R. Milkman (2019), 'Algorithmic control in platform food delivery work', *Socius*, **5**, 1–15.

Hatton, E. (2017), 'Mechanisms of invisibility: rethinking the concept of invisible work', *Work, Employment and Society*, **31** (2), 336–351.

Herod, A. (1997), 'From a geography of labor to a labor geography: labor's spatial fix and the geography of capitalism', *Antipode*, **29** (1), 1–31.

Herr, B. (2017), 'Riding in the gig-economy: an in-depth study of a branch in the app-based on-demand food delivery industry', Materialien zu Wirtschaftund Gesellschaft Working Paper-Reihe 169, Wien: Kammer für Arbeiter und Angestellte für Wien.

Hess, M. (2004), '"Spatial" relationships? Towards a reconceptualization of embeddedness', *Progress in Human Geography*, **28** (2), 165–186.

Howcroft, D. and B. Bergvall-Kåreborn (2019), 'A typology of crowdwork platforms', *Work, Employment and Society*, **33** (1), 21–38.

Huws, U. (2003), The making of a cybertariat: virtual work in a real world, New York: Monthly Review Press.

Huws, U., N. Spencer, D. S. Syrdal and K. Holts (2017), *Work in the European gig economy: research results from the UK, Sweden, Germany, Austria, the Netherlands, Switzerland and Italy*, Brussels: Foundation for European Progressive Studies, UNI Europa and Hertfordshire Business School.

Jesnes, K., A. Ilsøe and M. J. Hotvedt (2019), 'Collective agreements for platform workers? Examples from the Nordic countries', Nordic Future of Work Brief 3, Oslo: Fafo.

Johnston, H. (2020), 'Labour geographies of the platform economy: understanding collective organizing strategies in the context of digitally mediated work', *International Labour Review*, **159** (1), 25–45.

Johnston, H. and C. Land-Kazlauskas (2018), 'Organizing on-demand: representation, voice, and collective bargaining in the gig economy', Conditions of Work and Employment Series 94, Geneva: ILO.

Jonas, A. E. G. (1994), 'The scale politics of spatiality', *Environment and Planning D: Society and Space*, **12** (3), 257–264.

Joyce, S., D. Neumann, V. Trappmann and C. Umney (2020), 'A global struggle: worker protest in the platform economy', ETUI Policy Brief No. 2/2020, Brussels: ETUI.

Kässi, O. and V. Lehdonvirta (2018), 'Online labour index: measuring the online gig economy for policy and research', *Technological Forecasting and Social Change*, **137**, 241–248.

Kuek, S. C., C. M. Paradi-Guilford, T. Fayomi, S. Imaizumi and P. Ipeirotis (2015), *The global opportunity in online outsourcing*, Washington, DC: World Bank.

Lehdonvirta, V. (2016), 'Algorithms that divide and unite: delocalization, identity, and collective action in "microwork"', in J. Flecker (ed.), *Space, place and global digital work*, Basingstoke: Palgrave Macmillan, pp. 53–80.

Lehdonvirta, V., I. Hjorth, M. Graham and H. Barnard (2015), 'Online labour markets and the persistence of personal networks: evidence from workers in Southeast Asia', Paper presented at American Sociological Association Annual Meeting, Chicago, 22–25 August, accessed 15 December 2020 at http://vili.lehdonvirta.com/files/

Online%20labour%20markets%20and%20personal %20networks%20ASA%20 2015.pdf.

Maffie, M. D. (2020), 'The role of digital communities in organizing gig workers', *Industrial Relations: A Journal of Economy and Society*, **59** (1), 123–149.

Massey, D. (1995), 'The conceptualization of place', in D. Massey and P. Jess (eds), *A place in the world? Places, cultures and globalization*, Oxford: Oxford University Press, pp. 46–79.

Massey, D. (1999), 'Power-geometries and the politics of space-time', Hettner-Lecture 1998, Heidelberg: University of Heidelberg.

Morgan, G. and V. Pulignano (2020), 'Solidarity at work: concepts, levels and challenges', *Work, Employment and Society*, **34** (1), 18–34.

Pernicka, S. (ed.) (2015), *Horizontale Europäisierung im Feld der Arbeitsbeziehungen*, Wiesbaden: Springer.

Pongratz, H. J. (2018), 'Of crowds and talents: discursive constructions of global online labour', *New Technology, Work and Employment*, **33** (1), 58–73.

Robertson, R. (1994), 'Globalisation or glocalisation?', *Journal of International Communication*, **1** (1), 33–52.

Rosenblat, A. (2018), *Uberland: how algorithms are rewriting the rules of work*, Oakland, CA: University of California Press.

Shapiro, A. (2018), 'Between autonomy and control: strategies of arbitrage in the "on-demand" economy', *New Media and Society*, **20** (8), 2954–2971.

Soja, E. W. (1988), *Postmodern geographies: the reassertion of space in critical social theory*, London: Verso.

Swyngedouw, E. (1997), 'Neither global nor local: "glocalization" and the politics of scale', in K. Cox (ed.) *Spaces of globalization: reasserting the power of the local*, New York: Guilford Press, pp. 137–166.

Swyngedouw, E. (2004), 'Globalisation or "glocalisation"? Networks, territories and rescaling', *Cambridge Review of International Affairs*, **17** (1), 25–48.

Tarrow, S. (2001), 'Transnational politics: contention and institutions in international politics', *Annual Review of Political Science*, **4**, 1–20.

Tassinari, A. and V. Maccarrone (2020), 'Riders on the storm: workplace solidarity among gig economy couriers in Italy and the UK', *Work, Employment and Society*, **34** (1), 35–54.

Ticona, J. and A. Mateescu (2018), 'Trusted strangers: carework platforms' cultural entrepreneurship in the on-demand economy', *New Media and Society*, **20** (11), 4384–4404.

Ticona, J., A. Mateescu and A. Rosenblat (2018), 'Beyond disruption: how tech shapes labor across domestic work and ridehailing', *Data and Society*, accessed 15 December 2020 at https://datasociety.net/wp-content/uploads/2018/06/Data_Society_Beyond_Disruption_FINAL.pdf.

Vandaele, K. (2018), 'Will trade unions survive in the platform economy? Emerging patterns of platform workers' collective voice and representation in Europe', ETUI Working Paper 2018.05, Brussels: ETUI.

Veen, A., T. Barratt and C. Goods (2020), 'Platform-capital's "app-etite" for control: a labour process analysis of food-delivery work in Australia', *Work, Employment and Society*, **34** (3), 388–406.

Vida (2019), 'Weltweit erster KV für Fahrradboten abgeschlossen', accessed 15 December 2020 at www.vida.at/cms/S03/S03_4.8.a/1342616918551/kollektivvertrag/strasse/weltweit-erster-kv-fuer-fahrradboten-abgeschlossen.

Waters, F. and J. Woodcock (2017), 'Far from seamless: a workers' inquiry at Deliveroo', *Viewpoint Magazine*, accessed 15 December 2020 at www.viewpointmag.com/2017/09/20/far-seamless-workers-inquiry-deliveroo/.

Will-Zocholl, M., J. Flecker and P. Schörpf (2019), 'Zur realen Virtualität von Arbeit. Raumbezüge in digitalisierter Wissensarbeit', *AIS-Studien*, **12** (1), 36–54.

Wood, A. J. (2015), 'Networks of injustice and worker mobilisation at Walmart', *Industrial Relations Journal*, **46** (4), 259–274.

Wood, A. J., V. Lehdonvirta and M. Graham (2018), 'Workers of the internet unite? Online freelancer organisation among remote gig economy workers in six Asian and African countries', *New Technology, Work and Employment*, **33** (2), 95–112.

Wood, A. J., M. Graham, V. Lehdonvirta and I. Hjorth (2019), 'Networked but commodified: the (dis)embeddedness of digital labour in the gig economy', *Sociology*, **53** (5), 931–950.

PART II

Regulating platform work

8. Embedding platforms in contemporary labour law

Valerio De Stefano and Mathias Wouters

8.1 INTRODUCTION

The Centenary Declaration for the Future of Work by the International Labour Organization (ILO) in 2019 reaffirmed the continued relevance of the employment relationship 'as a means of providing certainty and legal protection to workers'.[1] This is, to some extent, striking because many new forms of work, especially platform work, often take place outside the boundaries of the employment relationship. Yet, at the same time, the reaffirmation is also understandable. The employment relationship is not only fundamental in accessing social protection, it is also a flexible instrument to govern businesses and facilitate innovation (Aloisi and De Stefano 2020). As such, if we do not move away from the 'employment relationship' as the backbone of labour law, we are left with the task of rethinking the scope of the employment relationship. Our aim here is to do so in the context of platform work by linking up discussions on 'Personal Work Relations' and the 'ABC tests' established in the United States (US).

After having discussed in the first part of this chapter the possibility of classifying platform operators as employers, the second part positions platforms as labour market intermediaries. It explains how the coverage of the Private Employment Agencies Convention 1997 (No. 181) and of the Home Work Convention 1997 (No. 177) is arguably broader than often assumed. This may provide regulators with an additional, second regulatory option in relation to platform operators which cannot be considered employers because they operate more decentralized platforms that leave workers with the ability to conduct a genuine business. Under these conditions, certain platform operators could still be considered private employment agencies, thereby having to abide by the rules applicable to labour market intermediation; whereas certain platform users could become designated as employers of crowdworkers, who are legally classified as homeworkers. Both Conventions are discussed because they need to be taken into account when construing the 'international governance system for digital labour platforms' for which the Global Commission on the Future of Work has called (ILO 2019, p. 44).

8.2 THE PLATFORM AS EMPLOYER

The issue of whether or not platform workers should be legally classified as employees or provided with a more limited set of protections has generated much scholarly debate. This discussion is fuelled by the many lawsuits in which platforms have to justify their use of self-employed workers. Since even a cursory survey of the case law on platform work will show that many platform workers remain in doubt about their employment status, the discussion continues to rage, thereby undermining the ILO Centenary Declaration's claim that the employment relationship is a means to provide certainty to workers. One potential outcome is to base the legal underpinnings of the employment relationship on the idea of 'Personal Work Relations' which can be put into practice through certain ABC tests.

8.2.1 Inconsistent Case Law

For platform workers, many of whom are arguably engaged under 'ambiguous employment relationships … that give rise to an actual and genuine doubt about the existence of an employment relationship' (ILO 2006, p. 73), even small legal details can make the difference. Consider Latin America, where multiple decisions have been rendered in relation to Uber. The Second Labour Court of Santiago, Chile, ruled that Uber drivers are indeed self-employed.[2] The labour courts in Montevideo, Uruguay, on the other hand, ruled in favour of the claimants: both the court of first instance and the appeal court ruled that Uber drivers are employees.[3] The courts in Uruguay justified their rulings based, among other things, on Uber's power to direct and sanction the driver even though the Chilean court had not considered Uber to have such abilities.

Moreover, if we focus on one country the courts may, likewise, not show unanimity. At least two Brazilian courts ruled in 2017 that Uber was the employer of drivers there. One court argued that diffuse control by the many Uber passengers was a strong method of control.[4] A subsequent court ruling had the judge first consider that the Uber driver could not be regarded as self-employed, after which it became plausible legally to conclude that the driver was indeed an employee.[5] Notwithstanding these claimants' successes, not all claimants have been as successful. Some of the lower courts have dismissed drivers' claims, influenced by the idea that Uber is a technology company and not a transportation company;[6] or that drivers have failed to prove subordination[7] or failed to establish a 'directive character in the performance of work'.[8] For now, the higher courts seem to have sided with Uber, for example because '[a]pplication drivers do not maintain a hierarchical relationship with the UBER company because their services are provided on an occasional basis, without pre-established hours and do not receive a fixed salary, which disfigures the employment relationship between the parties'.[9]

Although it is impossible for us to say what classification Uber drivers (ought to) receive under Brazilian law, the country's case law does illustrate the amount of ambiguity involved. This is an issue that also affects legal systems in the Global North. One court in Amsterdam ruled in July 2018 that Deliveroo bikers were self-employed; after which another court in Amsterdam – the same level in the judicial hierarchy – decided that they were employees.[10] The case is currently under appeal. The Australian Fair Work Commission twice had to rule on an unfair dismissal remedy for platform workers. First, the Commission decided that Uber drivers were not employed[11] before subsequently deciding that Foodora bikers, on the other hand, did not manage 'an independent operation' and were in fact employees.[12]

8.2.2 The Employment Relationship: An Endless Cycle of Prospective Changes

Because these rulings do not foster legal certainty, platform work can be considered an incubator that propels discussions on the boundaries of labour law. Such discussions are not novel and, in the past, took place in relation to parcel deliverers, newspaper boys and girls, door-to-door salespeople and so forth. Older research, for instance, has already highlighted that, for the purposes of determining an employment relationship, a 'business test' and an integration or 'organization test' can be applied. The first assesses if the worker has a business on his or her own account, whereas the latter intends to verify whether the worker performs duties that are integral to the business of the potential employer (ILO 1996a, p. 27). These tests were supposedly devised to move beyond the 'test of the right of control', focusing on the existence of control (or subordination) between workers and employers.

These repeated discussions on the boundaries of the personal scope of application illustrate how employees, workers and the self-employed are essentially 'human constructs and ought to be made (and unmade) on the basis of normative choices' (Countouris and De Stefano 2019, p. 19). In this day and age, platform work arrangements entice stakeholders to make new normative choices. One normative choice on display is that of 'Personal Work Relations', in which the aim is to move away 'from the Contract of Employment to the Personal Work Contract as a central organising category for the discussion of the law of the individual relationship' (Freedland 2006, p. 2). Contrary to the former, the latter, for instance, includes independent contractors that personally commit to execute work or labour (2006, p. 4).[13]

Such considerations were never limited to British law even though it might originally have been theorized by authors with a common law background there. This initial research culminated in a seminal book (Freedland and Kountouris 2011) that has, nevertheless, been subject to debate.[14] The ideas unmistakably have academic roots, aimed at expanding the personal scope of labour law to any 'work that he or she does or engages to do in person' in a 'relation' between a working person and another person or enterprise (Freedland and Kountouris 2013, p. 148). More recently the European Trade Union Confederation (ETUC) report on 'New trade union strategies for new forms of employment' has used this to develop more practical normative suggestions. The rather scholarly scope of application has been reformulated because '[u]ltimately, the idea of "personal work relation" can be used to define the personal scope of application of labour law as applicable to [1] any person that is engaged by another to provide labour, unless [2] that person is genuinely operating a business on her or his own account' (Countouris and De Stefano 2019, p. 7). The pattern is clear. A worker is covered (1), unless (2) proven otherwise.

This would extend the range of application of labour law with significant consequences for platform workers. Instead of having the workers prove that they have fixed working hours, are pressured to obey the directions given to them and are supervised by the platform or its users, this pattern would have the platform prove that the worker genuinely operates a business (Countouris 2019, p. 16). To this extent, the question remains what terminology legislators can use to define (1) persons that are engaged by another to provide labour; and (2) what persons are, nevertheless, considered genuinely to operate businesses on their own account. The ABC tests give an indication of how this form of classification mechanism can be put into action. European democracies should take up the challenge and reflect on the possibility to take such an 'employee, unless' style of approach. This is a process which is already underway in the Netherlands (Commissie Regulering van Werk 2020, pp. 70–71).

8.2.3 'ABC' Tests

The use of ABC tests has grabbed the attention of many American scholars. Some even refer to it as 'the most popular legal methodology for distinguishing between employees and independent contractors' (Pearce II and Silva 2018, pp. 1–2). Although it is perhaps permitted to refer to it as a legal methodology, in the sense that every kind of ABC test largely follows a similar pattern, one has to acknowledge the diversity among states' ABC tests. Therefore, it is preferable not to speak of *a test* but *tests* in order to do justice to the diversity among them.

One feature that arguably rests at the core of this 'methodology' is '[a]n initial presumption of employee status', which 'has been more implemented than any other form or feature of the ABC test' (Deknatel and Hoff-Downing 2015, p. 71). As such, instead of the worker having to prove that the relationship classifies as an employment relationship, it is the presumed employer which has to prove that the relationship with the worker cumulatively fulfils conditions A, B and C. If it does so, the worker can legitimately be considered an independent contractor. Consequently, ABC tests usually entail two steps: the first details who is presumed to be employed; while the second describes which individuals are, nevertheless, independent contractors under conditions A, B and C.

Lately this technique has obtained a lot of news coverage due to Assembly Bill (AB)5 in California. In this particular case, (1) the presumption reads: 'a person providing labor or services for remuneration shall be considered an employee'. Subsequently, (2) the exemption reads: 'unless the hiring entity demonstrates that all of the following conditions are satisfied':

> (A) The person is free from the control and direction of the hiring entity in connection with the performance of the work, both under the contract for the performance of the work and in fact.
> (B) The person performs work that is outside the usual course of the hiring entity's business.
> (C) The person is customarily engaged in an independently established trade, occupation, or business of the same nature as that involved in the work performed.[15]

As a result, ABC tests shift the risks predominantly away from the worker failing to prove employment status to the party with the most resources (Pinsof 2016, p. 368), the 'hiring entity', which might fail to prove the independent nature of the worker's activities (Seibert 2019). As a result, under a 'regular pattern', certain workers might be assisted by the law, making it easier for them to prove employment status.[16] Under a 'reversed pattern', however, one could see the opposite occurring in certain cases. AB5 clearly excludes physicians, lawyers, accountants, engineers, etc. from this codified ABC test, clarifying instead that the older 'Borello test' applies.[17] Pursuant to Borello, for instance, any engineer in service to another is presumed an employee, unless proven otherwise, using factors that are deemed relevant under the Borello jurisprudence. These are 'an almost endless number of factors that a court could consider in supporting a finding that the worker is an independent contractor' (Morgan 2018, p. 131). The 'reversed pattern' is, in other words, maintained but the criteria related to step (2) are redrafted to prove more easily that, for instance, engineers are self-employed.

In addition to the legal fine-tuning which could, for example, happen through industrial bargaining, this kind of 'reversed pattern' might also have other, unexpected benefits. A platform such as Uber makes use of opaque algorithms and incentives to influence workers (Rosenblat 2018, pp. 138–166). Having it prove that workers are independent could give an idea of how these algorithms work, which is subsequently important for collective bargaining purposes (De Stefano 2020). This 'new' approach might, therefore, have far-reaching effects that go beyond improving the clarity of the classification test itself and likely turning non-entrepreneurial workers, who are engaged in a platform's ordinary course of business, into employees (Cunningham-Parmeter 2019, p. 427). It might also force platforms to become more transparent.

For these reasons, we would argue that this 'reversed pattern' is well worth exploring. A broad debate should take place about the criteria that can identify persons who are genuinely operating a business on their own account. This does not necessarily require criteria A, B and C; two factors may suffice, or three quite different ones could be used.

8.3 THE PLATFORM AS INTERMEDIARY

Certain platforms grant their workforce a degree of autonomy that, irrespective of the classification mechanism used under national law, will almost certainly result in the designation of these workers as independent contractors. Yet, this does not necessarily mean that these activities are not embedded in the law of the labour market. For instance, the Private Employment Agencies Convention 1997 (No. 181) can be combined with the Home Work Convention 1996 (No. 177) to design a governance system that covers platform operators which are not employers in the usual sense of the word.

8.3.1 Employment Intermediation and Convention No. 181 Concerning Private Employment Agencies

Convention No. 181 can be used to regulate certain platforms. The Dutch ruling on Helpling – a platform which facilitates on-demand cleaning services – is, to our knowledge, the first court case on platform work that relies heavily on the rules governing private employment services. The court ruled that Helpling intermediates employment relationships between cleaners and households. Helpling had, therefore, to stop charging fees to the cleaners.[18] Given the circumstances at hand, it was fairly easy for the court to reach this conclusion because it was never in dispute that an employment relationship existed between Helpling's cleaners and the households. Once the court ruled that Helpling was neither an employer nor a temporary work agency (TWA), the contracts between the parties clearly established an 'employment contract for domestic help' between cleaner and household,[19] thus turning the platform into a modern-day employment intermediator. Had the platform's business model not relied on these 'employment contracts for domestic help', Dutch law would most likely not have applied the very same rules on labour brokerage.[20] That a platform's efforts to bring about an employment contract for casual household services are covered by national laws on private employment services, while similar efforts to bring about contracts for services essentially catering to the exact same household demand are not, is an issue that we have raised previously (De Stefano and Wouters 2019).

To be clear, from a legal point of view, scholars generally consider Convention No. 181 to cover only the intermediation of employment relationships (Blanpain 1999, p. 182; Countouris 2007, p. 158). However, this actually remains up for debate. Article 1(1)(a) of Convention No. 181 makes clear that a platform which provides 'services for matching offers of and applications for employment, without the private employment agency becoming a party to the employment relationships which may arise therefrom' is a private employment agency. It is the word 'employment' that makes it appear as if job-matching services must result in an employment relationship in order for the Convention to designate the service provider as a private employment agency. Yet, upon closer inspection, the word 'employment' does not necessitate an employment relationship.

The predecessor to Convention No. 181, the Fee-Charging Employment Agencies Convention (Revised) 1949 (No. 96), regulated agencies that conducted services 'as an intermediary for the purpose of procuring employment for a worker or supplying a worker for an employer'. TWAs at the time argued that they were not targeted by Convention No. 96 because they did not procure 'employment' for a worker by matching him or her with an employer. This argument was crucial in sustaining the growth of the temporary work industry and partly explains why TWAs decided to employ jobseekers directly so as to strengthen their argument that they were not prohibited/regulated (Freeman and Gonos 2005, p. 300).

Since many ILO member states wanted to enable the temporary work industry, the International Labour Office was asked to answer whether TWAs were regulated/prohibited by Convention No. 96. The Office answered affirmatively, amongst others because '[t]here would seem to be nothing inherent in the terms "employment" and "employer" to give the Convention so restrictive a meaning' (ILO 1966, p. 392). Even if the agency's user enterprise is not the legal employer, and neither is there any employment relationship between the agency worker and the user enterprise, Convention No. 96 still applies (ILO 2010, p. 52). Indeed, contrary to what is sometimes assumed, the ordinary meaning of 'employment', which should be used as the reference to interpret ILO Conventions,[21] does not require an employment contract/relationship.[22]

Those who want to dispute Convention No. 181's broad scope will refer to the second subclause of article 1(1)(a) which, of course, explicitly mentions 'employment relationships'. Yet this reference should also not necessarily be interpreted as limiting the broad, ordinary meaning of the word 'employment' in the first subclause. Based on publicly available working documents and the history that precedes Convention No. 181, it is completely plausible to argue that the second subclause aims to exclude the services of TWAs, that are separately defined in article 1(1)(b),[23] from also being considered 'services for matching offers of and applications for employment' under 1(1)(a). Hence TWAs that become a party to the employment relationships that may arise from job-matching services are only designated a private employment agency under 1(1)(b), not both 1(1)(a) and 1(1)(b).

This historically sensitive interpretation would, moreover, align with the views of the chairperson of the Committee on the Revision of Convention No. 96 and the views of the Committee's employer vice-chairperson, who both argued in favour of a 'living convention', as well as the worker vice-chairperson who wanted Convention No. 181 to be future proof (ILO 1997, pp. 267–281). If platforms intermediating contracts for personal services were covered by the regulation on private employment services, it would, furthermore, alleviate the pressure on established agencies that do apply the regulations (ING Economics Department 2018), at a time when even public employment agencies are rolling out their services to self-employed workers (European Public Employment Services 2019). What is more, certain ratifying states, such as Belgium, already have phrasing in their domestic laws that arguably extends the coverage of these regulations to the intermediation of contracts for personal services.

As illustrated by the Helpling case, this can be significant for local on-demand work. In addition to the articles of Convention No. 181 which prohibit fee-charging, prescribe complaint mechanisms and so forth, domestic regulations on private employment services could force these platforms to accept the codes of conduct which are issued by public authorities, are enforceable and reach beyond what the Convention demands. This would be a flexible solution that could cover those platforms intermediating between workers and enterprises or households, such as Care.com, Instacart and others, with the exception of those providing peer-to-peer services to pure end-market consumers, such as transportation network companies. The regulation on employment intermediation could additionally apply – at least in theory – to online platforms, both with regard to macro- and micro-tasks.

8.3.2 Online Platforms and Convention No. 177 Concerning Home Work

In 1990, the International Labour Office presented the 'geographical dispersal of the workforce on a global scale, known as "off-shoring"' as telework's 'most spectacular advantage to the employer' (ILO 1991, p. 25). A couple of decades later, online platforms present us with 'new offshoring institutions' that can serve this purpose (Lehdonvirta et al. 2019). The ILO Committee of Experts recently stated that:

> [I]nsofar as digital platform work or crowdwork is carried out at home or in a place other than the employers' premises on a regular basis and for remuneration, it could fulfil the conditions to be considered a form of regular home work, and as such could be covered by the provisions of the [Home Work] Convention. (ILO 2020, p. 235)

The Committee has, arguably, expressed itself rather carefully. According to our reading, ratifying member states may have little choice than to consider certain forms of crowdwork, by which we mean certain forms of work performed through online platforms, to constitute 'home work' covered under Convention No. 177.

Pursuant to article 1, a platform user that gives out 'home work in pursuance of his or her business activity' is an employer. 'In pursuance of his or her business activity' covers not only 'profit-making activities', but also 'the "business" of governmental or non-profit organizations' (ILO 1996b, p. 39). For instance, universities that use platform workers for research purposes are potentially covered.

This is on the condition that crowdworkers are performing 'home work', which means:

> [W]ork carried out by a person … (i) in his or her home or in other premises of his or her choice, other than the workplace of the employer; (ii) for remuneration; (iii) which results in a product or service as specified by the employer, irrespective of who provides the equipment, materials or other inputs used.

If the crowdworker's work fulfils these conditions, the crowdworker is, in principle, a homeworker covered by the Convention. This is the case unless, according to article 1(a), the homeworker 'has the degree of autonomy and of economic independence necessary to be considered an independent worker under national laws, regulations or court decisions'. The question is thereby not whether 'a presumed employee' has the degree of autonomy and of economic independence necessary to be considered an independent contractor, but whether the homeworker's degree of (economic) autonomy is sufficient to be considered an independent home-based worker. And it is, in this respect, completely reasonable to argue that certain crowdworkers should not be perceived as having a 'truly independent and entrepreneurial status' (ILO 1995, p. 37), and ought, as such, to be considered homeworkers covered under Convention No. 177.

The ordinary meaning of autonomy has to do with an 'individual's capacity for self-determination'[24] or 'self-governance'.[25] Autonomy should not be conflated with legal subordination, as illustrated by employees who can, nevertheless, experience significant autonomy (Hendrickx 2019, pp. 375–378). Therefore, if we conceive 'the degree of autonomy necessary' to be an independent worker in the sense of Convention No. 177 to presuppose at least a little 'autonomous agency in work' (Breen 2019, pp. 59–60), the result would arguably be that certain crowdworkers, predominantly on micro-tasks, cannot be considered to have the necessary degree of autonomy unless we want to disregard the ordinary meaning of autonomy completely. Crowdworkers that correspond to the 'basic definition' of homeworkers in article 1(a)(i–iii) and regularly provide services on a platform like Amazon Mechanical Turk do not have the degree of autonomy necessary to be considered an independent home-based worker. In fact, they have so little discretion when completing work orders that the service requester on the platform does not even care to know the identity of who is performing the task. They are, literally, a hired hand.

At the same time, apart from the extreme examples of clickworkers who are engaged without any personal/creative input, we have nevertheless to appreciate that many crowdworkers do feel autonomous and are able to exercise discretion (Wood et al. 2019, pp. 64–65). Therefore, a discussion should be had about the factors that signal a crowdworker's degree of (economic) autonomy, such as their ability, among others, truly to (re)negotiate piece rates and to provide direct services for the same client, i.e. off the platform. In relation to crowdworkers who cannot be considered independent home-based workers, Convention No. 177 would, among other things, demand that the 'national policy on home work' promote 'as far as possible, equality of treatment between' these non-autonomous crowdworkers and 'other wage earners'. Also, the respective responsibilities of platform users and online platforms would have to be determined by laws, regulations or court decisions.

8.4 AN INTERNATIONAL GOVERNANCE SYSTEM FOR PLATFORMS

A broad interpretation of Convention No. 177 and of the scope of Convention No. 181 provides us with two regulatory pinpoints for a system of the international governance of platforms. Pretty much all online platforms could, in principle, be covered through article 1(1)(a), or alternatively 1(1)(c), of Convention No. 181 on Private Employment Agencies; while Convention No. 177 concerning Home Work could apply to crowdworkers who classify as homeworkers and who are not considered independent home-based workers under national law. Furthermore, by recalling that unratified conventions are essentially equal to ILO recommendations (Maupain 2000, p. 375), and ILO member states' commitment in paragraph 22 of Recommendation No. 198,[26] the ILO arguably already has the necessary leads.

The issue that must be raised is what national law governs the relationships between crowdworkers, platforms and requesters (Cherry 2019). At first sight, one would expect the member state in which the crowdworker resides to determine at what point homeworkers/crowdworkers are independent home-based workers; and the member state whose law governs the legal status of platform operators to determine the rules applicable to the international intermediation of 'employment'. However, this may not turn out to be the case due to legal drafting, particularly choice-of-law clauses. Nor might this have desirable effects, considering how requesters in the Global North usually rely on workers in the Global South through platforms governed by common law legal systems that generally present less regulated environments for labour market intermediation.

In this respect, the Maritime Labour Convention (MLC) 2006 has often been cited as a source of inspiration for crowdworkers.[27] Similarly, ambitious discussions could be held on a 'Crowdworkers' Bill of Rights', like those had on the MLC's 'Seafarers' Bill of Rights' (Charbonneau and Vacotto 2019). Crowdworkers could, likewise, benefit from a designated mechanism for the implementation of such a bill. This mechanism would, however, require an ILO Convention on crowdwork, instead of an ILO Recommendation or Resolution, with complementary domestic regulation implementing the Convention. In practical terms, this mechanism might, furthermore, require ILO member states to monitor the requesters of crowdwork in their legal system, similar to how they exercise 'control over ships that fly [their] flag'.[28] This would be a real challenge considering how crowdwork's processes are less tangible, being based on computer chips instead of ships. Additionally, the successful implementation of this kind of bill would, seemingly, demand the utmost of the 'principle of solidarity' between ILO member states (Supiot 2020), for instance in relation to the taxation of crowdwork, as well as willingness from the employers' group to restrict telework's 'most spectacular advantage' (ILO 1991, p. 25), now facilitated through platforms.

These conditions seem unlikely to arise under a separate, designated standard-setting initiative. Perhaps the discussions on crowdwork would be less tense if they featured in a broader discussion on the revision of the instruments on homeworking: a revision that is timely in the light of the Covid-19 pandemic and policymakers' recent experiences with homeworking. This revision is, moreover, not necessarily problematic because such instruments have, in any case, mainly served as 'a benchmark and springboard' to gain recognition and rights for homeworkers through campaigning, organizing, etc. (Delaney et al. 2016, pp. 163–164). This is something that can be sustained by holding renewed discussions on homework that cover, *inter alia*, industrial homeworkers in global supply chains, teleworkers and their environmental impact, and crowdworkers engaged through online platforms.

8.5 CONCLUSION

Platform work has posed many challenges to domestic regulators. Some countries, such as Belgium and France, have responded by issuing rules specific to platform work while others might do so in the near future, such as Colombia and Argentina. Although these tailored measures can prove beneficial to formalize platform work and protect platform workers, one should not dismiss platform work's broader repercussions for the labour market. Digital labour platforms have ingrained themselves at the intersection between various pre-existing, and difficult to regulate, labour practices, including very short-term contracts, piece rate remuneration, off-shoring and unbridled employment intermediation.

Policymakers and social partners should be aware of how the broader regulatory frameworks that cover these labour practices can be used to chip away at the big, unregulated monolith that platform work presents. First, regarding the legal concept of the employment relationship, policymakers can build on the idea of Personal Work Relations and the ABC tests as a method for implementation. The ETUC Action Programme 2019–2023, for example, advocates 'for any person engaged in a "personal work relationship"' to have collective rights, including the right to strike, regardless of competition rules. Furthermore, the Dutch trade union FNV has acted on the Dutch Commission's proposal for an 'employee unless' approach. Both of these policies move in a similar direction, leading to a situation in which perhaps not all, but at least some, platform workers cannot be considered to be running a genuine business operation and should be provided with (almost) all, or at least some, labour protections. These endeavours should continue, taking into consideration the use of ABC tests in the US and ongoing litigation on how to apply AB5 to platforms in California. These US developments showcase the legitimacy of these policy proposals and the need to explore what prevents them from being implemented in a European context.

An additional issue that should be raised at country level is the option to explore a more modern take on employment intermediation, in which the intermediation of contracts for personal services is also covered by regulation. This could provide independent platform workers with certain guarantees against abusive behaviour by platforms and platform users, both enterprises and households.

Lastly, the ILO can make a valuable contribution to the regulation of platforms. It should explore the potential of Convention No. 181 and Convention

No. 177 in light of the call for an international governance system for online platforms. Additionally, although Covid-19 might raise many other issues, it also provides momentum to re-engage in discussions on Convention No. 177. One phenomenon that merits close attention is the offshoring of digital work through online platforms. Discussions on Convention No. 177 would present a real opportunity to uncover the 'invisible labour' taking place through online platforms and to appeal to member states' commitment to 'decent work for all' which, in this day and age, arguably extends to workers who generate wealth from abroad, like transnational teleworkers and crowdworkers.

NOTES

1. This chapter was also prepared within the framework of the Odysseus grant 'Employment rights and labour protection in the on-demand economy' granted by the *FWO Research Foundation – Flanders* made to Valerio De Stefano.
2. 2º Juzgado de Letras del Trabajo de Santiago 14 July 2015, No. O-1388-2015.
3. Juzgado Letrado del Trabajo de Montevideo de 6º Turno 11 November 2019, No. 77/2019; Tribunal de Apelaciones de Trabajo de Montevideo de 1° Turno 3 June 2020, No. 0002-003894/2019.
4. 33ª Vara do Trabalho de Belo Horizonte 13 February 2017, No. 0011359-34.2016.5.03.0112.
5. 42ª Vara do Trabalho de Belo Horizonte 12 June 2017, No. 0010801-18.2017.5.03.0180.
6. 37ª Vara do Trabalho de Belo Horizonte 30 January 2017, No. 0011863-62.2016.5.03.0137.
7. 42ª Vara do Trabalho de Belo Horizonte 20 July 2017, No. 0011231-35.2015.5.03.0181.
8. 12ª Vara do Trabalho de Belo Horizonte 30 May 2017, No. 0010044-43.2017.5.03.0012.
9. Superior Tribunal de Justiça 28 August 2019, No. 164.544 – MG (2019/0079952-0); also see Tribunal Superior do Trabalho 5 February 2020, No. TST-RR-1000123-89.2017.5.02.0038.
10. Rechtbank Amsterdam 23 July 2018, No. ECLI:NL:RBAMS:2018:5183; Rechtbank Amsterdam 15 January 2019, No. ECLI:NL:RBAMS:2019:198.
11. Fair Work Commission 21 December 2017, No. [2017] FWC 6610.
12. Fair Work Commission 16 November 2018, No. [2018] FWC 6836.
13. The textbook definition of an employment contract or relationship will mention that the worker has to be subordinated to, or perform the work under the control of, the employer. This is not necessarily the case for Personal Work Relations or a personal work contract.
14. In particular, the book symposium on Mark Freedland and Nicola Kountouris (2013).
15. Section 2750.3 Labor Code of California.
16. For example, section L7112-1 code du travail français.
17. Section 2750.3 Labor Code; *S. G. Borello & Sons, Inc.* v. *Department of Industrial Relations* (1989) 48 Cal.3d 341 (Borello).
18. Rechtbank Amsterdam 1 July 2019, No. ECLI:NL:RBAMS:2019:4546.

19. The contracts are based on a specific regulatory framework called 'Home services regulation' (Regeling dienstverlening aan huis). This regulation gives rise to a particular employment contract called 'Employment contract for domestic help' (Arbeidscontract hulp in huis).
20. The definitions clarify that the aim of employment intermediation is to 'establish a contract of employment under civil law' or an appointment as a civil servant. Section 1 Act on the allocation of workers by intermediaries, i.e. Wet allocatie arbeidskrachten door intermediairs.
21. Article 31 Vienna Convention on the Law of Treaties, 23 May 1969.
22. Employment means '[t]he quality, state, or condition of being employed; the condition of having a paying job'. Black's Law Dictionary (11th ed. 2019). 'The act of employing or being employed. The occupation, business, or profession to which one devotes his services, time, and attention, and which he depends upon for livelihood or profit.' Ballentine's Law Dictionary (3rd ed.).
23. '[S]ervices consisting of employing workers with a view to making them available to a third party, who may be a natural or legal person (referred to below as a 'user enterprise') which assigns their tasks and supervises the execution of these tasks.'
24. Black's Law Dictionary (11th ed. 2019).
25. The Wolters Kluwer Bouvier Law Dictionary Desk Edition.
26. 'Members should establish specific national mechanisms in order to ensure that employment relationships can be effectively identified within the framework of the transnational provision of services. Consideration should be given to developing systematic contact and exchange of information on the subject with other States.'
27. Like any other Convention, the MLC requires member states to act in order for the MLC to become operational. However, article 1 immediately obliges ratifying member states 'to give complete effect to its provisions in the manner set out in Article VI'. This approach differs from most Conventions because member states are usually given much more freedom in how to implement a Convention through all kinds of flexibility clauses. The MLC, instead, uses more enunciated phrasing and its structure makes it come across like a Maritime Labour Code. One of the big achievements of the Code is therefore that, despite this stringent language, it has achieved 97 ratifications, which is a big number compared to other 'recent' International Labour Standards, many of which are written in a far less stringent manner supposedly so as to make the instruments more 'ratifiable'.
28. Article 5(2) Maritime Labour Convention, 2006, as amended.

REFERENCES

Aloisi, A. and V. De Stefano (2020), 'Regulation and the future of work: the employment relationship as an innovation facilitator', *International Labour Review*, **159** (1), 47–69.

Blanpain, R. (1999), 'Belgian Report', in R. Blanpain (ed.), *Private employment agencies*, Bulletin of Comparative Labour Relations, **36**, 181–200.

Breen, K. (2019), 'Meaningful work and freedom: self-realization, autonomy, and non-domination in work', in R. Yeoman, C. Bailey, A. Madden and M. Thompson (eds), *The Oxford Handbook of Meaningful Work*, Oxford: Oxford University Press, pp. 51–72.

Charbonneau, A. and B. Vacotto (2019), 'La Convention du travail maritime, 2006: renouveau et source d'inspiration du droit international du travail', in G. P. Politakis, T. Kohiyama and T. Lieby (eds), *ILO100: Law for Social Justice*, Geneva: ILO, pp. 769–795.

Cherry, M. A. (2019), 'Regulatory options for conflicts of law and jurisdictional issues in the on-demand economy', *Conditions of Work and Employment Series 106*, Geneva: ILO.

Commissie Regulering van Werk (2020), 'In wat voor land willen wij werken? Naar een nieuw ontwerp voor de regulering van werk', accessed 6 October 2020 at www.reguleringvanwerk.nl/.

Countouris, N. (2007), *The changing law of the employment relationship: comparative analyses in the European context*, Aldershot: Ashgate.

Countouris, N. (2019), *Defining and regulating work relations for the future of work*, Geneva: ILO.

Countouris, N. and V. De Stefano (2019), *New trade union strategies for new forms of employment*, Brussels: ETUC.

Cunningham-Parmeter, K. (2019), 'Gig-dependence: finding the real independent contractors of platform work', *Northern Illinois University Law Review*, **39** (3), 379–427.

De Stefano, V. (2020), '"Negotiating the algorithm": automation, artificial intelligence and labor protection', *Comparative Labor Law and Policy Journal*, **41** (1).

De Stefano, V. and M. Wouters (2019), 'Should digital labour platforms be treated as private employment agencies?', *Foresight Brief*, Brussels: ETUI.

Deknatel, A. and L. Hoff-Downing (2015), 'ABC on the books and in the courts: an analysis of recent independent contractor and misclassification statutes', *University of Pennsylvania Journal of Law and Social Change*, **18** (1), 53–104.

Delaney, A., J. Tate and R. Burchielli (2016), 'Homeworkers organizing for recognition and rights: can international standards assist them?', in J. M. Jensen and N. Lichtenstein (eds), *The ILO from Geneva to the Pacific Rim: west meets east*, London: Palgrave Macmillan, pp. 159–179.

European Public Employment Services (2019), 'EU network of public employment services strategy to 2020 and beyond', accessed 6 October 2020 at www.pesnetwork.eu/download/pes-network-strategy-2020-and-beyond/.

Freedland, M. (2006), 'From the contract of employment to the personal work nexus', *Industrial Law Journal*, **35** (1), 1–29.

Freedland, M. and N. Kountouris (2011), *The legal construction of personal work relations*, New York: Oxford University Press.

Freedland, M. and N. Kountouris (2013), 'The legal construction of personal work relations as a continuing pursuit', *Jerusalem Review of Legal Studies*, **7** (1), 145–157.

Freeman, H. and G. Gonos (2005), 'Regulating the employment sharks: reconceptualizing the legal status of the commercial temp agency', *Working USA: The Journal of Labor and Society*, **8** (3), 293–314.

Hendrickx, F. (2019), 'From digits to robots: the privacy–autonomy nexus in new labor law machinery', *Comparative Labor Law and Policy Journal*, **40** (3), 365–387.

ILO (1966), 'Memorandum sent by the International Labour Office to the Ministry of Health and Social Affairs of Sweden', *ILO Official Bulletin*, **49** (3), 390–401.

ILO (1991), *Social protection of homeworkers*, Geneva: ILO.

ILO (1995), *Report V (2) Home work*, Geneva: ILO.

ILO (1996a), *Report VI (1) Contract labour*, Geneva: ILO.

ILO (1996b), *Report IV (2 A) Home work*, Geneva: ILO.

ILO (1997), 'Nineteenth sitting: report of the Committee on the Revision of Convention No. 96: submission, discussion and adoption', in ILO, *Record of proceedings: International Labour Conference, 85th Session*, Geneva: ILO, pp. 267–281.
ILO (2006), *Report V (1) The employment relationship*, Geneva: ILO.
ILO (2010), *Report III (Part 1B) General survey concerning employment instruments in light of the 2008 Declaration on Social Justice for a Fair Globalization*, Geneva: ILO.
ILO (2019), *Work for a brighter future: Global Commission on the Future of Work*, Geneva: ILO.
ILO (2020), *Promoting employment and decent work in a changing landscape*, Geneva: ILO.
ING Economics Department (2018), *Algorithms versus the temporary employment sector: is there a future for temporary employment agencies*, Amsterdam: ING Bank.
Lehdonvirta, V., O. Kässi, I. Hjorth, H. Barnard and M. Graham (2019), 'The global platform economy: a new offshoring institution enabling emerging-economy micro-providers', *Journal of Management*, **45** (2), 567–599.
Maupain, F. (2000), 'International Labor Organization recommendations and similar instruments', in D. Shelton (ed.), *Commitment and compliance: the role of non-binding norms in the international legal system*, Oxford: Oxford University Press, pp. 372–393.
Morgan, J. F. (2018), 'Clarifying the employee/independent contractor distinction: does the California Supreme Court's Dynamex decision do the job?', *Labor Law Journal*, **69** (3), 129–140.
Pearce II, J. A. and J. P. Silva (2018), 'The future of independent contractors and their status as non-employees: moving on from a common law standard', *Hastings Business Law Journal*, **14** (1), 1–36.
Pinsof, J. (2016), 'A new take on an old problem: employee misclassification in the modern gig-economy', *Michigan Telecommunications and Technology Law Review*, **22** (2), 341–371.
Rosenblat, A. (2018), *Uberland: how algorithms are rewriting the rules of work*, Berkeley, CA: University of California Press.
Seibert, B. (2019), 'Protecting the little guys: how to prevent the California Supreme Court's new "ABC" test from stunting cash-strapped startups', *Business, Entrepreneurship and the Law*, **12** (1), 181–202.
Supiot, A. (2020), 'The tasks ahead of the ILO at its centenary', *International Labour Review*, **159** (1), 117–136.
Wood, A. J., M. G. V. Lehdonvirta and I. Hjorth (2019), 'Good gig, bad gig: autonomy and algorithmic control in the global gig economy', *Work, Employment and Society*, **33** (1), 56–75.

9. The regulation of platform work in the European Union: Mapping the challenges

Sacha Garben

9.1 INTRODUCTION

The regulatory challenges or questions raised by the platform economy are manifold and 'span the entire map of the legal world, including work, tax, safety and health, quality and consumer protection, intellectual property, zoning, and anti-discrimination' (Lobel 2016).[1,2] One key legal question is the impact on labour law of the activities of platforms. This chapter analyses in greater detail the extent to which the various labour and employment regulations that have usually been designed with a traditional bilateral, standard, open-ended employment relationship in mind can, and should, be applied to the often atypical working arrangements used in the platform economy; and how policymakers and judiciaries across Europe have grappled with that set of problems.

The drivers, riders, cleaners, designers, translators, technicians and others working in the platform economy are often formally contracted as independent. Additionally, their working arrangements tend to exhibit features that are difficult to square with the traditional employment relationship, such as the use of own materials (the driver's car); autonomy concerning working hours (deciding to work by logging into a smartphone app); the short duration of the relationship (the translation of a single sentence (Milland 2017)); and the multilateral character of the relationship (the cleaner, the platform and the customer). At the same time, the worker may well be economically dependent on platform work, contractual independence can be constructed in rather artificial ways (a driver that works full time for a platform for several years but is formally contracted per journey) and the platform can exert significant control over the work and the person performing it, not in the least through advanced algorithms. This complex reality has challenged judges in many jurisdictions who have had to decide on claims brought by platform workers against the platforms, arguing that they should be treated not as independent contractors but as 'employees or workers'.

The additional protection that would usually result from such (re)qualification seems welcome from a social security and occupational health and safety (OHS) perspective, considering that the people working in this sector, in frequently precarious conditions, have a profile that is often vulnerable. It furthermore means that they would not be subject to the application of competition rules in the context of collective bargaining for minimum remuneration levels (or other employment standards). Although some consider that traditional labour protections are not suitable for the 'new' and 'innovative' aspects of platform work, and that the application of these protections would inhibit their dynamic development, many others argue that employment rules should apply at least in some form (Adams-Prassl and Risak 2016; Davidov 2017; Davies and Freedland 2006; Ratti 2017; Rogers 2016; Tolodi-Signes 2017).

The various legal solutions proposed to ensure such application vary widely, however. One key distinction is between approaches that consider it necessary to amend the current legal provisions to allow their application to platform work and those that would 'simply' apply the current legal provisions. A further difficulty is the cross-border dimension of digitalization, which puts into question whether a national legal system alone constitutes the most effective locus to regulate (labour in) the platform economy.

Countries across the world have been dealing with these questions, up until now mostly on an individual basis. In Europe, however, where generally the most important regulatory developments appear to have taken place, the institutions of the European Union (EU) are becoming increasingly involved. For the most part their recent activities have been, implicitly or explicitly, supportive of ensuring the labour protection of platform workers, both through the extension of protective standards in EU legislation and through judgments that suggest many platform workers will meet the definition of 'worker' in EU law. A remaining worry is that, where platform workers do not meet that definition, they may be not only locked out of crucial European labour rights but even actively banned by EU competition law from pursuing the protection of their interests (especially remuneration) through collective bargaining. For that reason, the Commission is currently considering adopting a Regulation to define EU competition law's scope of application to online platform workers, to enable an improvement of working conditions through collective bargaining agreements not only for employees but also, under some circumstances, for the solo self-employed (European Commission 2021a).

9.2 THE (REGULATORY) PROBLEM OF PLATFORM WORK

Platform work includes all labour provided through, on or mediated by platforms and features a wide array of standard and non-standard working relationships such as (versions of) casual work, dependent self-employment, informal work, piece-work, home work and crowdwork, in a wide range of sectors. As some of these transactions were previously conducted in the shadow economy, platform work has an important beneficial potential to transfer these transactions to the formal sector and subject them to the appropriate rules, which may also lead to a better protection of the workers concerned. However, the regulation of the activities of platforms, and its enforcement, has generally not been straightforward. This is due to the dynamics of the sector, the rule-avoiding behaviour of many platforms and the narrative – fostered by the platforms – that their activities are 'new' and 'unprecedented' features emerging from rapid technological change that should not be treated similarly to any existing economic activities. Furthermore, this difficulty results not least from some aspects of platform working not fitting easily into pre-established legal categories. This latter consideration applies particularly to employment law. The almost inevitably triangular[3] (or multilateral) nature of the arrangements, their often temporary nature, the sometimes relatively high measure of autonomy of the worker in terms of working place and time and the at times informal (citizen-to-citizen) nature of some of the activities, as well as the absence of a common workplace, all challenge the application of the concept of the standard, permanent, binary employment relationship which, in many countries, is the requirement for the application of employment and social protections.

This is problematic because platform work is often precarious, in the sense that it lacks job security and income stability, and it poses a range of both pre-existing and new OHS risks, both physical and psycho-social, as discussed in detail in the Introduction of this book. It is somehow ironic that, while these risks would make it all the more important for employment and OHS regulations to apply to platform work, it is precisely the cause of these risks – an atypical employment relationship – that means these protections are less likely to apply. Of course, it should be noted that these challenges as such do not seem to be unique to the platform economy (De Stefano 2016). The past few decades have seen an increase in the use of non-standard forms of work. Many of the working arrangements set up by the platforms coincide with, or closely resemble, forms of atypical work or a mixture thereof, sometimes with the only difference being that they make use of a digital tool. However, the extent to which some platforms seem to have built their entire business model deliberately around this precarity, and the extent to which technologies are used to compound this precariousness, does seem to lift the overall package of problems to a higher plane.

The precarious position of platform workers is further aggravated in that the specific features of platform work tend to hamper the collective organization of workers, and thus the defence of their rights and interests, as well as the development of social dialogue. Most workers on platforms do not know each other, there is a high turnover of workers, set working patterns may be lacking and workers may not consider the work they provide for/on/via the platform as their primary professional activity. Additionally, putting workers in direct competition with each other – through individual ratings and a competitive method of work allocation – is an operational feature of many platforms. These factors are not conducive to the solidarity and collaboration needed for effective unionization – while a consideration of such workers as 'self-employed' causes problems to unionization in legal terms. For all these reasons we should, at the same time, be careful not to overestimate the novelty of working arrangements in the platform economy and their accompanying risks, while also not underestimating the scale and depth of the problem.

9.3 THE REGULATION OF PLATFORM WORK IN EUROPEAN COUNTRIES

To meet the concerns set out in the previous section, commentators and regulators have proposed and adopted various different approaches. This chapter will only provide a short overview (for a recent extensive survey see Kilhoffer et al. 2020).

A first, default option is 'simply' to apply existing regulations to platform work. Often, this will entail a case-by-case determination, usually by national courts, whether the platform worker is an employee, self-employed or – in some countries, such as the United Kingdom (UK) – falls into a third category in-between. Depending on the (flexibility of the) test applicable to determine labour status, this may already encompass many platform workers as employees, or place them in an intermediate category, meaning that (most) employment and OHS rules would apply, at least in legal terms.[4] Active enforcement by the competent authorities and effective access to justice for workers are necessary for this approach to be effective, especially considering the systematic rule-avoiding behaviour of many platforms. For instance, in Spain, a Madrid court held that Deliveroo drivers are employees because they have to follow instructions and lack the autonomy characteristic of truly independent work,[5] in a case that was undertaken at the request of the General Treasury of Social Security and reportedly carried out thanks to the proactive work of the labour inspectorate.[6]

But even with such active enforcement, the case-by-case nature of the assessment means a certain measure of unpredictability for all parties involved. National courts will have to look carefully at the precise features of the platform and the working arrangement, 'fitting them' into the framework of the relevant legal provisions and case law – which some platforms have precisely been keen to avoid in the design of the (formal) contractual labour arrangements they use. This will lead to some platform workers being found to be workers while others are found to be self-employed. In the UK, a high-profile judgment of October 2016 by a London Employment Tribunal upheld the claim of Uber drivers who argued that they were workers – the case is now pending at the Supreme Court.[7] Subsequent cases followed the same trend; however, there was a 2018 High Court ruling in which Deliveroo drivers were considered self-employed. In the Netherlands, Deliveroo drivers were first considered self-employed by a district court but, in a subsequent judgment, were considered employees – a finding that was recently confirmed by the Amsterdam Court of Appeal.[8] In France, the *Cour de Cassation* held on 4 March 2020 in relation to an Uber driver that there had been a case of 'fictitious independent worker status'; and that an employment contract needed to be established based on the circumstance that the driver provided services under terms and conditions which placed him in a relationship of permanent legal subordination with regard to the principal.[9] In an earlier ruling, however, the *Cour de Cassation* had held that a courier with the platform Take Eat Easy would, on the basis of the circumstances, not be reclassed as an employee from an independent worker.[10] In Italy, the Employment Tribunal of Turin rejected in 2018 a claim from six riders of the platform Foodora that they be counted as employees, although this was overturned by the Turin Court of Appeal in February 2019.[11]

This first option thus leads to differing results within jurisdictions as well as between them. While Uber drivers have, in many national courts, been reclassed as workers instead of independents, a Brussels court in a 2019 judgment considered, conversely, that UberX drivers were self-employed.[12]

While this first scenario is the default one that will decide on labour status in the platform economy in the absence of any deliberate regulatory action, it is not necessarily a merely static and passive response. The courts can adapt to the specific features of platform work the tests of (self-)employment that they have often themselves developed, for instance by placing less emphasis on the ownership of key business assets (such as cars in the context of passenger transport) and more emphasis on de facto control mechanisms (such as the rating and pricing systems operated by the platforms). The discretion of the judiciary to do so depends on the national legal system and legislative framework but, in most of the jurisdictions mentioned above, we can detect a certain evolution in the way that the courts have approached the issue, in that they have tended to move away from a more formalistic approach to a broader one that attempts to capture the specificity of these emerging phenomena. While this may go a long way towards resolving the problem of the lack of employment and social protection for platform workers, one disadvantage is that it places courts in the slightly uncomfortable position of having to take policy decisions on the appropriate treatment of platform workers which, arguably, should instead be subject to democratic debate and political decision-making.

A second possibility is to take specific action to narrow the group of persons that will be considered 'self-employed'. This could most notably be achieved through the addition of an intermediate '(independent) worker' category. In the United States in particular, there has been a call for the introduction of a new 'independent worker' status (Harris and Krueger 2015). Advocates of that midway approach tend to call for legislative intervention to regulate relationships that do not easily fit into the employed/self-employed dichotomy (Ratti 2017). These independent workers 'would occupy a middle ground between the existing categories of employee and independent contractor; the latter typically [being] workers who provide goods and services to multiple businesses without the expectation of a lasting work relationship' (Harris and Krueger 2015). The idea would be that, based on a set of governing principles to guide the assignment of benefits and protections to independent workers, businesses would provide certain benefits and protections that employees currently receive without fully assuming the legal costs and risks of becoming an employer. Several jurisdictions, including the UK, already have such a third category and this approach would imply the expansion thereof. It has been argued that this expansion is necessary because:

> [I]n most countries the law requires some degree of subordination even for this intermediate group (or at least this is how legislation has been interpreted). This means that the intermediate group is much smaller than it should be, with many dependent workers still left outside the scope of (even partial) protection. Secondly, workers in the intermediate group usually receive only very minimal protections, especially related to social security and sometimes the ability to bargain collectively. In fact, large parts of labour law should apply when a worker is dependent on one client (who should be considered an employer for such purposes). (Davidov 2014)

A different way specifically to limit the group of self-employed people is through a rebuttable presumption of employment. This already exists, to some extent, in countries such as the Netherlands and Belgium. However, legal cases concerning the status of platform workers in these countries, as well as in the UK with its intermediate category, show that these mechanisms do not necessarily resolve the categorization difficulties and that, in the end, a case-by-case assessment (by the courts) is likely to remain necessary with all the legal uncertainty that this entails. What might be more effective is to introduce such a presumption of employment specifically for platform work, perhaps even in specific sectors, such as Portugal has done for the transport sector in Law 45/2018.[13] As discussed in Section 9.4, this may be an effective course of action for an EU initiative on online platform work.

A third regulatory response to the challenges posed by platform work is to provide specific protection for platform workers regardless of their employment status. This can be through state regulation, such as in France, whose Act of 8 August 2016 on work, modernization of social dialogue and securing of career paths provides: (a) that independent workers in an economically and technically dependent relationship with a platform can benefit from insurance for accidents at work which is the responsibility of the platform in question; (b) that these workers equally have a right to continuing professional training, for which the platform is responsible, and should at their request be provided by the platform with a validation of their working experience with it; (c) that they have the right to constitute a trade union, to be a member of a union and to have a union represent their interests; and (d) that they have the right to take collective action in defence of their interests.

The French legislation, while wide-ranging, does not address the specific issue of the labour status of platform workers. Up to now, only Portugal and Italy have adopted such specific legislative action on the employment position of platform workers. Portugal has adopted a presumption of employment for workers in the transport sector, as mentioned above, while Italy has adopted legislation providing labour rights similar to employment for platform workers in food delivery.[14] Perhaps most ambitiously, the Italian region of Lazio has adopted legislation concerning the remuneration, health and safety and social protection of all types of platform work regardless of the employment status of the worker.[15]

Regulation can, of course, also be done by the stakeholders themselves. In Denmark, a first collective agreement was signed between Hilfr, the Danish digital platform for cleaning services, and the United Federation of Danish Workers. This guarantees the same conditions as elsewhere on the Danish labour market. The framework for the agreement was created by the Danish government's aim to ensure fair competition by creating the same rules for all (for example in relation to taxation). Since the Hilfr collective agreement, others have been concluded in Denmark, Sweden and Norway (Jesnes et al. 2019). It has been found, on the basis of initial experiences with the Hilfr agreement, that collective agreements can indeed lift wages and working conditions in the platform economy (Ilsøe 2020). It is testimony to the strength of Nordic social dialogue that this approach is succeeding in the face of a range of characteristics of platform work that can hamper collective bargaining. Moreover, the validity of this approach depends on the interpretation of the *FNV* case law, as discussed in Section 9.5.

Finally, a 'weaker' form of self-regulation has taken place in Germany where, in 2017, eight Germany-based platforms[16] signed a Code of Conduct in which they agreed to conclude local wage standards as a factor in setting prices on their platforms. Together with the German Crowdsourcing Association and the German Metalworkers' Union (IG Metall), they have also set up an office tasked with resolving disputes between crowdworkers, clients and crowdsourcing platforms as well as with overseeing the enforcement of the Code of Conduct.

9.4 THE EUROPEAN UNION LEGISLATOR AND PLATFORM WORK

The EU shares competence with its member states on a range of employment issues, in accordance with the provisions of the Social Policy Title in the Treaty on the Functioning of the European Union (TFEU). While this competence is, in principle, limited to setting minimum standards and should not prejudice member states' responsibility for the fundamental organization of their social security systems, the EU's social *acquis* should not be underestimated since it consists of a rich body of law on issues such as non-standard employment, working time and OHS.

In its Communication on a European agenda for the collaborative economy of 2 June 2016,[17] the European Commission set out the conditions under which it considers that an employment relationship exists for the purposes of applying these EU social law provisions. It considers that the definition of 'worker' laid down by the EU Court of Justice (CJEU), as applied in the context of the free movement of workers, also guides the application of EU labour law. This requires that 'the essential feature of an employment relationship is that for a certain period of time a person performs services for and under the direction of another person in return for which he receives remuneration'. Whether an employment relationship exists or not has to be established on the basis of a case-by-case assessment, considering the reality of the relationship and looking cumulatively at the existence of a subordination link, the nature of work and the presence of remuneration.

The EU has, furthermore, adopted two measures in the context of the European Pillar of Social Rights that may have an impact on platform workers' social and employment rights.[18] Firstly, it has adopted Directive 2019/1152 on transparent and predictable working conditions in the EU, a revision of the Written Statement Directive 91/533/EEC. This reinforces the rights already contained in that older Directive about the information a worker is entitled to receive in their employment contract and, in addition, defines a number of core labour standards for all workers, particularly regarding the protection of atypical and casual forms of employment. It also lays down a maximum duration of probation of six months (where a probation period is foreseen); the right to reference hours, in which working hours may vary under very flexible contracts to allow some predictability of working time; the right to request a new form of employment (and the employer's obligation to respond); the right to training; and the right to a reasonable notice period in the case of dismissal/early termination of the contract and the right to adequate redress in such cases.[19] Article 1(2) provides that the Directive applies to 'every worker in the Union who has an employment contract or employment relationship as defined by the law, collective agreements or practice in force in each Member State with consideration to the case-law of the Court of Justice'.

While the reference to an employment contract under national law may suggest that platform workers are, once again, in the throes of uncertainty concerning their labour status, the reference to 'every worker' and 'the case law of the Court of Justice' gives strong reason to assume that, when called upon to interpret the Directive's scope of application, the Court will apply its broad definition of 'worker' regardless of whether the person concerned qualifies as an employee under national law. In principle, the Directive only applies to contracts of more than three working hours a week, but this criterion does not apply to contracts without a guaranteed amount of paid work such as zero hours contracts and many platform-based working arrangements.[20]

Secondly, the EU has adopted Council Recommendation of 8 November 2019 on access to social protection for workers and the self-employed,[21] hoping to tackle the problem of up to one-half of people in non-standard work and self-employment being at risk of not having sufficient access to social protection and/or employment services across the EU. The Recommendation urges member states to provide similar social protection rights for similar work regardless of labour status and to ensure the transferability of acquired social protection rights but, as is inherent in this type of legal instrument, it does not legally bind the member states to do so. The Commission initially considered a Directive on this issue but lowered the ambition to the adoption of a non-binding Recommendation which, nevertheless, serves to highlight the importance of this issue to European and national policymakers.

The European Parliament's position, in general terms, has been that fair working conditions and adequate legal and social protection should be assured for all workers in the platform economy regardless of their status. In its Resolution of 19 February on a European Pillar of Social Rights, Parliament called on the Commission to broaden the Written Statement Directive to cover all forms of employment and to include relevant existing minimum standards for work intermediated by digital platforms and other instances of dependent self-employment, emphasizing that:

> [T]hose employed as well as those genuinely self-employed who are engaged through online platforms should have analogous rights as in the rest of the economy and be protected through participation in social security and health insurance schemes; Member States should ensure proper surveillance of the terms and conditions of the employment relationship or service contract, preventing abuses of dominant positions by the platforms. (Point 5(c))

In the beginning of 2021, the Commission launched the first phase consultation of social partners under Article 154 TFEU on possible action addressing the challenges related to working conditions in platform work (European Commission 2021b). It recognizes that 'questions around employment status of people working through platforms impact their working conditions' and considers the possibility of an EU legislative initiative on the basis of Article 153 TFEU.

In light of the discussion above, it would arguably be most appropriate for the EU-level action to strengthen current national efforts (mostly judicial) to adapt the classification of employment to the specific context of online platform work, by laying down a rebuttable presumption of employment for work provided via online platforms. It could for instance provide that 'Member States shall ensure online platform workers all the rights and protections under the relevant national law applicable to persons with an employment contract', while, by way of derogation providing:

> Member States may decide to disapply the relevant provisions of national law to those online platform workers whose relationship to the platform clearly does not feature the essential characteristics of a work relationship and who are to be regarded as self-employed in light of, in particular, their full autonomy in terms of the pricing, organisation and execution of the work in question.

An additional element that could point towards the finding of an employment relationship could be the 'core business' test as applied recently by the Amsterdam Court of Appeal: when the work provided by the online platform worker constitutes part of the core business of the online platform in light of its overall functioning and purpose and the role of the work provided therein.

9.5 THE EUROPEAN UNION COURT OF JUSTICE AND PLATFORM WORK

In a landmark judgment concerning the platform economy, the CJEU examined the nature of the activities of the platform company Uber.[22] The central question was whether Uber's activities were to be classed as 'information society services' under EU law, in which case market access should be granted and restrictions on its operation should have been notified and could only be accepted in limited circumstances; or whether they instead constituted 'transportation services' which fall outside the scope of the EU rules in question and could, therefore, in principle be freely regulated by member states.

In its judgment, the Court considered that the intermediation service provided by Uber is based on the selection of non-professional drivers using their own vehicle to whom the company provides an app without which: (a) those drivers would not be led to provide transport services; and (b) persons who wished to make an urban journey would not use the services provided by those drivers. In addition, Uber exercises decisive influence over the conditions under which that service is provided by those drivers. Uber determines at least the maximum fare, receives that amount from the client before paying part of it to the non-professional driver of the vehicle and exercises a certain control over the quality of the vehicles, the drivers and their conduct which can, in some circumstances, result in their exclusion. Therefore, it was 'inherently linked to a transport service' and, accordingly, must be classed as 'a service in the field of transport' which can be freely regulated by the member states.[23] While the judgment does not concern the labour status of Uber drivers, the CJEU's considerations concerning the measure of control of the platform can be expected to be relevant for future labour law cases at national and EU level in the future.[24] Indeed, in the Airbnb and Star Taxi cases, the CJEU found that the activities of the platforms did constitute 'information society services' in light of the limited 'level of control' of the platform over the services provided (their pricing, quality and method of delivery).[25]

Less positive from the perspective of the labour protection of platform workers is an older judgment of the Court in the context of collective bargaining and competition law, *FNV*. In this case it held that the prohibition on agreements between undertakings set out in Article 101 TFEU applied to the minimum fees set down in collective agreements between self-employed service providers and their contractor, unless the self-employment actually constituted 'false self-employment'.[26] The Court gave further indications as to who could be considered falsely self-employed for these purposes:

> [A] service provider can lose his status of an independent trader, and hence of an undertaking, if he does not determine independently his own conduct on the market, but is entirely dependent on his principal, because he does not bear any of the financial or commercial risks arising out of the latter's activity and operates as an auxiliary within the principal's undertaking … On the other hand, the term 'employee' for the purpose of EU law must itself be defined according to objective criteria that characterise the employment relationship, taking into consideration the rights and responsibilities of the persons concerned. In that connection, it is settled

> case-law that the essential feature of that relationship is that for a certain period of time one person performs services for and under the direction of another person in return for which he receives remuneration … From that point of view, the Court has previously held that the classification of a 'self-employed person' under national law does not prevent that person being classified as an employee within the meaning of EU law if his independence is merely notional, thereby disguising an employment relationship … It follows that the status of 'worker' within the meaning of EU law is not affected by the fact that a person has been hired as a self-employed person under national law, for tax, administrative or organisational reasons, as long as that persons acts under the direction of his employer as regards, in particular, his freedom to choose the time, place and content of his work … does not share in the employer's commercial risks … and, for the duration of that relationship, forms an integral part of that employer's undertaking, so forming an economic unit with that undertaking.[27]

The precise implications of this ruling for platform workers are not entirely clear. It suggests that those platform workers that would be considered self-employed would not be allowed to enter into collective agreements with the platform to set minimum payment standards. As regards employment conditions other than fees/payment, these could be permitted because they could fall within the *de minimis* exception to the application of Article 101 TFEU. The 2014 De Minimis Notice excludes from the scope of EU competition law those agreements between parties with an aggregate market share equal to or below 5 per cent and an annual turnover equal to or below €40 million because they are considered to have no effect on trade. The *de minimis* exception, however, does not apply to so-called 'hardcore' restrictions such as price fixing which a collective agreement on remuneration would constitute from the perspective of competition law in the case of the (genuinely) self-employed.

Thus it is once again crucial to determine whether platform workers are considered self-employed from the perspective of EU competition law or, conversely, that they constitute workers. Being held as the latter implies that they can therefore benefit from the *Albany* doctrine under which the Court has excluded from the application of Article 101 TFEU collective agreements which result from 'collective negotiations between organisations representing employers and workers' and have the purpose of 'jointly adopt[ing] measures to improve conditions of work and employment'.

Two different interpretations of *FNV* have been offered in this respect. Schiek and Gideon have argued that, in *FNV*, the Court:

> [D]eveloped a specific notion of worker for the purposes of competition law, which determines whether micro-entrepreneurs qualify as undertakings, which are subject to competition law or as workers which are excluded from competition law coverage when bargaining collectively. The Albany exclusion from Article 101 TFEU is thus not accepted for those micro entrepreneurs who cannot be considered as falsely self-employed under this definition. These micro entrepreneurs would still

> be subject to the control of competition authorities when organising collectively to improve their working conditions. (Schiek and Gideon 2018, p. 8)

This would imply a more restrictive interpretation of the notion of worker than the one generally applicable under EU law, to the detriment of (some) platform workers. A textual, narrow reading of the *FNV* judgment indeed lends some support to that interpretation, but it would not align with the spirit of the CJEU's more recent *Uber* judgment discussed above. On the other hand, Klebe and Heuschmid (2016) consider that the *FNV* judgment means that 'the (solo) self-employed who are in a similar situation to regular employees' are not subject to competition law.

The judgment seems to leave sufficient room for the CJEU to consider, when specifically asked whether platform workers on a certain platform are to be treated as 'undertakings' in the sense of Article 101 TFEU or instead as 'workers' in the sense of the Albany exception, that the general definition of worker under EU law applies. Furthermore, there also seems to be room for a future consideration whether, in line with its more recent *Uber* judgment, the degree of control of the platform over the activities of the people providing work for it would be determinative in that respect. This would suggest, for instance, that *Uber* drivers would not be considered 'undertakings' but 'workers' and, therefore, would be free to conclude a collective agreement with *Uber* setting down some minimum pay standards. Alternatively, the Court could consider accepting as protection against social dumping a justification in the context of Article 101 TFEU, in line with the proposal made by Advocate General Wahl in *FNV*, or it could more explicitly adapt its definition of worker for the purpose of competition law to focus on the aspect of economic (and not organizational) dependency (Schiek and Gideon 2018). Whatever option is chosen, it seems very desirable that the Court delivers, in a new judgment, a clear statement about the position of platform workers under competition law and that it takes a cautious, protective stance so as to include in the scope of Article 101 TFEU only those platform workers that are clearly self-employed and economically independent. Until then, the chilling effect of the *FNV* ruling will act as a further hurdle to any collective bargaining on remuneration in the context of platform work – compounding the problem that this is an area where collective bargaining is already inherently difficult, as described above.

The European Commission has recently recognized the importance of this issue, with Margrethe Vestager, Commissioner for Competition, stating that 'the competition rules are not there to stop workers forming a union. So we need to make clear that those who need to can negotiate collectively, without fear of breaking the competition rules' (Vestager 2020). Accordingly, the Commission has launched an initiative aiming to define EU competition law's scope of application, so as to enable an improvement of working conditions through collective bargaining agreements not only for employees, but also, under some circumstances, for the solo self-employed (European Commission 2021a). The options range from exempting the solo self-employed providing their own labour through online platforms to exempting also solo self-employment in the offline economy. The Commission plans to adopt a Regulation in the fourth quarter of 2021.

9.6 CONCLUSION

Regulators and courts at national and EU level are only just starting to grapple with platform work and the disruption it entails. At national level, regulators are mulling over the various different policy options. As discussed, these include: (a) applying the existing definition of employment to platform work with the result that some fall within, and some outside, its scope; (b) enlarging the scope of application specifically to include platform work in employment/ social protection; (c) taking specific (self-)regulatory measures to offer special protections to platform workers. While courts are becoming increasingly adept at resolving the issues on a case-by-case basis, by adapting their national 'tests' of employment status to the specificity of online platform work, there remains a measure of unpredictability in the absence of a clarified legislative framework. There is furthermore an issue with the effective access of online platform workers to such labour protection even where it does apply in theory, as the current opaque legal situation, and the practices of most platforms that play into that uncertainty, forces workers to take legal action in court which they are unlikely to do. This is where the possible EU initiative could provide a tangible added value, by strengthening the position of online platform workers and aiding national courts in finding an employment relationship in the right circumstances, through laying down a presumption of employment for work provided in the online platform economy and providing enhanced enforcement options. Similarly, the current Commission consultation exploring options to make sure that competition law does not harm the possibility of collective bargaining for online platform workers is very welcome. It seems fair to say that these workers already face an uphill battle in terms of social protection and labour conditions and standards. For EU law to add another hurdle would not be in the spirit of the European Pillar of Social Rights or the EU's commitment to a fair society and a social market economy.

NOTES

1. The author wishes to thank Mathias Wouters for useful comments on an earlier draft of this chapter. The views expressed are entirely personal.

2. For a comprehensive discussion of the full range of the implications of EU law for the platform economy, see Hatzopoulos (2018).
3. In the sense that there are three parties in play: the person performing the work; the person receiving the work; and the platform through which the work is provided.
4. The practical reality is more complicated as workers may generally not be aware of all their rights, let alone in the 'uncertain' context of platform work, and furthermore may be hesitant to enforce them.
5. Autos nº. 510/18, Juzgado de lo Social nº 19 de Madrid – sentencia Nº. – 188/2019.
6. More generally the situation in Spain has been quite unpredictable. As reported in the 2020 study for the European Commission, seven cases appeared before the Spanish courts in 2018 and 2019 concerning the employment status of food couriers for Deliveroo, Take Eat Easy and Glovo. The courts classed the couriers as employees in four cases but as self-employed in the other three (European Commission 2020, p. 115).
7. *Uber BV and others (Appellants)* v. *Aslam and others (Respondents)*, Case ID: UKSC 2019/0029.
8. Judgment of 16 February 2021, ECLI:NL:GHAMS:2021:392.
9. Arrêt n°374 du 4 mars 2020 (19-13.316) – Cour de cassation – Chambre sociale – ECLI:FR:CCAS:2020:SO00374.
10. Arrêt n°1737 du 28 novembre 2018 (17-20.079) – Cour de cassation – Chambre sociale – ECLI:FR:CCASS:2018:SO01737.
11. Court of Appeal of Turin, Judgment no. 26 of 4 February 2019, partially reforming judgment no. 778/2018 by the Court of First Instance.
12. A/18/02920, Tribunal de l'entreprise francophone de Bruxelles.
13. Lei n.º 45/2018–N.º15410.08.2018, pp. 3972–3980.
14. L. 2 novembre 2019, n. 128, Conversione in legge, con modificazioni, del decreto-legge 3 settembre 2019, n. 101, recante disposizioni urgenti per la tutela del lavoro e per la risoluzione di crisi aziendali.
15. Giunta Regionale: deliberazione N.308 norme per la tutela e la sicurezza dei lavoratori digitali, 2018n.40,20.03.2019.
16. Nine platforms have been part of the Code since 2019.
17. COM(2016)0356.
18. 2017, OJ C428/10. The European Pillar of Social Rights is a solemn inter-institutional proclamation by the Commission, Parliament and Council which (re)affirms the commitment of the EU and the member states to the various social rights and principles set out therein. It also constitutes a sort of action plan, providing the framework for further legislative and soft law initiatives at EU level to enhance the protection of the social standards set out in the Pillar (Garben 2019).
19. Article 18: 'Member States shall take the necessary measures to prohibit the dismissal or its equivalent and all preparations for dismissal of workers, on the grounds that they have exercised the rights provided for in this Directive.' In accordance with recital 43, equivalent detriment to dismissal includes 'an on-demand worker no longer being assigned work' – a practice that is familiar as regards some platforms.
20. Article 1(4) of the Directive.
21. OJ C 387, 15.11.2019, pp. 1–8.
22. ECLI:EU:C:2017:981.
23. In a recent judgment, the Court took a different approach as regards the activity of the platform Airbnb, considering it did have to be classed as an 'information

society service' under Directive 2000/31. This, however, does not have any implications for the issue of the labour protection of platform workers. Case C-390/18, *X v YA, Airbnb Ireland UC, Hôtelière Turenne SAS, Association pour un hébergement et un tourisme professionnels (AhTop), Valhotel*, ECLI:EU:C:2019:1112.
24. See also Schiek and Gideon (2018).
25. Case C-390/18, Criminal proceedings against X, ECLI:EU:C:2019:1112.
26. Case C-413/13, *FNV Kunsten Informatie en Media* v. *Staat der Nederlanden*, ECLI:EU:C:2014:2411.
27. Paras 33–36 of the judgment.

REFERENCES

Adams-Prassl, J. and M. Risak (2016), 'Uber, Taskrabbit, & Co: platforms as employers? Rethinking the legal analysis of crowdwork', *Comparative Labour Law and Policy Journal*, **37** (3), 619–649.

Davidov, G. (2014), 'Setting labour law's coverage: between universalism and selectivity', *Oxford Journal of Legal Studies*, **34** (3), 543–566.

Davidov, G. (2017), 'The status of Uber drivers: a purposive approach', *Spanish Labour Law and Employment Relations Journal*, **6** (1–2), 6–15.

Davies, P. and M. Freedland (2006), 'The complexities of the employing enterprise', in G. Davidov and B. Langile (eds), *Boundaries and frontiers of labour law*, Oxford: Hart Publishing, pp. 273–294.

De Stefano, V. (2016), 'The rise of the "just-in-time workforce": on-demand work, crowdwork and labour protection in the "gig-economy"', ILO Conditions of Work and Employment Series No. 71, Geneva: ILO.

European Commission (2020), 'Competition: the European Commission launches a process to address the issue of collective bargaining for the self-employed', Press Release, 30 June, accessed 12 October 2020 at https://ec.europa.eu/commission/presscorner/detail/en/IP_20_1237.

European Commission (2021a), 'Inception impact assessment, ares(2021)102652', available at: https://ec.europa.eu/info/law/better-regulation/have-your-say/initiatives/12483-Collective-bargaining-agreements-for-self-employed-scope-of-application-EU-competition-rules.

European Commission (2021b), 'First phase consultation of social partners under Article 154 TFEU on possible action addressing the challenges related to working conditions in platform work', C(2021) 1127 final.

Garben, S. (2019), 'The European Pillar of Social Rights: an assessment of its meaning and significance', *Cambridge Yearbook of European Legal Studies*, **21**, 101–127.

Harris, S. and A. Krueger (2015), 'A proposal for modernizing labour laws for twenty-first century work: the "independent worker"', Discussion Paper 2015-10, Washington, DC: Hamilton Project.

Hatzopoulos, V. (2018), *The collaborative economy and EU law*, Oxford: Hart Publishing.

Ilsøe, A. (2020), 'The Hilfr agreement: negotiating the platform economy in Denmark', FAOS Research Paper 176, Copenhagen: University of Copenhagen.

Jesnes, K., A. Ilsøe and M. Hotvedt (2019), 'Collective agreements for platform workers? Examples from the Nordic countries', Nordic Future of Work Brief 3, Oslo: Fafo.

Kilhoffer, Z., W. P. De Groen, K. Lenaerts, I. Smits, H. Hauben, W. Waeyaert, E. Giacumacatos, J. P. Lhernould and S. Robin-Olivier (2020), *Study to gather evidence on the working conditions of platform workers*, VT/2018/032, Brussels: European Commission.

Klebe, T. and J. Heuschmid (2016), 'Collective regulation of contingent work: from traditional forms of contingent work to crowdwork: a German perspective', in E. Ales, O. Deinert and J. Kenner (eds), *Core and contingent work in the European Union: a comparative analysis*, Oxford: Hart Publishing, pp. 176–197.

Lobel, O. (2016), 'The law of the platform', Legal Studies Paper Series No. 16-212, San Diego, CA: University of San Diego School of Law.

Milland, K. (2017), 'Slave to the keyboard: the broken promises of the gig economy', *Transfer*, **23** (2), 229–231.

Ratti, L. (2017), 'Online platforms and crowdwork in Europe: a two-step approach to expanding agency work provisions?', *Comparative Labour Law and Policy Journal*, **38** (3), 477–511.

Rogers, B. (2016), 'Employment rights in the platform economy: getting back to basics', *Harvard Law and Policy Review*, **10** (2), 479–520.

Schiek, D. and A. Gideon (2018), 'Outsmarting the gig-economy through collective bargaining: EU competition law as a barrier to smart cities?', *International Review of Law, Computers and Technology*, **32** (2–3), 275–294.

Tolodi-Signes, A. (2017), 'The "gig-economy": employee, self-employed or the need for a special employment regulation? *Transfer*, **23** (2), 193–205.

Vestager, M. (2020), 'Keeping the EU competitive in a green and digital world', Speech to the College of Europe, Bruges, 2 March, accessed 12 October 2020 at https://ec.europa.eu/commission/commissioners/2019-2024/vestager/announcements/keeping-eu-competitive-green-and-digital-world_en.

10. Workers, platforms and the state: The struggle over digital labour platform regulation

Sai Englert, Mark Graham, Sandra Fredman, Darcy du Toit, Adam Badger, Richard Heeks and Jean-Paul Van Belle

10.1 INTRODUCTION

Emerging from Cohen and Sundararajan's 'Self-regulation and innovation in the peer-to-peer sharing economy' (2017), the question of regulating digital labour platforms has taken up much space within public, policy and academic debate. The discussion has focused on platforms' and states' (in)ability to regulate this growing sector effectively, the role of potentially industry-wide standards and the relationship between platforms and analogue industries. Self-regulation, an approach championed by the platforms themselves and their supporters, centres on the ability of these companies to police themselves, while also emphasizing the supposed disciplining effect of market forces on their potential excesses.

Governments, the argument goes, are ill-equipped to understand these new forms of organization of work, service provision and rapidly changing technology (Allen and Berg 2014; Cohen and Sundararajan 2017; Koopman et al. 2015). Attempts at applying existing labour, tax or licensing law to them would stifle their growth, undermine their ability to provide work and services and sound the death knell for the industry before it had had the chance to achieve its full potential. This chapter, however, argues that 'self-regulation' is best understood not as an actual attempt by platforms independently to establish best practice in the sector but as an active process of reshaping existing laws and regulations by these companies.

After giving an overview of the self-regulation literature and identifying the problems with its key claims, for both workers and clients, the chapter argues that platforms' approach to regulation can best be understood in two ways. First, they move in and – not unlike the preferred strategy of settler colonial regimes – lay down 'facts on the ground' (McKee 2017, p. 107). They ignore existing regulations, arguing that these do not concern them given their nature as technological connecting services. Second, once platforms have achieved a significant presence in a city or country, which limits lawmakers' ability to shut down or control their operations, they engage extensively with state and local government to reform existing regulations to their advantage. The picture that emerges is one in which platforms use their leverage as job creators and (often popular) service providers to reshape unfavourable legal frameworks. Far from self-regulating in isolation from the law, platforms have consciously participated in processes of deregulation and re-regulation.

The chapter then moves on to discuss a number of responses to this behaviour which aim to develop alternative forms of regulation for digital labour platforms. It touches upon industrial and legal processes, some of which are ongoing, through which workers and their advocates seek to impose labour rights on companies in the sector, as well as the attempts by trade unions and other collectives to develop extra-legal standards for the sector. In this regard, it especially highlights the *Fairwork* project. The chapter argues that, in the absence of robust state action, these various regulatory initiatives from below, led by workers and their supporters, are attempting to respond to platforms unilaterally imposing regulations from above.

10.2 SELF-REGULATION AGAINST THE STATE

Much of the self-regulation literature takes platforms' claims about themselves at face value. Platforms, so the argument goes, should be considered primarily as technology-innovating firms whose business is not being taxi or hotel companies but connecting service providers with service users. Allen and Berg (2014, p. 2), for example, write: 'the sharing economy describes a rise of new business models ("platforms") that uproot traditional markets, break down industry categories, and maximise the use of scarce resources'. Issues emerge when these platforms come into contact with the state, which misunderstands their true nature and tries to impose the same regulations on – say – Ola, Lyft or Uber as they do on their local taxi companies. By doing so, it is locking these new innovative entities into an old and inappropriate shell. Instead, the state should step away and let the platforms self-regulate. In the words of Cohen and Sundararajan (2017, p. 116):

> [R]egulatory barriers may slow the growth of employment that involves individuals providing goods, services, labor, and capital through peer-to-peer platforms … [T]he resolution of these challenges must include self-regulatory approaches. *Self-regulation* is not the same as deregulation or no regulation. Rather, it is the reallocation of regulatory responsibility to parties other than the government.

The first aspect highlighted by supporters of platform self-regulation are the feedback tools set up within apps and websites that allow both 'peers' in the exchange to rate one another. This, they argue, helps to undermine the long-lasting problem of information asymmetries in the marketplace, referring to the challenges faced when a buyer and a seller come to the proverbial market and are confronted with the problem that they know nothing about one another, the quality of the goods or services on sale or the trustworthiness of the parties involved (Allen and Berg 2014). It is for this reason that consumer protection laws, licensing and other forms of 'soft regulation', such as watchdogs and ombuds offices, have emerged in an attempt to enforce certain standards on the otherwise free market (Ducci 2018).

It is also this process that the self-regulation literature wants to consign to the dustbin of history. '[T]he internet,' Koopman et al. argue (2015, pp. 540–541), 'largely solves this problem by providing consumers with robust search and monitoring tools to find more and better choices', lowering 'both search costs and transaction costs associated with commercial interactions'. As our actors go to market, they are immediately able to find out all the necessary information about each other and make informed decisions about which trading partner to choose. Stemler (2017, p. 683) concurs and adds that these 'reputation systems create incentives for participants to self-police'. The market, through the development of technology, has therefore generated a mechanism to address one of its fundamental limitations and, in the process, has not only made the need for state regulation obsolete but has surpassed it.

A second aspect of the self-regulation literature focuses on the disciplining power of competition. Given the need for companies to capture as much market share as possible while also undermining their competitors' ability to do so, the idea is that the market serves as a great equalizer which forces companies to regulate naturally in order not to be outdone by more ethical competitors. Furthermore, by undermining state regulations platforms 'have opened traditionally cartelized industries to new competition' (Koopman et al. 2015, p. 543), thereby acting as an accelerator for the natural tendency of the market to regulate. This brings it into areas hitherto held back from the liberating power of the invisible hand. The platforms are then seen as more effective tools for the propagation of a free and uncontrolled, equilibrating market.

The third aspect repeatedly celebrated in the literature is the platforms' ability to set their own terms of service and enforce them. Allen and Berg (2014, p. 23) exalt it as a new form of 'civil-society governance', while Johal and Zon (2015, p. 19) consider it will allow the more effective enforcement of a number of the responsibilities of the state, such as fiscal regulation. The argument is two-fold. Not only are platforms – spurred on by competition and feedback – able to respond more efficiently to issues emerging in the course of their business transactions than states, but they are also better able to police the implementation of and respect for the rules they develop. This is achieved first and foremost through sanctions – and ultimately deactivation – of both service users and providers who fail to live up to the standards set by the platforms. This realization leads Cohen and Sundararajan (2017, p. 130) to celebrate platforms' 'tremendous potential enforcement capabilities as regulatory entities' because 'they control the channels for demand for their drivers, and as digital platforms, disconnecting a driver involves minimal transaction costs for the companies'. No thought is given to the effects of this absolute control on the workers, however.

This emphasis on the efficiency of platform self-regulation and the superiority of its development, adaptability and enforcement of regulatory measures leads some authors to theorize about the desirability of platforms expanding their field of action beyond the limits of their own companies or industry. Cohen and Sundararajan (2017, p. 133) muse that, as self-regulating platforms 'establish a track record of credibility and enforcement and gain legitimacy as partners in regulation, they can then be called on to help invent self-regulatory solutions to societal issues [such as discrimination] that are especially difficult to address by centralized governmental intervention'.

A brave new world indeed.

10.3 EVADING AND OBSCURING: THE LIMITS OF THE SELF-REGULATION PARADIGM

As might be expected, there is a gap between the theory and the practice of self-regulation. Looking in turn at feedback mechanisms, working conditions and wages reveals a more complicated, and considerably less positive, image of the consequences of platform self-regulation. It suggests that, given the lack of regulation, the disciplining power of competition is driving both labour and service standards down, leaving workers particularly vulnerable.

The most celebrated aspect of platforms' ability to self-regulate, as discussed above, is the feedback mechanisms they put in place. However, it is important to point out that there are a number of shortcomings built into these. Reputational mechanisms can be manipulated. The writing of fake reviews, either for oneself or through friends and colleagues, is a well-known mechanism with which users of websites ranging from Just Eat to Amazon are familiar. As Stemler (2017, p. 699) points out, the practice can also involve material rewards in exchange for good reviews.

In addition, platforms might have incentives to encourage better scores in order to encourage greater engagement with the platform. Airbnb's content policy, for example, states that it has 'the ability to remove a review, "in whole or part", for any reason at its "sole discretion"' (Stemler 2017, p. 700), raising at least the possibility of platforms tampering with reviews. Feedback mechanisms tend to be undermined by a number of compounded biases that range from the participation only of the highly satisfied or the extremely dissatisfied to confirmation biases through which those with higher pre-existing scores tend to be rewarded by users because they already assume a certain quality of service independent of their own experience (Stemler 2017).

Moreover, it appears that feedback mechanisms have a tendency to facilitate discrimination. This matters especially given the high proportion of black, Asian and migrant digital labour platform workers (van Doorn 2017, p. 907). Leong and Belzer (2017, p. 1292) found that the visibility of names and profile pictures of black customers led repeatedly to higher rates of rejection and cancellation, as well as poorer ratings in both Airbnb and Uber.

Platforms, then, raise barriers to demonstrating, challenging and holding guilty parties to account for discriminatory behaviour. If sexist Uber drivers or users systematically give poor scores to their female counterparts, how can the latter seek redress? Similarly, if alternative data points (such as location or name) become proxies for race, how can the targets of discrimination demonstrate intent? Far from levelling the playing field, the feedback mechanism facilitates discriminatory behaviour by offering protection to the perpetrators: 'online databases, despite implying neutrality, actually capture and perpetuate the biases of their creators and administrators' (Leong and Belzer 2017, p. 1291).

It is more constructive to understand feedback mechanisms as providing users with the illusion of engagement and agency, thereby strengthening identification with the platform, while serving as a tool to control and discipline labour. The app itself serves to track and monitor worker productivity (De Stefano 2016), whereas the rating mechanisms mobilize clients in the disciplining of platform workers who are striving to keep their scores high in order to avoid deactivation.

Another key aspect of platform self-regulation has been their ability to avoid existing labour regulations by refusing to acknowledge any employment relationship between themselves and their workers. The latter are systematically identified by platforms as self-employed service providers or independent contractors and therefore kept out of the reach of existing labour laws and regulations (Drahokoupil and Fabo 2016; Ducci 2018; Fabo et al. 2017; Graham et al. 2017; van Doorn 2017). In doing so, minimum wages, sick and holiday pay, pensions and even health and safety standards can easily be withheld.

The issue is, therefore, how to determine when platform workers are genuinely self-employed, using online services to increase their reach when looking for clients, and when platforms are misusing the employment classification to avoid regulation. Ducci (2018) solves the question not by arguing over the definition of the employment relationship but by focusing on the issue of control. If workers have no control over the setting of rates for each job, for example, with these being imposed by the platform, he argues that control over the work process remains in the hands of the platform and therefore makes the service provider a platform worker who should be treated as such.

It appears that this approach is also gaining traction in the courts. In the case of *Uber* v. *Aslam*, the majority found in favour of Uber drivers who claimed that they were in fact employed by the platform and not independent drivers because:

> First, Uber in effect requires drivers to accept trips and not to cancel trips, enforcing these requirements by logging off drivers who breach them. Second, Uber subjects its drivers through the rating system to what is in effect a performance management or disciplinary procedure. Third, Uber imposes numerous conditions on drivers, including the choice of acceptable vehicles. It also instructs drivers on how to do their work and controls them in the performance of their duties. Uber interviews and recruits drivers; and controls the key information, including the passenger's surname, contact details and intended destination, and excludes the driver from it. (Fredman and Du Toit 2019, p. 268)

In this case, the court considered that it was the nature of the relationship rather than the detail of the contractual agreement that was important in order to establish the relationship between workers and the platform. However, while the legal victory in the courts was important, it was also limited. Uber has been able to halt the process through appeals – most recently to the United Kingdom Supreme Court – thus kicking the issue further into the long grass.

Finally, platform self-regulation often takes the shape of mobilizing market mechanisms to drive down wages and undermine working conditions even further. For example, online platforms, such as Upwork or Amazon Mechanical Turk, allow individuals and companies to put out jobs for any worker worldwide to bid for and execute, putting isolated workers behind their screens in competition with one another. This generates massive pressures to deliver their services as quickly as possible for the lowest possible price which, in turn, leads to 'some platforms pay[ing] the equivalent of less than €1 per hour, a low rate of pay that might decline yet further as the growing number of workers without special skills from all around the world registers on the platforms' (Fabo et al. 2017, p. 170).

Online platforms also reinforce international divisions of labour, with the majority of jobs being offered by actors in the Global North but carried out in the Global South (Graham et al. 2017), facilitating the use of cheap labour by platforms and further increasing downwards pressures on pay and conditions. This increased competition also leads workers to accept that they shoulder the costs associated with carrying out jobs, further depreciating their wages (Irani 2015). These issues are not limited to online work. Platforms, like Uber for example, also expect workers to take on the entirety of the costs associated with car rental, petrol prices, insurance and repairs, increasing their margins in the process. They are, after all, independent entrepreneurs using an app, not company employees.

In addition, not only are online platform workers located in different countries and therefore not always able effectively to challenge the central platform, but most platforms 'are registered in countries different to where their workers are physically located' (Fabo et al. 2017, p. 171), creating issues in attempting to hold them to account. Workers in South Africa, for example, who legally challenged Uber over their employment relationship, discovered that they were in fact hired by a company based in the Netherlands and had therefore taken the wrong party to court (Woodcock and Graham 2019).

The cases discussed here illustrate the reality of self-regulation in the sector for platform workers. They also demonstrate the way in which the market tends towards equalizing pay and labour conditions and standardizing the organization of the work process. However, far from conforming to the optimistic picture painted in the literature discussed above of a process through which everyone – platforms, workers and clients – benefits, the reality is quite different: platforms benefit from the exploitation of workers locked in isolated and individualized competition for limited jobs. The self-regulating mechanisms of the platforms also bar them from appealing to existing labour laws and regulations by denying that such a relationship exists. To make matters worse, even if (or when) workers decide to challenge platforms over the quality of their employment, a number of issues confront them.

10.4 SELF-REGULATION, DEREGULATION, RE-REGULATION

It is important, however, to point out that while platforms sing the praises of their self-regulatory abilities, this stance has been rather partial. Many laws are crucial for the smooth running of platform operations. McKee (2018, p. 110) reminds his readers that 'intellectual property, which guarantees these platforms control over their software', is crucial for the apps and websites to remain in control of their markets. Similarly 'the assumption that people can offer "their" apartments for rent through Airbnb presupposes the existence of a system of property law that determines what certain people can do with regard to certain things.' Far from operating in a lawless and self-regulated environment, platforms have positioned themselves in such a way as to profit as much as possible from those laws that ensure their ability to do business while aggressively campaigning to undermine the ones that limit their market capture or profit margin.

In fact, as Finck (2017, p. 5) has argued, a 'crucial point to note is that the regulatory disruption created by platforms is not an accidental effect of the platform economy but rather a constituent characteristic thereof'. Indeed, it is the platforms' disruptive power that has allowed the largest players in the field not so much to evade regulation but to participate in a process of redrawing the regulatory boundaries to their advantage. While it will not be possible to give a full overview here of these processes, it is useful to provide a couple of examples that illustrate the point. What is striking throughout these cases is the preparedness of lawmakers to facilitate the platforms' desire to avoid existing regulation. Their approach can be explained by a number of factors, not least ideological commitments to market 'freedom', echoing many of the debates surrounding the state's participation in the rolling back of its own involvement in production and service provision over the last 40 years (see for example Harvey 2006), and the easy (if often cosmetic) fix they represent in tackling growing unemployment numbers (Taylor et al. 2017).

In the case of taxi services, the state of California created a new category of Transportation Network Companies, which devolved the implementation of regulation to the platforms themselves (Finck 2017, p. 12). The government of Quebec, on the other hand, agreed to lift the ban on Uber services as well as the requirement that their drivers should obtain the same licences as taxi drivers, in exchange for the company's promise to perform thorough background checks on drivers and ensure that all vehicles were safe and insured (McKee 2018, p. 38). The state of Massachusetts went further by 'officially delegating to the platforms the power to issue a public licence' (McKee 2018, pp. 40–41).

While it might appear that these are negative outcomes for the platforms, subjecting them to unwanted regulation, the picture is more complicated. In many cases, the platforms have effectively captured parts of the regulatory mechanism in exchange for fees paid to the state (McKee 2018, p. 36). In addition, it appears that these new laws have little relation to their regulatory effectiveness but represent a (temporary) settlement in the confrontation between platforms and existing legislation (Wyman 2018). The lack of effective implementation of the agreements by platforms appears to bear this out.

Far from casting platforms as effective self-regulators, the evidence points to their ability to undermine existing regulations and encourage the establishment of new ones to their advantage. The issue at hand is, therefore, not one of 'self-regulation' at all but instead of who should regulate whom and to what end. The narrative surrounding self-regulation is a strikingly slippery one. Claims of freedom *from regulation* by platforms (Section 10.2) shift rapidly to freedom *to regulate* workers and consumers (Section 10.3) and, as illustrated in this section, to *deregulate* and *re-regulate* the state.

10.5 ALTERNATIVES FROM BELOW TO SELF-REGULATION FROM ABOVE

Faced with the dual challenge of a lack of rights at work on the one hand and limited state protection on the other, workers on digital labour platforms and their allies have responded in a number of different ways in the attempt to improve their pay and conditions. These responses can broadly be classified in three categories: industrial action; legal challenges; and extra-legal standard-setting initiatives. It is worth pointing out that, while it is helpful to discuss each of these approaches in turn, the divisions between them are not as solid as might appear. Indeed, labour struggles and legal challenges have more often than not taken place in concert, while standard-setting initiatives can only be fruitful if taken up as part of broader campaigns. It is perhaps useful to think of all three fields of action as different aspects of a constellation of regulatory interventions from below.

Around the world workers have started organizing and fighting back, despite confident claims by certain commentators and more established unions alike that the sector's reliance on isolated individuals without formal collective meeting places, and on platforms' internal regulations, would hamper worker organization (Woodcock and Graham 2019). These initiatives have taken different forms, ranging from online networks of crowdworkers like Turkernation, where Amazon Mechanical Turk workers can rate clients and leave comments for one another, to the development of unions for platform workers – particularly among delivery and taxi services – across the world. The use of WhatsApp groups and Facebook chats, as well as the newly launched online tools mentioned above, have replaced the staff room and the canteen of the traditional workplace (Woodcock and Graham 2019). The focus of these campaigns has ranged from better conditions for platform workers – mainly better pay and lower platform fees – to demands that accurate employment relations be acknowledged. Kilhoffer et al. (2017) provide a useful summary of labour initiatives and the key themes within their organizing efforts, while Du Toit (2019) underscores the importance of emerging platform worker cooperatives.

Alongside the emerging forms of worker organization, a growing number of legal challenges have pitted workers against platforms, overwhelmingly focusing on the definition of the existing employment relationship. Whether in South Africa, New York, London, Brazil or Switzerland, Uber drivers have argued that they should be classified as employees rather than as self-employed taxi service providers (Fiveash 2017; Haynes 2017; Rathi 2017). These legal challenges have met with fierce opposition from the platforms. Appeals are dragging out the process and, in some cases, halting the implementation of court decisions or overturning them. It is important to point out that traditional union federations in western Europe have played an important role in these legal challenges (Joyce et al. 2020). While dependency on these tactics represents an organizational weakness of the labour movement – unable to enforce labour rights at the point of production – they also illustrate the ability of the workers' movement to push for regulatory solutions of its own. Just as the platforms are moving to enforce regulations that serve their interests, so too are platform workers entering the institutional fray.

Other regulatory challenges have attempted to develop extra-legal standard-setting initiatives. Bringing together academics, activists, lawyers and labour organizers, and collaborating with platform workers themselves, these projects highlight the serious structural issues within the sector that need to be addressed, primarily regarding labour rights and working conditions, and reflect on new and better adapted approaches to implementing them.

For example, in Germany, the Fair Crowd Work project, set up by a number of unions and spearheaded by IG Metall, aims to provide platform workers with advice and support. Its website notes that it 'collects information about crowd work, app-based work, and other "platform-based work" from the perspective of workers and unions. Uniquely, the site offers ratings of working conditions on different online labour platforms based on surveys with workers' (IG Metall 2017). More recently, Silberman and Johnston (2020), who were involved with this initiative, have pointed out that Article 40 of the General Data Protection Regulation (GDPR) can serve as a helpful framework for workers to improve their conditions, across different platforms and regardless of their employment status. They argue that, while not resolving the issue of implementation *per se*, GDPR offers avenues for workers to demand access (and corrections) to information ranging from their ratings by customers to termination or pay being withheld without explanation by platforms.

Similarly, the 'Frankfurt declaration on platform-based work' (Fair Crowd Work 2016) aimed to launch a 'network of European and North American unions, labor confederations, and worker organizations … for transnational cooperation … to ensure fair working conditions and worker participation in governance in the growing world of digital labor platforms'. The network focused on minimum wages, accurate employment relations, the implementation of relevant national laws and social benefits and the development of representative mechanisms for platform workers. An important aspect of this work has been the establishment of an ombuds office for platform workers in Germany, setting a precedent for dispute resolution in this sector.

Another example of such an approach is Fairwork,[1] a project bringing together academics in the United Kingdom, South Africa, India and Germany, which has developed five areas on which to judge the quality of working conditions in digital labour platforms: fair pay; fair contracts; fair management;

fair conditions; and fair representation. Taking its cue from the living wage campaign and the Fairtrade project (Graham and Woodcock 2018, pp. 246, 248), Fairwork aims to mobilize consumer choice and intra-company competition to push platforms to regulate labour conditions in the industry while simultaneously providing the labour movement with greater information and a number of tools for action.

Judged on their compliance with the five areas of fair work, platforms are allocated a score out of ten after evidence has been gathered by the researchers, submitted by the platforms themselves and collected through worker interviews (see Fairwork 2019; Fairwork 2020a; Fairwork 2020b). Fairwork's grading mechanism was developed in concert with a host of stakeholders in the different countries where it is active including platforms, lawyers, unions and International Labour Organization representatives, as well as platform workers themselves. The aim of the grading process is to develop an international framework to quantify the quality of work in digital labour platforms and offer comparisons between different platforms within and between the countries surveyed.

In turn, the project hopes that this can be used by workers and organizers in their own campaigns. Comparisons between platforms open important avenues for potential worker activism and demands to equalize conditions upwards against the sector's tendency to do the opposite (see Section 10.3). It is striking, for example, that Uber offers better conditions in South Africa than it does in India (Fairwork 2019, p. 20). So, while it pays minimum wages, including costs, in both countries, only in South Africa does the company mitigate work-related risks by providing workers with a panic button linked to a private security firm to respond to car-jackings and assaults, or provide workers with a mechanism to communicate with a platform representative, on the app and in person.

In addition, comparison between companies that deliver the same services can mobilize existing competition between platforms and push them to take action to avoid being outdone by the competition in the field of ethical consumption. Indeed, the important differences in fairness scores which emerge between platforms offering the same services – different taxi services in India, or different delivery or cleaning services in South Africa – creates a pressure on the platforms that perform less well to catch up while also creating opportunities for workers and clients to make demands from the platforms directly to match – at the very least – industry standards. This is a potential opportunity for workers to mobilize competition, so celebrated as a disciplining factor in the self-regulation literature, against the platforms.

Moreover, the project has already achieved certain – if limited – success. In Fairwork's first year of activity in South Africa, NoSweat 'implemented minimum wage, health and safety, and grievance policies' under its influence and the delivery platform Bottles 'committed to supporting the emergence of independent, collective worker representation' (Fairwork 2019, p. 24). In its second year of activity, this time in Germany, Fairwork obtained that Zenjob 'formally indicate its willingness to encourage workers to form a collective body and engage in negotiations with it', while '[b]oth Zenjob and InStaff have … incorporat[ed] into their terms of service the anti-discrimination and anti-harassment guidelines recommended by the federal and regional agencies' (Fairwork 2020b, p. 18). While there can be no doubt that these good intentions will need worker vigilance to move from policy to implementation, they lay out a positive basis on which to do so.

What these three different aspects of regulation from below have in common is that they emerged out of the existing demands of platform workers and the attempt to develop – and impose – new regulatory frameworks. While competition and consumer choice are celebrated in the literature as market mechanisms leading to self-regulation, these projects illustrate a different possibility: collective regulation, often through conflict and confrontation, led by platform workers' demands rather than unilateral self-regulation imposed by the platforms.

10.6 CONCLUSION

Arguments in favour of self-regulation by digital labour platforms make a number of assumptions that are not borne out in reality. Most importantly, the claim that the market and the platforms can not only develop better standards than the state, but that they can do so for the collective benefits of shareholders, workers and consumers, has been shown to have serious limitations.

Instead, this chapter has argued that platforms have not sought to self-regulate in isolation from the state but have instead mobilized existing laws where they found them useful while leveraging their economic power and popularity to reshape those that hampered their business model. Much like any other company, platforms are led, first and foremost, by a desire to increase their market share, limit their costs and increase their profit margin. This process has led them to undermine labour regulations and working conditions as well as aggressively (and largely successfully) to attempt to reshape laws and lay hold on the levers of policy implementation.

Finally, the chapter has pointed to emerging forms of regulatory struggles from below, led by platform workers themselves, and amplified by a constellation of supporters. Whether through industrial action, legal challenges or the development of alternative standard-setting mechanisms, a wide array of strategies have and continue to be developed. Their success, however, does not depend on the goodwill of the platforms but on the ability of a different kind of regulation to emerge – and impose itself – from below.

This goes to the heart of the problem. While the alternatives discussed in Section 10.5 highlight the growing number of alternatives from below, the lack of a concrete, mass transformation of the sector is equally noticeable. As it stands, these initiatives have created a series of alternative readings of the law, underlined structural problems in the sector and pointed to different solutions to the struggles workers face. However, despite discrete victories – in the courts or through specific industrial disputes – the balance of forces remains everywhere to the advantage of the platforms.

As Ducci (2018, p. 299) points out, similar issues surrounding the lack of regulation and the ability of new industries to impose new (and roll back old) labour practices, to the detriment of workers, emerged in the past following rapid technological and economic changes. It was the combined development of labour militancy on the one hand and legal reform on the other that gave birth to the regulatory frameworks which platforms are seeking to destabilize. Reforms are currently being proposed by the different initiatives discussed in this chapter, but only through the development of sustained workers' struggle can these be imposed.

NOTE

1. This chapter's authors have all – at one time or another – worked for and/or on the Fairwork project.

REFERENCES

Allen, D. and C. Berg (2014), *The sharing economy: how over-regulation could destroy an economic revolution*, Melbourne: Institute of Public Affairs.

Cohen, M. and A. Sundararajan (2017), 'Self-regulation and innovation in the peer-to-peer sharing economy', *University of Chicago Law Review Online*, **82** (1), 116–133.

De Stefano, V. (2016), 'The rise of the "just-in-time workforce": on-demand work, crowd work and labour protection in the "gig-economy"', Conditions of Work and Employment Series 71, Geneva: ILO, accessed 8 January 2021 at www.ilo.org/wcmsp5/groups/public/---ed_protect/---protrav/---travail/documents/publication/wcms_443267.pdf.

Drahokoupil, J. and B. Fabo (2016), 'The platform economy and the disruption of the employment relationship', Policy Brief 5/2016, Brussels: ETUI.

Du Toit, D. (2019), 'Platform work and social justice', *Industrial Law Journal*, **40** (1), 1–11.

Ducci, F. (2018), 'Competition law and policy issues in the sharing economy', in S. Tremblay-Huet, D. McKee, F. Makela and T. Scassa (eds), *Law and the 'sharing economy': regulating online market platforms*, Ottawa: University of Ottawa Press, pp. 295–318.

Fabo, B., J. Karanovic and K. Dukova (2017), 'In search of an adequate European policy response to the platform economy', *Transfer*, **23** (2), 163–175.

Fair Crowd Work (2016), 'The Frankfurt declaration on platform-based work', *Fair Crowd Work*, accessed 8 January 2021 at http://faircrowd.work/unions-forcrowdworkers/frankfurt-declaration/.

Fairwork (2019), 'The five pillars of Fairwork: labour standards in the platform economy', Oxford: Oxford Internet Institute, accessed 8 January 2021 at https://fair.work/wp-content/uploads/sites/97/2019/10/Fairwork-Y1-Report.pdf.

Fairwork (2020a), 'Fairwork Germany ratings 2020: labour standards in the platform economy', accessed 8 January 2021 at https://fair.work/wp-content/uploads/sites/97/2020/05/Germany-English-report.pdf.

Fairwork (2020b), 'Fairwork South African ratings 2020: labour standards in the gig economy', accessed 8 January 2021 at https://fair.work/wp-content/uploads/sites/97/2020/04/Fairwork-South-Africa-2020-report.pdf.

Finck, M. (2017), 'Digital co-regulation: designing a supranational legal framework for the platform economy', LSE Legal Studies Working Papers 15/2017, accessed 8 January 2021 at http://dx.doi.org/10.2139/ssrn.2990043.

Fiveash, K. (2017), 'Uber driver is employee and must be treated as such, rules Swiss agency', *ArsTechnica*, 1 June, accessed 8 January 2021 at https://arstechnica.com/tech-policy/2017/01/uber-driver-is-employee-rules-swiss-agency/.

Fredman, S. and D. Du Toit (2019), 'One small step towards decent work: Uber v Aslam in the Court of Appeal', *Industrial Law Journal*, **48** (2), 260–277.

Graham, M. and J. Woodcock (2018), 'Towards a fairer platform economy: introducing the Fairwork Foundation', *Alternate Routes*, **29**, 242–253.

Graham, M., I. Hjorth and V. Lehdonvirta (2017), 'Digital labour and development: impacts of global digital labour platforms and the gig economy on worker livelihoods', *Transfer*, **23** (2), 135–162.

Harvey, D. (2006), 'Neo-liberalism as creative destruction', *Geografiska Annaler, Series B*, **88** (2), 145–158.

Haynes, B. (2017), 'Brazil judge rules Uber drivers are employees, deserve benefits', *Reuters*, 14 February, accessed 8 January 2021 at www.reuters.com/article/us-uber-tech-brazil-labor/brazil-judge-rules-uber-drivers-are-employees-deserve-benefits-idUSKBN15T2OC.

IG Metall (2017), *Fair crowd work*, accessed 28 June 2020 at http://faircrowd.work.

Irani, L. (2015), 'Difference and dependence among digital workers: the case of Amazon Mechanical Turk', *South Atlantic Quarterly*, **114** (1), 225–234.

Johal, S. and N. Zon (2015), 'Policymaking for the sharing economy: beyond whack-a-mole', Mowat Research 106, Toronto: Mowat Centre, accessed 8 January 2021 at https://munkschool.utoronto.ca/mowatcentre/wp-content/uploads/publications/106_policymaking_for_the_sharing_economy.pdf.

Joyce, S., D. Neumann, V. Trappmann and C. Umney (2020), 'A global struggle: worker protest in the platform economy', Policy Brief 2/2020, Brussels: ETUI.

Kilhoffer, Z., K. Lenaerts and M. Beblavý (2017), 'The platform economy and industrial relations: applying the old framework to the new reality', CEPS Research Report 2017/12, Brussels: CEPS.

Koopman, C., M. Mitchel and A. Thierer (2015), 'The sharing economy and consumer protection regulation: the case for policy change', *Journal of Business, Entrepreneurship and the Law*, **8** (2), 529–547.

Leong, N. and A. Belzer (2017), 'The new public accommodations: race discrimination in the platform economy', *Georgetown Law Journal*, **105** (1271), 1293–1295.

McKee, D. (2017), 'Neoliberalism and the legality of peer platform markets', *Environmental Innovation and Societal Transitions*, **23** (1), 105–133.

McKee, D. (2018), 'Peer platform markets and licensing regimes', in D. McKee, F. Makela, T. Scassa and S. Tremblay-Huet (eds), *Law and the 'sharing economy': regulating online market platforms*, Ottawa: University of Ottawa Press, pp. 17–54.

Rathi, A. (2017), 'A UK court has upheld the ruling that Uber drivers have workers' rights – and deserve benefits', *Quartz*, 10 November, accessed 8 January 2021 at https://qz.com/1126154/uber-employment-case-a-uk-court-rules-drivers-should-get-worker-benefits/.

Silberman, M. and H. Johnston (2020), 'Using GDPR to improve legal clarity and working conditions on digital labour platforms' Working Paper 2020/05, Brussels: ETUI.

Stemler, A. (2017), 'Feedback loop failure: implications for the self-regulation of the sharing economy', *Minnesota Journal of Law, Science and Technology*, **18** (2), 673–712.

Taylor, M., G. Marsh, D. Nicol and P. Broadbent (2017), 'Good work: the Taylor review of modern working practices', accessed 8 January 2021 at https://assets.publishing.service.gov.uk/government/uploads/system/uploads/attachment_data/file/627671/good-work-taylor-review-modern-working-practices-rg.pdf.

van Doorn, N. (2017), 'Platform labor: on the gendered and racialized exploitation of low-income service work in the "on-demand" economy', *Information, Communication, and Society*, **20** (6), 898–914.

Woodcock, J. and M. Graham (2019), *The gig economy: a critical introduction*, Cambridge: Polity Press.

Wyman, K. (2018), 'The novelty of TNC regulation', in N. Davidson, M. Finck and J. Infranca (eds), *Cambridge handbook on law and regulation of the sharing economy*, Cambridge: Cambridge University Press, pp. 129–140.

11. Trade union responses to platform work: An evolving tension between mainstream and grassroots approaches

Simon Joyce and Mark Stuart

11.1 INTRODUCTION

Platform work has presented major challenges to trade unions including: a geographically dispersed workforce; novel algorithmic management methods; lack of employment rights and social protections; and legal restrictions on collective bargaining for workers (mis)classified as self-employed.[1] Predictions that these difficulties would render the unionization of platform workers impossible have proved mistaken, however. Platform worker organization has grown rapidly, with impressive displays of energy, creativity and determination. Indeed, some commentators see platform worker organization as 'a new power in the labour movement' (Vandaele 2021; see also Cant and Woodcock 2020).

The great irony of these developments is that this energy and creativity has largely bypassed established, mainstream unions. Indeed, it has often taken place outside any union organization, in loose networks of workers coordinated in town squares or online. Where platform workers have joined unions, these have often been radical grassroots unions. The 'organizational creativity' (Vandaele 2021) of grassroots unions means they have led the way in standing up to platform companies, despite being much smaller than their mainstream counterparts. This remarkable upsurge of worker organizing outside the ranks of mainstream unions poses a number of important questions. First, are established forms of worker organizing ineffective in mobilizing and representing platform workers? Second, can the established solutions on which mainstream unions rely, such as institutionalized collective bargaining and litigation, effectively address the problems of platform workers? Third, have the new grassroots organizers found alternative solutions to these problems?

The distinction between 'mainstream' and 'grassroots' unions has become common but is rarely defined. Following definitions used by the Leeds Index of Platform Labour Protest (Trappmann et al. 2020), we define mainstream unions as established, typically longstanding, politically moderate (social democratic) and usually affiliated to a national union confederation. By contrast, grassroots unions are usually newer, politically more radical and not part of a national confederation. Furthermore, mainstream unions tend to be more hierarchically organized while grassroots unions usually have a more horizontal structure (Cini and Goldman 2020; Chesta et al. 2019; Mrozowicki and Maciejewska 2017). Of course, there are exceptions: some grassroots unions are longstanding or affiliated to union confederations while some established unions are politically radical. Nevertheless, the broad distinction is adequate for current purposes.

The argument of this chapter is that the response of mainstream unions to the spread of platform work has been uneven at best. Platform worker activists complain that established unions 'have their own agenda' or simply do not 'get' platform work. Where mainstream unions have attempted to organize platform workers, gains from orthodox collective bargaining have, so far, been minimal. By contrast, grassroots unions have developed direct action methods combining street and online organizing to produce high-profile protests better suited to challenging platforms. Grassroots unions have contributed significantly to raising the public and political profile of issues in platform work and have made significant gains for workers in some cases. What remains unclear, though, is the extent to which the gains of grassroots unions can be converted into lasting improvements or sustained organization.

The coverage of this chapter is limited in two important but unavoidable respects. First, we mainly examine union organization of in-person platform workers, especially in delivery: couriers, taxi drivers and other transport work. These location-based platforms are where worker protests have been most widespread and visible, and where the workers involved are most likely to be unionized. While platform workers in other industries – including online platform work – do organize collectively, these are much less likely to be unionized and therefore fall outside the specific concerns of this chapter. Second, and relatedly, we focus mainly on the Global North. Platform worker organization and protest in the Global South is significantly less likely to be unionized. If a shortage of research on non-union worker organization presents difficulties in understanding the dynamics of worker struggles in the Global North, it is a fundamentally more serious problem in the Global South (Atzeni 2020). Further research in this area is urgently needed.

The chapter is set out as follows. First, we examine mainstream union responses to platform work and then review ones from grassroots unions. Finally, we assess the relative contributions of each and argue that grassroots unions have developed methods from which mainstream unions could learn.

We support our analysis with empirical evidence from a wide range of relevant literature and from our own research.[2]

11.2 MAINSTREAM UNIONS: THE LIMITS OF INSTITUTIONALISM

The response of mainstream unions to platform work has been uneven. Evidence from the Leeds Index of Platform Labour Protest suggests the involvement of mainstream unions in platform worker organizing has been low and patchy. Mainstream unions were involved in only 19 per cent of platform worker protests between January 2017 and June 2020 (Bessa et al. forthcoming). Mainstream union involvement was also concentrated in the Global North: some 34 per cent of platform worker protests in Europe involved mainstream unions, but only 11 per cent in North America. Figures for the Global South as a whole are, however, even lower. In Latin America, 16 per cent of platform worker protests involved mainstream unions, but this fell to only 10 per cent in Asia and 7 per cent in Africa.

The shortcomings of mainstream unions in relation to platform work can be seen as an extension of their wider difficulties in dealing with precarious work. Too often, mainstream unions adopt 'exclusive' approaches, defending core members in secure work but excluding precarious workers (Doellgast et al. 2018). A similar reticence is evident in relation to platform work. In interviews with us, mainstream union officers expressed uncertainty about who platform workers are, where they might be found and how or even whether they might be organized.[3] Union officers told us that the employment practices and organizational forms of platform work do not fit established, institutionalized organizing models. Insofar as platform workers do not fit these models, mainstream unions find it difficult to organize them. The difficulties cited by mainstream union officers include the lack of a fixed workplace, difficulties identifying an employer and a lack of employment rights. In the European Union (EU), competition law rules out union representation for self-employed workers on the questionable grounds that such organization comprises a cartel to fix prices (De Stefano 2017). Similarly, in the United States (US), collective bargaining procedures under the National Labor Relations Act apply only to employees (Dubal 2017). While the challenges of organizing workers outside the institutionalized core are real enough, the relative inactivity of many mainstream unions in the platform work sector begs the question as to whether alternative methods might be more applicable.

Nevertheless, some mainstream unions have attempted to organize platform workers. In Denmark, the 2018 agreement between mainstream union *3F* and platform-based domestic cleaning company *Hilfr* was greeted as 'groundbreaking' (Uni Global Union 2018); a 'landmark achievement' (Aloisi 2019); and a 'model for other sectors' (Hale 2018). The initial enthusiasm hid a number of problems, however. The agreement set minimum hourly pay for self-employed workers – known as 'Freelance Hilfrs' – while more experienced workers could become employees – 'Super Hilfrs' – and achieve higher rates of pay. By March 2020, there were only 36 Super Hilfrs alongside 180 Freelance Hilfrs who remained self-employed on lower pay, albeit underpinned by an agreed hourly minimum (Ilsøe 2020).

The differential coverage of employed and self-employed workers was only one way that the *Hilfr* agreement diverged from typical Danish collective bargaining. Other differences included: the agreement was signed at company, not industrial, level; employee status was entirely voluntary; and disputes were excluded from labour court coverage. Although the union saw the *Hilfr* agreement as a 'staircase' agreement, anticipating future improvement, it obviously left a lot of ground to be made up. A further peculiarity of the *Hilfr* agreement was its unusual origin in a high-level initiative between government, major unions and employer organizations (Ilsøe 2020). Furthermore, *Hilfr* promoted social responsibility as a feature of its business model and consequently sought a collective agreement despite low levels of union membership: a rare approach among platform companies. The unusual nature and origins of the *Hilfr* agreement significantly limit its overall applicability as a blueprint for other unions.

In a further blow, the *Hilfr* agreement was struck down in August 2020 by the Danish Competition and Consumer Authority (DCCA) on the basis of the EU competition law mentioned previously. Specifically, the DCCA ruled that minimum hourly pay rates 'may limit competition between the Freelance Hilfrs' (DCCA 2020). Limiting competition between workers is, of course, exactly the purpose of collective agreements on wages. Consequently, this ruling rendered a central function of unions unlawful for platform workers. For some commentators the ruling is a misapplication of EU competition law and therefore open to challenge (Countouris and De Stefano 2020). Certainly, the use of EU law to ensure that domestic cleaners compete against each other seems particularly cruel. Whatever the final outcome, this case highlights the dependence of mainstream union approaches on sympathetic legal and institutional settings and the difficulties they face when conditions are less conducive.

Similar difficulties arise in the 2019 agreement between the British union GMB and platform courier company Hermes, which controversially retains self-employment while allowing drivers to opt in voluntarily to a 'self-employed plus' (SE+) contract. The SE+ contract, which also applies to new starters, includes minimum hourly pay at least equal to the United Kingdom (UK) minimum wage, paid holiday and bonuses for hitting delivery targets (Rolf et al. forthcoming). The agreement also includes union representation for SE+ drivers, with negotiation and consultation based on a ‘Partnership Principles Agreement’ committing to ‘co-operative’ relations. As of mid-2020, some 3000 *Hermes* drivers were on the SE+ contract, with around 10 000 on standard self-employed terms – far larger than the *Hilfr* agreement.

While workers report some improvements under the agreement, significant difficulties remain. Previously, the GMB had won a legal ruling that Hermes drivers should be classed as workers, occupying a place in UK law between self-employed and employee. Hermes planned to appeal, and further litigation seemed likely, but this ceased when the agreement was signed. Under the agreement, SE+ drivers have more protections than the self-employed (minimum hourly pay and paid holidays) but fewer than the full legal entitlement for workers which would extend to pension enrolment, health and safety regulation, paid rest breaks and collective consultation over redundancy. In other words, the collective agreement conceded issues previously won in court. Defending itself from critics – including grassroots unions – the GMB emphasized the expense and uncertainty of legal proceedings. It remains to be seen how much the agreement might be improved in future.

Where platform workers are already classed as employees, mainstream unions have been able to extend industrial agreements to include platform workers. Swedish union *Unionen* has agreements with a small number of companies that describe themselves as ‘platforms’ but which are effectively temporary work agencies supplying mostly white-collar or professional workers to client businesses and who are consequently covered by an existing sectoral agreement (Jesnes et al. 2019). Moreover, the workers concerned are legally employees. This arrangement is therefore somewhat different from the usual model of platform work (Soderqvist and Bernhardtz 2019). Recent agreements with Just Eat in Denmark and Foodora in Sweden similarly gain purchase when riders are employees although, in Denmark, employees conduct only around 10 per cent of Just Eat’s deliveries (Havstein 2021). Again, while such cases provide positive examples for unions, this model is difficult to apply where platform workers are not employees.

Legal employee status has also enabled mainstream unions to establish works councils on platforms in a few European locations, including Foodora in Vienna; Foodora and Deliveroo in several cities in western Germany; and *Laconsegna* in Italy (Vandaele 2021). It is unclear, though, how many of these bodies are still functioning. For instance, Deliveroo undermined works council arrangements by reducing its workforce below the minimum threshold before withdrawing from Germany altogether.

In a few cases, employee status has helped mainstream unions organize more militant action. In 2019, Norwegian union *Fellesforbundet* organized a five-week strike of Foodora riders, winning a collective agreement with increased pay; annual rises; expenses payments for bicycles, phones, uniforms and winter gear; and provision for early retirement (Jesnes et al. 2019; see also Chapter 16). Strike tactics included methods seen in grassroots unions including 'critical mass' ride-outs, a city centre focus for strike activity, soup kitchens, free bicycle repairs for the public and energetic use of social media. The strike gained significant public support and spread from Oslo to Trondheim with the number of strikers doubling to over 200 (ITF 2020).

Where platform workers lack legal employee status, some mainstream unions have adopted what are effectively non-union approaches. For instance, IG Metall pioneered the online provision of information to platform workers in a model that contributed to the development of the Fairwork Foundation (see Chapter 10). In the US, unions have adopted the 'worker association' model, which permits organization outside the usual framework of US collective bargaining, often focusing instead on lobbying city- or state-level government for labour rights and greater regulation of platforms. As previously, these efforts have had mixed results.

In 2014, Teamsters Local 117 in Seattle set up the App-Based Drivers Association – later, the Drivers Union – to organize drivers on Uber, Lyft and similar platforms (Stott 2014). In 2015, a Seattle city statute gave independent contractors the right to form unions and collectively bargain, although this law was challenged by business interests (Wiessner 2018). In California, Gig Workers Rising (GWR) is a worker association linked to the Service Employees International Union (SEIU). GWR campaigned alongside unions – both mainstream and grassroots – for the passage of Assembly Bill 5 (AB5), the 2019 California state law which reclassified platform workers as employees.[4] Similarly, in New York, the Independent Drivers Guild (IDG), set up in 2016 by the International Association of Machinists, took part in the campaign for the increased regulation of app-based taxi services in New York City which led to caps on car and driver numbers and a minimum wage for drivers (Brooks 2018).

There have also been problems, however. In 2019, GWR provoked controversy when SEIU officials were discovered to have met executives from Uber and Lyft, seemingly to discuss the possibility of an agreement retaining independent contractor status for drivers (Scheiber 2019). Similarly, the IDG has drawn criticism due to its agreement with Uber which, though kept secret, is thought to have accepted independent contractor status for drivers in exchange for limited concessions and, most controversially, direct funding from the company (Scheiber 2017; for a more critical view, see DeManuelle-Hall 2019). In key instances, then, the adoption of non-union methods, while potentially offering a pragmatic solution to the problems of organizing platform workers, has seen mainstream US unions move towards problematic concessions to platform companies, including over the crucial issue of employment status.

Given the importance of employment status, a key strategy of mainstream unions has been to challenge the (mis)classification of platform workers. Reclassification can bring important benefits such as minimum wages, paid leave or even healthcare (Forde et al. 2017) as well as collective rights to union representation and collective bargaining. Legal actions are more common in the Global North than the Global South, reflecting employment status having a far greater material impact in the former than the latter (Bessa et al. forthcoming). Furthermore, evidence from the Leeds Index shows that although mainstream and grassroots unions generally combine legal action and other methods in their campaigns, mainstream unions are considerably more likely to engage in legal action than grassroots unions and less likely to organize strikes or demonstrations (Joyce et al. 2020b). These differences suggest that mainstream unions and grassroots unions adopt different strategic orientations when it comes to selecting methods for pursuing grievances.

The evidence to date suggests that mainstream unions have been reticent about organizing platform workers. Where platform workers have legal employee status and can be incorporated into existing collective bargaining arrangements, some progress has been made. There are few countries where this is the case, however. Where employee status is lacking, representation by mainstream unions is rare and collective agreements tend to be weak and at risk of legal annulment. While mainstream unions have defended such arrangements as better than nothing, it is not surprising that these deals have attracted considerable criticism from grassroots unions. Faced with the difficulties of organizing platform workers through established channels, mainstream unions show a strategic preference for legal action over strikes or demonstrations. In the next section, we examine the contrasting approach of grassroots unions to platform work which can broadly be characterized as a strategy of militant direct action.

11.3 GRASSROOTS UNION RESPONSES: THE LOGIC OF MOBILIZATION

Two things stand out about grassroots union responses to platform work: their willingness to organize workers who are not legally employees; and the sheer energy and dynamism of these often small organizations. A growing research literature highlights the 'organizational creativity' (Vandaele 2021) of grassroots unions and draws contrasts with mainstream unions (for instance, Aloisi and Gramano 2019; Bronowicka and Ivanova 2020; Chinguno 2019; Cini and Goldmann 2020; Chesta et al. 2019; Mrozowicki and Maciejewska 2017; Panimbang et al. 2020; Tassinari and Maccarrone 2020; Vicente 2019). Typically, grassroots unions combine established methods such as strikes and demonstrations with newly innovated ones including log-offs, extensive online organizing and legal challenges that augment rather than substitute for mobilization. Grassroots platform worker unions are found in all regions but are most active in Europe and feature in platform worker protests with roughly the same frequency as mainstream unions (Bessa et al. forthcoming) – a remarkable finding given the huge disparities in size and resources.

In general, self-activity and self-organization feature prominently in all forms of platform worker organization – both union and non-union. Indeed, trade unions of any type are actually absent in most cases of platform worker protest (Bessa et al. forthcoming). Most platform worker protests are organized by loose networks of workers in town squares or online groups. Sometimes, workers give themselves a collective name or may approach a union – mainstream or grassroots – or even form their own union. This complex of overlapping forms and dynamics highlights the difficulties in uncritically applying to platform work the familiar conceptual frameworks that are drawn from institutionalized industrial relations in the Global North. It also underlines the importance of analysing platform worker organization in terms of class-based dynamics (Atzeni 2020). Bottom-up organizing and the relative absence of mainstream unions help to explain the prevalence of grassroots unions in this sector (cf. Cini and Goldman 2020). Platform workers may form new grassroots unions – for instance, Rideshare Drivers United in California or Riders Union Bologna in Italy – or look to existing ones, such as the Independent Workers Union of Great Britain (IWGB) or *Freie ArbeiterInnen-Union* in Germany. While this section focuses on grassroots organizations that describe themselves as unions, it is important to understand that much of their dynamic character comes from processes of worker self-organization which, to date, are significantly under-researched.

Strikes organized by grassroots unions tend to be energetic and high profile. For location-based platform workers such as delivery riders, gathering in central locations close to popular restaurants is part of the working day that transfers readily to collective action. Consequently, strikes by these workers often feature gatherings in town squares with crowds of brightly uniformed workers augmented by flags, flares and music (for instance Riders Union Bologna 2021; Scott 2019). Grassroots unions have a capacity for engaging media attention, increasing the public profile of strikes and demonstrations and highlighting issues in platform work and insecure employment more generally.

Grassroots unions have also expanded the received models of strike action by integrating tactics from other social movements. One obvious example is the 'critical mass' style ride-out, where large numbers of cyclists or scooter riders ride slowly around city streets in a loud and colourful demonstration, slowing or blocking traffic and stopping to protest outside the offices of platform companies, their clients or regulatory authorities (Cant 2020; Chesta et al. 2019; Tassinari and Maccarrone 2020). Platform workers have also augmented traditional strike methods with log-offs, whereby platform workers turn off the app and make themselves unavailable for work. This tactic can have a significant impact on platform operations, as Callum Cant (2020: 116) describes:

> After just half an hour the app was in meltdown. Restaurant workers later told union reps that orders had been delivered three hours late, and that order volume had collapsed by over 50 per cent. Food was stacking up in the kitchens and no one was turning up to deliver it.

Food delivery is a just-in-time industry which gives workers a degree of structural power (Vandaele 2020, 2021); something commonly ignored in accounts that recognize only the difficulties of organizing platform workers.

Grassroots unions have even been able to coordinate international strike action. In Europe, food delivery workers in different countries communicate regularly and have, at times, coordinated action (Bronowicka and Ivanova 2020). Perhaps most impressively, grassroots unions of Uber drivers coordinated international strike action to coincide with the date in May 2019 when Uber shares were first traded on stock exchanges (Dubal 2019). Originating in California, strikes and other protests took place across the US and in Chile, Brazil, UK, Australia, Japan and India.

A notable but under-researched feature of grassroots strikes of platforms is that they often involve workers who are not union members. For mainstream unions, and for most academic observers, strikes are strongly associated with union membership. While there is little research on this issue – studies of grassroots unions tend not to examine membership – it is evident in some accounts that grassroots unions promote participation in strike action in terms of mobilizing as many workers as possible rather than seeing strikes narrowly in terms of organizational membership. For instance, Cant (2020, p. 123) notes union membership being concentrated among cyclists but that 'participation in the strike had been more even' across both cyclists and scooter riders. Furthermore, the strike committee comprised both union members and non-members (2020, p. 116). Cant (2020, p. 127) goes on to record strikes developing out of WhatsApp discussions. Evidence from Italy also suggests that grassroots unions encourage participation in strike action beyond union membership (Chesta et al. 2019; Cini and Goldmann 2020). Further research would be beneficial but what evidence there is does suggest that the strategy of grassroots unions is focused on wider mobilization.

A further innovation of grassroots unions is the widespread use of online organizing. Research shows this to be an almost universally important method of grassroots platform worker organizing, found across work on location-based platforms such as food delivery and ride-sharing apps as well as fully online workers on platforms like Upwork or Amazon Mechanical Turk (Anwar and

Graham 2020; Maffie 2020; Wood et al. 2018; Zyskowski and Milland 2018). Grassroots union activists often report involvement in online groups prior to joining a union. For instance, London Uber driver and union organizer Yaseen Aslam was a member of a WhatsApp group of 50–60 drivers even before he started driving for the platform (Aslam and Woodcock 2020). While these groups often pre-figure union organization, grassroots unions have been notably more successful in integrating online methods into their organizing activity than have most mainstream unions.

While grassroots unions mobilize workers who are not members, they are nevertheless concerned to recruit and increase membership. Again, research in this area is scarce. Where membership figures are available, these may include non-platform workers. For instance, the IWGB registered a membership of 4623 in 2019 (Certification Office 2020) but this number includes members from non-platform industries. While many grassroots unions remain small, some have achieved a relatively significant size. The New York Taxi Workers Alliance, whose membership includes non-platform drivers, claims around 21 000 members (NYTWA 2021). In California, Rideshare Drivers United similarly claims 19 000 members but these are all platform-based drivers (Moore 2020). Grassroots unions' capacity to mobilize workers beyond their membership makes comparisons with mainstream unions complicated, however.

Grassroots unions are, like mainstream unions, also active in challenging legal frameworks around platform work. One approach is to challenge platforms directly through litigation aimed at changing the legal status of platform workers (Aloisi and Gramano 2019; Dubal 2017). Unions such as the IWGB have been extremely active in this regard, often pressing cases that mainstream unions have been unwilling to take up. Grassroots unions have been additionally active in campaigns to change statute law on employment status; for instance, the long-running campaign over AB5 in California. Another approach has been to campaign for changes to the regulation of local markets for particular platform services. Most notably, this includes the re-regulation of taxi services to encompass platforms such as Uber by restricting the numbers of drivers and cars on the road; changing driver licensing requirements; extending minimum wage regulations to include ride-share drivers; or establishing minimum standards for the fair treatment of platform workers (Aslam and Woodcock 2020; Chesta et al. 2019).

What makes grassroots campaigns for legal reform different from mainstream union approaches, however, is that they tend to be integrated into wider campaigns of collective action. This reintegration of 'legal enactment' with wider collective campaigns (Webb and Webb 1902) contrasts with the more familiar approach of mainstream unions where legal action is often a 'cut-price' (Hyman 1989, p. 58) alternative to direct action such as strikes.

The gains made by grassroots unions are more difficult to track than institutionalized negotiations with annual pay rounds, detailed recording of outcomes and the circulation of minutes. Nevertheless, gains certainly have been made. In addition to the legal and regulatory reforms noted above, grassroots unions have won increases in pay or the withdrawal of pay reductions (Tassinari and Maccarrone 2020); restrictions on the recruitment of new workers to the platform, thereby holding up pay rates (Cant 2020); and modifications to algorithmic management regimes (Bronowicka and Ivanova 2020). Moreover, the establishment of persistent union organization among a workforce widely considered unorganizable should be considered a notable achievement in itself.

Despite these gains, however, difficulties remain. Grassroots-style unions have historically experienced difficulties in sustaining organization in the long term (Darlington 2013) and that is reflected in the ups and downs of membership and activity noted in recent research (for instance Cant 2020; Tassinari and Maccarrone 2020). In part, these difficulties reflect the transitory platform workforce. Grassroots unions also face the considerable capacity of platform companies to counter-mobilize through their control over the app and the deep pockets of their venture capital funders. Furthermore, the institutionalization of collective bargaining that lays behind the more conservative approach of mainstream unions also contributes a stability to union organization that grassroots unions in non-institutionalized settings struggle to maintain (cf. Darlington 2013). Nevertheless, despite these challenges, the efforts of grassroots unions have shown beyond question that more determined union responses to platform work are entirely achievable.

11.4 CONCLUSION

Contrary to expectations, platform workers have developed new ways of organizing and new methods of struggle and have revisited and refreshed older, more familiar forms. These methods are developing in 'tension' (Vandaele 2020) with established, mainstream methods. We have outlined some ways in which this tension is developing, and it is now possible to answer – at least in outline – the questions set out at the beginning of this chapter.

To date, established forms of worker organizing have proved somewhat ineffective in mobilizing and representing platform workers. In particular, the response of mainstream unions has been at best uneven. Most obviously, there is a continuing problem that many mainstream unions simply ignore platform workers, repeating an established pattern whereby mainstream unions focus on core membership at the expense of emerging industries (Doellgast et al. 2018). Some mainstream unions have made efforts to organize platform workers. As shown above, however, mainstream unions have difficulties organizing workers who are not legally defined as employees. Yet the experience of grassroots unions shows that established methods such as strikes can be highly effective in platform work. Consequently, the difficulties encountered by mainstream unions in mobilizing and representing platform workers stem less from the methods themselves than from the formulaic way that mainstream unions apply them to platform work. Improvements in mainstream union responses to platform work ought, therefore, to include both a reorientation on platform work as an organizing priority and the adaptation of methods to the particular challenges of the sector.

The established solutions on which mainstream unions rely, such as institutionalized collective bargaining and litigation, have a similarly mixed record in terms of effectively addressing the problems of platform workers. Certainly, the highly institutionalized forms of collective bargaining prevalent in the Global North have had little impact so far. This is due to the lack of employee status that is key to accessing legal frameworks of employment rights and protections including institutionalized collective bargaining (for non-institutional understandings of collective bargaining, see especially Webb and Webb 1902; and cf. Aloisi and Gramano 2019). Consequently, the strategic orientation on institutional access to collective bargaining typical of mainstream unions necessarily leads to difficulties when workers are not legally defined employees, as is almost always the case in platform work.

By contrast, grassroots unions commonly adopt a strategy of direct action where non-institutionalized or informal collective bargaining is used to translate pressure from collective action into gains for workers. Similarly, a difference in strategic orientation can be identified in relation to the litigation in which grassroots unions also engage: where mainstream unions tend to use it as an alternative to collective action, grassroots unions are more likely to use it to strengthen campaigns of collective action. Thus, the effectiveness of particular methods differs depending on the strategic orientation of the unions using them.

Finally, have the new grassroots organizers found alternative solutions to these problems? The key innovation of grassroots unions to date has been to develop ways of organizing workers who fall outside the standard institutionalized methods that have become the default strategic orientation of mainstream unions in the Global North.

Grassroots unions have built memberships of workers who are usually not legally categorized as employees. They have organized strikes outside usual institutional frameworks and have augmented traditional strike methods with digital methods such as collective log-offs and the widespread use of online organizing. They have successfully incorporated protest methods from social movements, such as the critical mass ride-out, and have developed a capacity for social media campaigning significantly in advance of that of most mainstream unions. Furthermore, they have shown how the method of 'legal enactment' can be incorporated into activist campaigns. In short, they have shown themselves adept at operating in non-institutionalized spaces of worker organization, in marked contrast to most mainstream unions.

This is not to say that all problems have been solved. Considerable challenges remain for mainstream and grassroots unions alike. What is important for current debates is that, in an industry of highly precarious work and workers, where the difficulties of union organization are well known and where the methods of mainstream unions have once more been found wanting, new and radical grassroots unions have made significant strides in developing alternative methods. A key question for the future of trade unions – not only in platform work – is therefore whether, and how fast, mainstream unions can learn from grassroots unions and both adopt and adapt those methods in respect of the challenges which undoubtedly lie ahead.

NOTES

1. This work was supported by the Economic and Social Research Council (grant number ES/S012532/1) as part of the Digital Futures at Work Research Centre (Digit).
2. The research includes: 48 interviews with platform work stakeholders in Europe (see Forde et al. 2017), including mainstream and grassroots activists and officers; a case study of platform workers in the UK; attendance at conferences and seminars of platform worker unions (mainstream and grassroots); and monitoring online media coverage of platform worker protests (see Bessa et al. forthcoming; Joyce et al. 2020a; Joyce et al. 2020b; Trappmann et al. 2020).
3. Interviews conducted with mainstream union officers in Bulgaria, Denmark, Germany, Italy, Poland and the UK (see Forde et al. 2017); other evidence from sources cited above.
4. In November 2020, AB5 was overturned after platform companies spent more than $200 million campaigning to preserve the independent contractor status of their workforce (De Stefano 2020).

REFERENCES

Aloisi, A. (2019), 'At the table, not on the menu: non-standard workers and collective bargaining in the platform economy', *EUIdeas*, 25 June, accessed 17 February 2021 at https://euideas.eui.eu/2019/06/25/at-the-table-not-on-the-menu-non-standard-workers-and-collective-bargaining-in-the-platform-economy/.

Aloisi, A. and E. Gramano (2019), 'Workers without workplaces and unions without unity: non-standard forms of employment, platform work and collective bargaining', in V. Pulignano and F. Hendrickx (eds), 'Employment relations for the 21st century', Bulletin of Comparative Labour Relations 107, Alphen aan den Rijn: Wolters Kluwer.

Anwar, M. A. and M. Graham (2020), 'Hidden transcripts of the gig economy: labour agency and the new art of resistance among African gig workers', *Environment and Planning A: Economy and Space*, **52** (7), 1269–1291.

Aslam, Y. and J. Woodcock (2020), 'A history of Uber organizing in the UK', *South Atlantic Quarterly*, **119** (2), 412–421.
Atzeni, M. (2020), 'Worker organisation in precarious times: abandoning trade union fetishism, rediscovering class', *Global Labour Journal*, **11** (3), 311–314.
Bessa, I., S. Joyce, D. Neumann, M. Stuart, V. Trappmann and C. Umney (forthcoming), 'Worker protest in the platform economy', ILO Working Paper, Geneva: ILO.
Bronowicka, J. and M. Ivanova (2020), 'Resisting the algorithmic boss: guessing, gaming, reframing and contesting rules in app-based management', accessed 17 February 2021 at www.ssrn.com/abstract=3624087.
Brooks, C. (2018), 'How New York taxi workers took on Uber and won', *Labor Notes*, **474**, 23 August, accessed 17 February 2021 at www.labornotes.org/2018/08/how-new-york-taxi-workers-took-uber-and-won.
Cant, C. (2020), *Riding for Deliveroo: resistance in the new economy*, Cambridge: Polity Press.
Cant, C. and J. Woodcock (2020), 'Fast food shutdown: from disorganisation to action in the service sector', *Capital and Class*, **44** (4), 513–521.
Certification Office (2020), 'Independent workers union of Great Britain: 2019 annual returns', accessed 17 February 2021 at www.gov.uk/government/publications/independent-workers-union-of-great-britain-iwgb-annual-returns.
Chesta, R. E., L. Zamponi and C. Caciagli (2019), 'Labour activism and social movement unionism in the gig economy: food delivery workers struggles in Italy', *Partecipazione & Conflitto*, **12** (3), 819–844.
Chinguno, C. (2019), 'Power dynamics in the gig/share economy: Uber and Bolt taxi platforms in Johannesburg, South Africa', *Labour, Capital and Society*, **49** (2), 30–65.
Cini, L. and B. Goldmann (2020), 'The worker capabilities approach: insights from worker mobilizations in Italian logistics and food delivery', *Work, Employment and Society*, accessed 17 February 2021 at https://doi.org/10.1177/0950017020952670.
Countouris, N. and V. De Stefano (2020), 'Collective-bargaining rights for platform workers', *Social Europe*, 6 October, accessed 17 February 2021 at www.socialeurope.eu/collective-bargaining-rights-for-platform-workers.
Darlington, R. (2013), *Radical unionism: the rise and fall of revolutionary syndicalism*, Chicago, IL: Haymarket Books.
DCCA (2020), 'Commitment decision on the use of a minimum hourly fee', *Danish Competition and Consumer Authority News*, 26 August, accessed 12 February at www.en.kfst.dk/nyheder/kfst/english/decisions/20200826-commitment-decision-on-the-use-of-a-minimum-hourly-fee-hilfr/.
De Stefano, V. (2017), 'Non-standard work and limits on freedom of association: a human rights-based approach', *Industrial Law Journal*, **46** (2), 185–207.
De Stefano, V. (2020), '"I now pronounce you contractor": Prop22, labour platforms and legislative doublespeak', *UK Labour Law Blog*, 13 November, accessed 12 February 2021 at https://uklabourlawblog.com/2020/11/13/i-now-pronounce-you-contractor-prop22-labour-platforms-and-legislative-doublespeak-by-valerio-de-stefano.
DeManualle-Hall, J. (2019), 'Strike by drivers disrupts Uber launch', *Labor Notes*, **483**, 31 May, accessed 16 February 2021 at https://labornotes.org/2019/05/strike-drivers-disrupts-uber-launch.
Doellgast, V., N. Lillie and V. Pulignano (2018), *Reconstructing solidarity: labour unions, precarious work, and the politics of institutional change in Europe*, Oxford: Oxford University Press.

Dubal, V. (2017), 'Winning the battle, losing the war? Assessing the impact of misclassification litigation on workers in the gig economy', *Wisconsin Law Review*, **4**, 739–802.

Dubal, V. (2019), 'Why the Uber strike was a triumph', *Slate*, 10 May, accessed 12 February 2021 at https://slate.com/technology/2019/05/uber-strike-victory-drivers-network.html.

Forde, C., M. Stuart, S. Joyce, L. Oliver, D. Valizade, G. Alberti, K. Hardy, V. Trappmann, C. Umney and C. Carson (2017), *The social protection of workers in the platform economy*, Brussels: European Parliament.

Hale, J. (2018), 'In Denmark, a historic collective agreement is turning the "bogus self-employed" into "workers with rights"', *Equal Times*, 4 July, accessed 12 February at www.equaltimes.org/in-denmark-a-historic-collective?lang=en#.YCZ6DHnLdPb.

Havstein, A. (2021), 'Nu kan svenskere også bestille takeaway med god samvittighed', *Fagbladet 3F*, 30 January, accessed 17 February 2021 at https://fagbladet3f.dk/artikel/nu-kan-svenskere-ogsaa-bestille-takeaway-med-god-samvittighed.

Hyman, R. (1989), *Strikes*, 4th edition, Basingstoke: Macmillan.

Ilsøe, A. (2020), 'The Hilfr agreement: negotiating the platform economy in Denmark', Research paper 176, Copenhagen: FAOS.

ITF (2020), 'Union win! Historic agreement for food delivery workers', *ITF Global*, 7 October, accessed 17 February 2021 at www.itfglobal.org/en/news/union-win-historic-agreement-food-delivery-workers.

Jesnes, K., A. Ilsøe and M. J. Hotvedt (2019), 'Collective agreements for platform workers? Examples from the Nordic countries', Nordic Future of Work Brief 3, Oslo: FAFO.

Joyce, S., D. Neumann, V. Trappmann and C. Umney (2020a), 'A global struggle: worker protest in the platform economy', ETUI Policy Brief 2, Brussels: ETUI.

Joyce, S., M. Stuart, V. Trappmann, I. Bessa, C. Umney and D. Neumann (2020b), 'A global struggle: worker protest in the platform economy', 72nd Annual meeting of the Labor and Employment Relations Association conference, 13–16 June.

Maffie, M. D. (2020), 'The role of digital communities in organizing gig workers', *Industrial Relations: A Journal of Economy and Society*, **59** (1), 123–149.

Moore, N. (2020), '"Ride hailing", Platform Workers Forum: global perspectives on organizing and policy', ILR School, Cornell University, 12–13 November.

Mrozowicki, A. and M. Maciejewska (2017), '"The practice anticipates our reflections": radical unions in Poland', *Transfer*, **23** (1), 67–77.

NYTWA (2021), 'Our history', New York Taxi Workers Alliance, accessed 17 February 2021 at www.nytwa.org/mission-history.

Panimbang, F., S. Arifin, S. Riyadi and D. S. Utami (2020), 'Resisting exploitation by algorithms: drivers' contestation of app-based transport in Indonesia', Trade Unions in Transformation 4.0, Berlin: Friedrich-Ebert-Stiftung.

Riders Union Bologna (2021), accessed 17 February 2021 at www.facebook.com/ridersunionbologna/.

Rolf, S., J. O'Reilly and M. Meryon (forthcoming), 'The rise of privatised social and employment protection in the platform economy: evidence from the UK courier sector'.

Scheiber, N. (2017), 'Uber has a union of sorts, but faces doubts on its autonomy', *New York Times*, 12 May, accessed 12 February 2021 at www.nytimes.com/2017/05/12/business/economy/uber-drivers-union.html.

Scheiber, N. (2019), 'Debate over Uber and Lyft drivers' rights in California has split labor', *New York Times*, 29 June, accessed 12 February 2021 at www.nytimes.com/2019/06/29/business/economy/uber-lyft-drivers-unions.html.

Scott, K. (2019), 'Medical couriers at the Doctors Laboratory vote in favour of pay deal', *Employee Benefits*, 21 June, accessed 12 February 2021 at https://employeebenefits.co.uk/couriers-tdl-pay-deal/

Soderqvist, C. F. and V. Bernhardtz (2019), 'Labor platforms with Unions: discussing the law and economics of a Swedish collective bargaining, framework used to regulate gig work', Working Paper 57, Örebro: Swedish Entrepreneurship Forum.

Stott, R. (2014), 'Driven to organize: Seattle-area Uber drivers form association', *Associations Now*, 20 May, accessed 12 February 2021 at https://associationsnow.com/2014/05/driven-organize-seattle-area-uber-drivers-form-association/.

Tassinari, A. and V. Maccarrone (2020), 'Riders on the storm: workplace solidarity among gig economy couriers in Italy and the UK', *Work, Employment and Society*, **34** (1), 35–54.

Trappmann, V., I. Bessa, S. Joyce, D. Neumann, M. Stuart and C. Umney (2020), 'Global labour unrest on platforms: the case of food delivery workers', Trade Unions in Transformation 4.0, Berlin: Friedrich-Ebert-Stiftung.

Uni Global Union (2018), '3F reaches groundbreaking collective agreement with platform company Hilfr', *Uni Europa News*, 18 September, accessed 12 February 2021 at www.uniglobalunion.org/news/3f-reaches-groundbreaking-collective-agreement-platform-company-hilfr.

Vandaele, K. (2020), 'From street protest to "improvisational unionism": platform-based food delivery couriers in Belgium and the Netherlands', Trade Unions in Transformation 4.0, Berlin: Friedrich-Ebert-Stiftung.

Vandaele, K. (2021), 'Collective resistance and organizational creativity amongst Europe's platform workers: a new power in the labour movement?', in *Work and labour relations in global platform capitalism*, Cheltenham, UK and Northampton, MA, USA: Edward Elgar Publishing and Geneva: ILO.

Vicente, M. (2019), 'Collective relations in the gig economy', *E-Journal of International and Comparative Labour Studies*, **8** (1), 84–93.

Webb, S. and B. Webb (1902), *Industrial democracy*, London: Longmans.

Wiessner, D. (2018), 'US court revives challenge to Seattle's Uber, Lyft union law', *Reuters*, 11 May, accessed 12 February 2021 at www.reuters.com/article/us-uber-seattle-unions/u-s-court-revives-challenge-to-seattles-uber-lyft-union-law-idUSKBN1IC27C.

Wood, A. J., V. Lehdonvirta and M. Graham (2018), 'Workers of the internet unite? Online freelancer organisation among remote gig economy workers in six Asian and African countries', *New Technology, Work and Employment*, **33** (2), 95–112.

Zyskowski, K. and K. Milland (2018), 'A crowded future: working against abstraction on Turker Nation', *Catalyst: Feminism, Theory, Technoscience*, **4** (2), 1–30.

PART III

Case studies across the globe: Online labour platforms

12. The uneven potential of online platform work for human development at the global margins

Mark Graham, Vili Lehdonvirta, Alex J. Wood, Helena Barnard, Isis Hjorth and David Peter Simon

12.1 INTRODUCTION

Increased access to the internet has led to the emergence of a new world of work. Online platform work is becoming increasingly important to workers living in low- and middle-income countries. Our multi-year and multi-method research project investigated the potential for this new economic practice to benefit human development by providing basic capabilities and potentially improving wellbeing and health. Our findings highlight in particular that online platform work brings some benefits, such as improved earning opportunities, stimulating work and increased autonomy, for workers in south-east Asia and sub-Saharan Africa. However, it also represents a risk to the health and wellbeing of these workers as a result of social isolation, overwork and insecurity. Moreover, the above benefits are spread unevenly due to high levels of inequality being inherent to this form of work organization. Additionally, access to these benefits may be blocked by discrimination and predatory intermediaries. Online labour platforms also operate outside the regulatory and normative frameworks that could provide workers with protections or generate tax revenues to fund development more widely.

This chapter highlights the ways in which the observed risks materialize in this novel labour market. Data come from 152 face-to-face semi-structured interviews with workers and stakeholders. We interviewed 27 stakeholders, comprising government and non-governmental organization officials and representatives of remote online work platforms, and 125 workers: 45 in south-east Asia (16 in the Philippines; eight in Malaysia; and 21 in Vietnam) and 80 in sub-Saharan Africa (38 in Kenya; 23 in Nigeria; and 19 in South Africa). Interviews were conducted during seven months of fieldwork in south-east Asia and sub-Saharan Africa during 2014–2015. A second source of data comes from a survey of 679 online workers also located in south-east Asia and sub-Saharan Africa.[1] The chapter investigates whether online platform work has the potential to benefit human development at the world's economic margins.[2]

12.2 ONLINE LABOUR PLATFORMS

Online labour platforms mediate digitally delivered services with clients and workers situated in different locations – potentially on different sides of the planet – with clients often residing in high-income countries and workers in low- and middle- income countries.[3] For example, transaction data provided to us by one major platform demonstrates that, in 2013, clients based in the United States were major purchasers of digital labour located in the Philippines, India, China and Bangladesh. The major international platforms operating in this sector have millions of registered clients and offer a similarly broad range of freelance services such as: programming; web design; graphic design; transcription; translation; writing; online marketing; lead generation; personal assistance; data entry; customer service; and online research.

The history of online platform work is deeply enmeshed with the history of the internet and international business process outsourcing. The idea that individual workers can meaningfully participate in the global economy through a combination of the internet and outsourcing began in the 1980s with the concept of offshore outsourcing taking root in the modern business enterprise (Davis-Blake and Broschak 2009; Manning et al. 2017; Sako 2005). American companies like General Electric and American Express were early adopters, moving business processes from the United States to India (Booth 2013). These early relationships developed in offshore outsourcing practices ultimately influence today's geographic diffusion of online platform work. While sub-Saharan Africa has historically faced severe barriers to entering into global competition – namely infrastructural issues and connectivity challenges – the relatively recent expansion of fibre optic connectivity around the continent has opened opportunities for the entire region (Graham et al. 2015). In this context, online work platforms – platforms that model and manage the relationship between independent contractors and clients – present new avenues particularly for small to medium business owners willing to hire workers in previously untapped outsourcing locations.

At their core, digital labour platforms act as a matchmaking service. They coordinate buyers and sellers of temporary contract work, regulating the individualized, temporary relationship between buyers and sellers in a similar fashion to traditional labour intermediaries. 'Online platform work' itself refers to contingent paid work that is allocated, transacted and delivered by way of internet platforms without an explicit or implicit contract for long-term employment. In this environment, workers undertake a number of more or less overlapping tasks rather than having a single job role. For instance, a job on an online work platform could exist in the form of a temporary graphic design project, which itself might refer to a bundle of Photoshop tasks. These tasks could be any number of different things, including more mundane tasks such as data entry.

There are arguably three types of platform: those based upon what we term 'double auction'; single-side buyer/seller-posted tasks; and labour management mechanisms. The largest platforms are double auctions, while there are some medium-sized single-side buyer/seller-posted platforms. Labour management platforms, however, tend to be relatively small. The key difference is how tasks are sourced and delivered to workers. In double auction mechanisms, for example, both the worker and the 'client' can post tasks and suggest prices. In a single-side market, though, only one party – either the worker or the client – can post tasks and demand prices.

Meanwhile, in a labour management situation, the platform company regulates the allocation of tasks through administrative rules. These companies often offer a specific service, such as transcription or data entry. Algorithmic controls – including ratings and reputation scores and automated tests – play a prominent role on all types of platform. A reputation rating is a score that a worker receives from a client after completing a task and is an especially important feature of the double auction and single-side platforms. These ratings are combined with work history (number of completed jobs, hours worked and total earnings) and test scores in order to rank workers. Platforms algorithmically filter tasks towards the highest-ranked workers, meaning that those with the best overall reputation are more likely to receive more work. Scores and reputation, therefore, act as a powerful system of labour control and engender both rewards and risks for workers (see Wood et al. 2019a).

12.3 ONLINE PLATFORM WORK AND THE PROMISE OF DEVELOPMENT

Major international institutions suggest that the growth of online platform work will enable the frictionless entry of workers in low- and middle-income countries into a global marketplace (Kuek et al. 2015). Economists have previously suggested that access to internet-based marketplaces would permit a kind of 'virtual migration' that offers economic benefits akin to physical migration (Horton 2010). Policymakers expect that regions like sub-Saharan Africa and south-east Asia, in particular, will capitalize on this digitally mediated work opportunity. New sources of work are especially needed as the youth to adult unemployment rate hits historic peaks (UNDP 2015) while average wages remain significantly lower in emerging economies than in developed economies (ILO 2015). The growth of online platform work is, therefore, seen as a relatively welcome phenomenon among economic development experts, and the world's largest global development network is promoting its potential to aid human development (UNDP 2015).[4] Some governments agree, seeking to support the practice to advance human development. For instance, Nigeria has pursued a new initiative entitled 'Microwork for Job Creation', namely to help workers make money and build skills. In the words of an official in Nigeria's Ministry of Communications Technology and Digital Economy, the vision is for it 'to really be … leveraging ICT to engage, create, and develop our people. To create wealth for them and develop them'.

There is, so far, little evidence on whether online platform work can enhance human development for workers.

Human development refers to a perspective in which people are both the ends of and the means to growth. Therefore, development in this sense is understood as the extent to which people are provided with the basic 'capabilities' to do and be the things that people have reason to value (Sen 1999). Paid work has an important role in the provision of such capabilities and thus potentially improving wellbeing and health. The manner in which it does so is more complex than simply creating more jobs, as the degree to which jobs are beneficial is dependent upon their quality (Mahmood et al. 2014; Monteith and Giesbert 2017; UNDP 2015). However, job quality, or the degree to which jobs furnish basic capabilities (Green 2006), cannot be evaluated through subjective satisfaction as people adapt their preferences to poor employment conditions and circumstances which distorts their ability to meaningfully evaluate their situation (Burchell et al. 2014; Cohen 1990; Egdell and Beck 2020).

Drawing upon economic, sociological and psychological research, there is thus a solid theoretical and empirical understanding of job characteristics which, at least in high-income countries, contribute to wellbeing and thus human development (Burchell et al. 2014; Green 2006; Green et al. 2016; Kalleberg 2011). Below we look at the relationship between online platform work and several work-related basic capabilities that have been highlighted in the job quality literature (Green 2006; Green et al. 2016; Kalleberg 2011; Monteith and Giesbert 2017). Doing so enables us to highlight the risks alongside the rewards of online platform work in order to reveal the complex and sometimes problematic reality of the 'new world of work' in south-east Asia and sub-Saharan Africa.

12.4 BENEFITS OF ONLINE PLATFORM WORK FOR HUMAN DEVELOPMENT

Economic research has tended to focus upon pay, equating good jobs with high-paying ones. Although the job quality approach argues for a wider conception of the capabilities provided by paid work, material rewards (pay and fringe benefits) are, nevertheless, recognized as an important element of job quality as this is how most workers receive the majority of their income. A basic requirement for a job or combination of jobs is, therefore, remuneration which is at least sufficient to satisfy a worker's (and their dependent family's) basic material needs (Green 2006; Kalleberg 2011).

Online platform work was an important source of income for most of our interview informants and the main source of income for many. Likewise, 73 per cent of our survey respondents reported that online platform work was an important source of income for their households, and 61 per cent reported that it was their main occupation. Our interview informants mostly gained an income which was sufficient to allow them to avoid material hardship. This is supported by our survey data, with the mean weekly income from this work, among those paid in United States dollars (N = 610), being \$165 (σ = \$209) (seven outliers of \$2000 and over were excluded as improbably large although not wholly impossible given the unpredictability of online platform work). A geographic difference was apparent, with south-east Asian workers' earnings averaging \$181 (N = 304) compared to \$150 for those in sub-Saharan Africa (N = 307). This relates to south-east Asian respondents being more likely to carry out less routine tasks which, therefore, required more specific skills than their African counterparts.

Tasks that included more specific skills were ones such as programming, website design, graphic design and translation while more routine services included microwork and data entry, customer service, online research, search engine optimization (SEO) and writing, personal assistance and non-technical customer support. Across all respondents, those carrying out work that required more specific skills earned on average \$44 a week more than those doing more routine work. This suggests that the ability of national education systems to provide people with specific technical skills plays an important role in maximizing the earnings potential of online platform work. Moreover, some successful workers explained that, as a result of online platform work, they could afford to save some income and pay for private health insurance. As Angel from the Philippines described: 'For me, it's a high paying job because now I was able to afford an apartment, pay my own bills, my own internet connection, my own cable, paying for our own food, for my kids' milk. That's all on my own.'

In contrast to the well-paying online work opportunities experienced by workers like Angel, local labour markets frequently offered such workers only low-paying jobs or none at all.

The avoidance of material deprivation is only one benefit of paid work. The potential for self-fulfilment, self-realization and identity-making is unique to human productive activities but requires that the tasks that comprise such activities are stimulating (Felstead et al. 2016; Green 2006). An important element of experiencing a task as stimulating is variety since, without diversity, all tasks become routine, repetitive and thus boring. Seventy-two per cent of our survey respondents felt able to choose and change the order in which they undertook online tasks, 74 per cent were able to choose or change their methods of work and 62 per cent agreed that their job involved unforeseen problem solving. Our interviews also demonstrated that many workers found their work stimulating, with 57 per cent of survey respondents agreeing that their job involved solving complex tasks. Moreover, our interview informants also explained how they had the necessary space to complete work independently. This type of autonomy is typically not afforded in other work contexts, such as business process outsourcing centres focused on customer support. For example, Victor, a worker from Nigeria, explained: 'You have freedom of choice. Who you want to work with, when you want to work and how you want to work.'

12.5 RISKS OF ONLINE PLATFORM WORK

Growing awareness of online platform work, increasing global connectivity and the lack of good jobs on local labour markets is causing many new workers to join online labour platforms. This growing supply of workers is not necessarily matched by equal increases in the demand for their work, resulting in underemployment and downward pressures on pay rates. Table 12.1 shows that, on one major platform, there are many times more workers yet to find any work than those who have found work, although the numbers exclude successful workers who have made their profiles invisible to public searches.

This oversupply meant that while the income that workers gained from this work tended to be relatively good compared to what was available locally, it was difficult to significantly increase pay rates for individual jobs. This situation contributed to the overwork discussed below, as in-demand workers tended to respond to the difficulty in raising their hourly rate by increasing their hours. As Simon (Kenya; data entry) explained: 'A client [is] paying me $3.50 an hour. I'm so broke, this is someone who's ready to give me the money, so why don't you want 18 hours in one day.'

Table 12.1 *Labour oversupply on one major online labour platform*

Country	Potential workforce	Successful workers	Oversupply
Global	1 775 500	198 900	1 576 600
Philippines	221 100	32 800	188 300
Malaysia	11 900	500	11 400
Vietnam	7700	1000	6700
Kenya	21 700	1500	20 200
Nigeria	7000	200	6800
South Africa	10 200	800	9400

Note: Shown are the countries studied in this chapter. Data from 7 April 2016; not seasonally adjusted. Potential workforce refers to total searchable worker profiles. Successful workers are those with at least one hour billed or $1 earned. Oversupply = potential workforce minus successful workers.

On the other hand less in-demand workers were forced to spend longer doing the unpaid work of searching and applying for paid work, which further reduced their potential to earn (our average survey respondent reported spending 16 hours per week looking for work, or 39 per cent of the total time spent on online platform work). As the chief executive officer of one major platform explained to us:

> There are 7.1 billion people on the planet, there are 2.4 billion people on the internet … They're what I call 'PHDs'; poor, hungry, driven … They're willing to work on any sort of job, right, a lot harder than maybe you or I are, for less money … It's highly competitive and it changes dramatically as the internet gets turned on in various countries … And those [routine job] rates are going down because the more [workers there are], when you're talking about unskilled jobs there's almost no floor as to where those actual prices go.

Moreover, while online platform work could provide workers with decent earning opportunities that were superior to what was available in the local labour market, these rewards were very unevenly distributed. As discussed above, workers providing services that required more specific skills earned significantly more than those doing more routine tasks but, even among survey respondents doing routine work, we find massive income disparities: the income ratio between the 10th and 90th percentile is 1:19 (N = 328; excluding respondents earning nothing or above $2000 in the last week). This suggests that other forms of worker power besides skills are also important in determining income.

Our interviews suggest this other form of power is reputation, both in the conventional sense of good standing and networks among potential clients and in the sense of the symbolic power which accrues to some workers as a result of client ratings. Work flowed to those workers who had managed to maintain a strong reputation over a long period and were thus known by clients and highly ranked by platform algorithms. A key theme of the interviews was the difficulties faced by workers who lacked strong reputations. Due to labour oversupply they had low incomes, in some cases below their countries' monthly minimum wages, and a handful survived around the global poverty line of $58 per month. Workers who were struggling financially described how, for example, their pay was not enough to 'actually survive' (Helen, South Africa, writing and virtual assistant).

12.5.1 Job Insecurity

Health is a key basic capability. There is a wealth of evidence that job insecurity is associated with lower physical and mental wellbeing (Cheng and Chan 2008; Wood and Burchell 2018). Platforms do little to mitigate this insecurity, with some of the largest platforms boasting that workers are on-demand and can be fired at any time. For example, one platform advised clients on managing 'on-demand' labour that: 'All users have the freedom to end a contract at any time … Ending a contract without warning can be interpreted as firing.'

Framing workers as freelancers and independent contractors places them outside national employment institutions. As a result of this lack of employment protection many online platform workers experience job insecurity. Nearly half of the workers surveyed (44 per cent) felt easily replaceable. As Amanda, a South African worker, explained: 'There's a lot of people out there, if they're not satisfied with you, they are going to try somebody else … So, they can replace you. This is one of those jobs that you can be replaced.'

In some cases, workers could quickly find another project if they had a strong platform reputation, providing income security if not job security, but this was not the case for all workers.

12.5.2 Social Isolation

Several reports highlight the benefit of being able to work from home when undertaking online platform work (Kuek et al. 2015). Indeed, we find that workers value this flexibility. However, many successful workers have little option but to work from home, which can lead to social isolation. Social contact is an important mental health benefit of paid work (Wood and Burchell 2018). Our research shows that some workers feel detached from others with limited opportunities to interact with people outside their family. In fact, 77 per cent of survey respondents say they rarely or never communicate face to face with other people who use platforms. As Sarah, a South African worker, put it: 'It can get very lonely and it can get very isolating … When you work at a company you can just have coffee with someone. That element is missing.'

This problem is compounded with most clients being in different time zones (usually in north America and western Europe) and, consequently, many workers find themselves working unsocial hours like evenings, nights and weekends.

12.5.3 Overwork

Given how many people work from home, it is not surprising that online platform work can both help and harm work–life balance. Overwork is not uncommon: both in terms of long hours and working at a high intensity. Fifty-four per cent of survey respondents said they worked at a very high speed. Sixty per cent worked to tight deadlines and 22 per cent experienced pain as a result of their work. Interviews with workers indicated that this intensity is driven by low pay rates, employment insecurity and bidding for jobs. The primary way that workers increase their earnings is through working more hours. Justine, from the Philippines, explained how: 'You tend to overwork. You have a hard time separating your personal life from work. Basically, even if it's your rest day, your phone and computer would be there. It's about work. You can't help it but go check on it.'

Others said they would frequently work over 70 or 80 hours a week for $3.50 per hour, and sometimes throughout the night when tight deadlines arose. Sharon, a Kenyan, said that she puts in 'like 40 [hours a week] or even more depending on the magnitude of work … That's just at night … I work [locally] during the day so when I go back home, I put in five to six hours … then over the weekend … In total it will be like 70.'

12.5.4 Opacity and Taxation

Survey responses suggested that workers rarely know their clients, with 8 per cent admitting they do not even know the name of the person who hired them. Moreover, 71 per cent indicated that they would like to know more about the person for whom they are working. This lack of connection is important as it alludes to a loss of a shared understanding. This informality and the associated uncertainty over legal responsibilities is also reflected in that only 18 per cent of our survey respondents suggested that they had paid income tax on their online earnings in the past year. This is corroborated by our interview data. Moreover, clients were located in high-income countries, as were the headquarters of nearly all the platforms and the vast majority of their subsidiaries. The limited tax revenues generated by this 'normatively disembedded work' (Wood et al. 2019b) raises further questions regarding the potential for online platform work to benefit wider human development in low- and middle-income countries.

12.5.5 Discrimination

An additional barrier to the provision of basic capabilities via platform work is the potential for discrimination. Many stakeholders suggest that qualified people in Nairobi will have the same access to clients as their peers in New York City thanks to online work platforms (Kuek et al. 2015). In principle, platforms welcome any qualified worker regardless of gender, origin or other attributes. However, workers at the global margins often feel that they are being discriminated against in practice because of their countries of origin – sometimes subtly, sometimes blatantly. For example, Moses is a 26-year-old translator from the Nairobi slums who uses a digital work platform. He often changes the geographical location listed on his profile, explaining how the platform makes it hard given that he is Kenyan. He says that identifying clients in platform-based work forces people to realign their profile constantly to fit the job description. As a result, many of Moses's clients believe that he is based in Australia, just as they are unaware that he is a college dropout rather than a successful translator. The pronounced discriminatory practices with which Moses is met means that he is constantly creating a persona in order to be able to win tasks. Moses and others like him say that bowing down to discriminatory practices on digital work platforms is a commonplace necessity: 'You have to create a certain identity that is not you. If you want to survive online, you have to do that.'

Joyce from the Philippines explained: 'I would get a bounced email saying that only workers in the United States are allowed. Yes, so if you put your address there, they would not hire you because you're from the Philippines. That's how racist some companies are.'

Clients may assume that workers from low- and middle-income countries provide less valuable work than workers from high-income countries, unless the worker has evidence such as testimonials from previous clients that attest to the high quality of their work. This causes workers from low- and middle-income countries, who are just starting out and lack extensive work histories, to earn disproportionately little.

12.5.6 Intermediation

The positive impact of online labour platforms can also be constrained by the existence of complex subcontracting chains. Even though digital labour platforms can cut out intermediaries, the opposite can also be true. Because of the heavy role that reputational feedback scores play in online platform work, work can flow to intermediaries who already have a high score. These then re-outsource that work, keeping a part of the client's fee for themselves. The existence of a large pool of potential workers online (Table 12.1) helps this practice to continue and allows the intermediaries to pay low rates. From a client perspective, intermediaries can add value to the process by, for example, breaking larger projects into more manageable tasks, providing project management and taking responsibility for timely delivery to the client. Experienced intermediaries may also be better at picking workers than inexperienced clients. However, from a workers' perspective, intermediaries can complicate the flow of information from clients to workers, potentially hindering skill development.

The experience of Maya, a 26-year-old Malaysian woman studying for her master's degree, is illustrative. Maya's digital platform work activities are very much driven by her passion for writing. Maya applies for writing-related jobs on all five online work platforms with which she has signed up. She has written more than 74 articles and has had more than 15 different clients on the largest online work platforms. Initially, it was difficult for Maya to find suitable writing jobs. She eventually learned about SEO tasks, which seemed like a good starting point. SEO jobs tend to produce text that is read by computers rather than humans. Even though Maya's work consists of writing for machines rather than humans, the jobs are still ultimately based on writing tasks.

Some of her clients are actually other contractors who have developed strong reputations characterized by high positive feedback scores, making them able to attract a much larger number of tasks at much higher rates than Maya. Maya knows this because she often finds that jobs she has unsuccessfully applied for are reintroduced to the market by another digital platform worker. For example, she once applied for an SEO writing task, suggesting a price of $25 rather than the listed suggestion of $50. But the job went to another contractor (who had a higher positive feedback rating) who requested a price of $75. This contractor then offered the job to Maya for just $7.50, an amount below her minimum wage. She accepts tasks from intermediaries but not without thinking that this is an inherently unfair set-up. Because she lacks any direct contact with her clients, it is also difficult for her to understand the needs of clients: this makes the writing process more challenging.

12.6 CONCLUSION

This chapter has empirically investigated the potential for online platform work to benefit human development in south-east Asia and sub-Saharan Africa. Despite the optimistic claims of some actors within the development community regarding the potential for online platform work to benefit human development, we find that the reality is far more complicated and uneven. While this form of work can provide individuals in low- and middle-income countries with opportunities to earn incomes above what is available via their local labour markets, such opportunities are spread far from equally. High levels of inequality are seemingly inherent to this form of work organization, with work flowing to those with the strongest platform reputations and profiles. This situation is aggravated by the platforms being saturated with an oversupply of labour, meaning that many workers are unable to earn decent incomes.

Workers who struggle to attract clients also risk being taken advantage of by those with stronger reputations. By virtue of their in-demand profiles, such workers hoover up more work than they can complete themselves and then re-outsource this work to desperate workers at the bottom of the subcontracting chain, sometimes paying the subcontracted worker a small fraction of what they will be paid by the client. While these subcontracting activities can add value to the service provided to the client, they can also act as a barrier to skill development and block workers at the bottom of the subcontracting chain from developing their own platform reputations. Workers at the global margins also feel that they face barriers to greater participation in this global labour market due to discrimination by clients who tend to be located in rich countries.

The uneven benefits of online platform work continue to be evident as we take a wider perspective on human development via a multidimensional understanding of job quality that moves beyond earning opportunities. We find that, while online platform workers often experience the work as stimulating, and value its autonomy and flexibility, they also suffer from job insecurity, work intensification and social isolation.[5] Moreover, this work is disembedded from those labour market and welfare institutions that could otherwise provide workers with greater protection and rights, not the least of which is the generation of tax revenues. That this activity goes largely untaxed further weakens the potential for this new economic phenomenon to benefit wider development at the global margins. Maximizing the impact of online platform work for human development will require the embedding of platforms within normative, legal and collective institutions which ensure workers are protected against discrimination, unscrupulous intermediaries, poverty pay, insecurity, social isolation and work intensification.[6]

ACKNOWLEDGEMENTS

We would like to thank the grantor, the International Development Research Centre (107384-001) and the workers who patiently and generously participated in the research. We would like to give additional thanks to Tess Onaji-Benson for fieldwork assistance in Nigeria and David Sutcliffe for ongoing support in Oxford. We also gratefully acknowledge Ellie Marshall, Ilinca Barsan and Liinus Hietaniemi for their help with the collection of the survey data, and Kaveh Azarhoosh for his assistance.

NOTES

1. This chapter draws on our earlier report (Graham et al. 2017). However, some statistics differ slightly due to being recalculated following further data collection. Some of the findings are elaborated on and discussed in more detail in Wood et al. (2019a, 2019b).
2. By economic margins we refer to economically disadvantaged countries; that is, those falling in the categories of low and middle income. These countries might also be broadly known under the inadequate umbrella term 'the developing world'.
3. While using the term 'online' to refer to the role of the internet in the organization of work, this does not imply that the work is immaterial or taking place in some kind of ethereal cyberspace (see Graham 2013).
4. For example, the Rockefeller Foundation initiative 'Digital Jobs Africa'.
5. See Wood et al. (2019a) for more detail. See also Anwar and Graham (2019, 2020).
6. This point is theoretically developed in Wood et al. (2019b) (see also Graham et al. 2020), while the potential for the collective organization of these workers is empirically discussed at length by Lehdonvirta (2016), Wood et al. (2018) and Wood and Lehdonvirta (2021).

REFERENCES

Anwar, M. A. and M. Graham (2019), 'Hidden transcripts of the gig economy: labour agency and the new art of resistance among African gig workers', *Environment and Planning A: Economy and Space*, **52** (7), 1269–1291.

Anwar, M. A. and M. Graham (2020), 'Between a rock and a hard place: freedom, flexibility, precarity and vulnerability in the gig economy in Africa', *Competition and Change*, **25** (2), 237–258.

Booth, T. (2013), 'Special report: Outsourcing and offshoring: Here, there and everywhere', *The Economist*, 17 January.

Burchell, B. J., K. Sehnbruch, A. Piasna and N. Agloni (2014), 'The quality of employment and decent work, methodologies, and ongoing debates', *Cambridge Journal of Economics*, **38** (2), 459–477.

Cheng, G. H. L. and D. K. S. Chan (2008), 'Who suffers more from job insecurity? A meta-analytic review', *Applied Psychology*, **57** (2), 272–303.

Cohen, G. (1990), 'Equality of what? On welfare, goods and capabilities', *Recherches Économiques de Louvain /Louvain Economic Review*, **56** (3–4), 357–382.

Davis-Blake, A. and J. P. Broschak (2009), 'Outsourcing and the changing nature of work', *Annual Review of Sociology*, **35**, 321–340.

Egdell, V. and V. Beck (2020), 'A capability approach to understand the scarring effects of unemployment and job insecurity: developing the research agenda', *Work, Employment and Society*, **34** (5), 937–948.

Felstead, A., D. Gallie, F. Green and G. Henseke (2016), 'The determinants of skills use and work pressure: a longitudinal analysis', *Economic and Industrial Democracy*, **40** (3), 730–754.

Graham, M. (2013), 'Geography/internet: ethereal alternate dimensions of cyberspace or grounded augmented realities?' *The Geographical Journal*, **179** (2), 177–182.

Graham, M., C. Andersen and L. Mann (2015), 'Geographical imagination and technological connectivity in East Africa', *Transactions of the Institute of British Geographers*, **40** (3), 334–349.

Graham M., V. Lehdonvirta, A. J. Wood, H. Barnard, I. Hjorth and D. P. Simon (2017), *The risks and rewards of online gig work at the global margins*, Oxford: Oxford Internet Institute.

Graham, M., J. Woodcock, R. Heeks, P. Mungai, J.-P. Van Belle, D. du Toit, S. Fredman, A. Osiki, A. van der Spuy and S. Silberman (2020), 'The Fairwork Foundation: strategies for improving platform work in a global context', *Geoforum*, **112,** 100–103.

Green, F. (2006), *Demanding work: the paradox of job quality in the affluent economy*, Princeton, NJ: Princeton University Press.

Green, F., A. Felstead, D. Gallie and H. Inanc (2016), 'Job-related well-being through the great recession', *Journal of Happiness Studies*, **17**, 389–411.

Horton, J. J. (2010), 'Online labour markets', in A. Saberi (ed.), *Internet and Network Economics, 6th International Workshop, WINE 2010, Stanford, CA, USA, December 13–17, 2010*, Berlin: Springer, pp. 515–522.

ILO (2015), *Global wage report 2014/15*, Geneva: International Labour Office.

Kalleberg A. L. (2011), *Good jobs, bad jobs: the rise of polarized and precarious employment systems in the United States, 1970s–2000s*, New York: Russell Sage Foundation.

Kuek, S. C., C. Paradi-Guilford, T. Fayomi, S. Imaizumi, P. Ipeirotis, P. Pina and M. Singh (2015), *The global opportunity in online outsourcing*, Washington, DC: World Bank Group.

Lehdonvirta, V. (2016), 'Algorithms that divide and unite: delocalisation, identity and collective action in "microwork"', in J. Flecker (ed.), *Space, place and global digital work*, London: Palgrave Macmillan, pp. 53–80.

Mahmood, M., W. Lee, M. Mamertino, E. Ernst, C. Viegelahn, E. Bourmpoula and M. Giovanzana (2014), 'Employment patterns and their link with economic development', in ILO (ed.), *World of work report 2014: developing with jobs*, Geneva: International Labour Office, pp. 33–50.

Manning, S., M. M. Larsen and C. G. Kannothra (2017), 'Global sourcing of business processes: history, effects, and future trends', in G. L. Clark, M. P. Feldman, M. S. Gertler and D. Wojcik (eds), *The new Oxford handbook of economic geography*, Oxford: Oxford University Press, pp. 407–435.

Monteith, W. and L. Giesbert (2017), '"When the stomach is full we look for respect": perceptions of "good work" in the urban informal sectors of three developing countries', *Work, Employment and Society*, **31** (5), 816–833.

Sako, M. (2005), *Outsourcing and offshoring: key trends and issues*, Oxford: Said Business School.

Sen, A. (1999), *Development as freedom*, Oxford: Oxford University Press.

UNDP (2015), *Human development report 2015*, New York: United Nations Development Programme.

Wood, A. J. and B. J. Burchell (2018), 'Unemployment and well-being', in A. Lewis (ed.), *The Cambridge handbook of psychology and economic behaviour*, Cambridge: Cambridge University Press, pp. 234–259.

Wood, A. J. and V. Lehdonvirta (2021), 'Antagonism beyond employment: how the "subordinated agency" of labour platforms generates conflict in the remote gig economy' *Socio-Economic Review*, (ePub, ahead of print), available at https://ssrn.com/abstract=3820645.

Wood, A. J., M. Graham, V. Lehdonvirta and I. Hjorth (2019a), 'Good gig, bad gig: autonomy and algorithmic control in the global gig economy', *Work, Employment and Society*, **33** (1), 56–75.

Wood, A. J., M. Graham, V. Lehdonvirta and I. Hjorth (2019b), 'Networked but commodified: the (dis)embeddedness of digital labour in the gig economy', *Sociology*, **53** (5), 931–950.

Wood, A. J., V. Lehdonvirta and M. Graham (2018), 'Workers of the internet unite? Online freelancer organisation among remote gig economy workers in six Asian and African countries', *New Technology, Work and Employment*, **33** (2), 95–112.

13. From outsourcing to crowdsourcing: Assessing the implications for Indian workers of different outsourcing strategies

Janine Berg, Uma Rani and Nora Gobel

13.1 INTRODUCTION

The development of information and communications technologies (ICT) in the 1980s was believed to offer a new 'development paradigm' for developing countries such as India by creating new markets and employment opportunities in knowledge-intensive services. Since then, India has embarked on an ambitious effort to integrate ICT development into its national policies, offering both a range of IT-enabled services from software development and research and development services at one end of the skill spectrum to back office data entry and call centre work at the other.

These trends have accelerated since the mid-2000s as digital technologies such as cloud computing have enabled firms to 'crowdsource' work via digital labour platforms ('platforms') to individual workers located throughout the globe. Indeed, crowdsourcing is the latest manifestation in a decades-long shift of outsourcing across borders. The one critical difference is that outsourcing is not to enterprises but rather to individuals. Indians are active on platforms, comprising one-quarter of workers on the Oxford Internet Institute's Online Labour Index, which tracks projects and workers on the five largest English-language online labour platforms.[1]

This chapter analyses the potential implications for Indian workers of work that is crowdsourced through labour platforms. It asks whether the opportunities from online outsourcing represent an improvement for Indian workers. The analysis draws on a 2017 International Labour Office (ILO) survey of 2350 crowdworkers working on five leading micro-task platforms, of which 343 respondents were from India, providing information on their motivation for undertaking this work and their working conditions. It also compares workers performing platform work with similar types of 'offline' work, either in business processing centres or in other companies, using data from India's National Sample Survey for 2017. Despite the opportunities of platform work, also documented in Chapter 12, this chapter argues that, compared with work in business processing centres, as elucidated in the literature, the outsourcing of work through digital labour platforms has been less beneficial for Indian workers.

13.2 BUSINESS PROCESS OUTSOURCING: ASSESSING WORKERS' EXPERIENCE

Over the past several decades, business process outsourcing (BPO) has become a central organizational strategy of lead enterprises located throughout the Global North. Companies use BPO to reduce labour costs by accessing a cheaper yet skilled labour force outside their country of origin to perform service work (Hirschheim and Dibbern 2009). The BPO industry is comprised of two main types of activities:

1. Call centres that use voice-based, information technology-enabled services to perform inbound customer service, outbound sales and technical support.
2. IT-enabled back office services that range from basic data entry (e.g. credit card or payroll processing) to professional business services such as accounting, human resource administration or IT services.

These outsourced activities may be 'in-house', which are sometimes referred to as 'captive units' in that, even though they have been offshored, they remain part of the lead firm, or activities may be subcontracted to a third party. The differences in activities performed, but also whether they are captive or subcontracted, have a bearing on job quality and the career prospects of workers. Owing to their higher skill requirements, technical activities in business services are typically of higher quality than call centre jobs, particularly in terms of task complexity, variety and job discretion or autonomy as well as pay and career prospects (Messenger and Ghosheh 2010; D'Cruz and Noronha 2007).

India is a leading destination for BPO not only due to its advantage in its use of the English language but also due to its institutional compatibility in terms of having similar legal and accounting systems (Basant and Rani 2004). The BPO industry emerged in force with the economic liberalization of the economy in 1991 which encouraged the inflow of financial capital but also knowledge and technology transfer (D'Cruz and Noronha 2010). In addition, the government gave specific support to the industry via incentives, concessions and subsidies, as well as the creation of technology parks supported by industrial policy and legal frameworks. The National Association of Software and Services

Companies, which is the trade association of the IT and BPO industry in India, spearheaded many of these initiatives. Since their inception in 1988, they have grown to 2800 members including leading IT multinationals from across the globe.[2] In 2017, 1.1 million people worked in the Indian BPO industry,[3] offering outsourcing services to more than 65 countries. Throughout its 30 year history, growth in the BPO industry has been sustained, growing by about 9 per cent in 2019, amounting to aggregate revenues of $181 billion. Exports in the IT-BPO industry grew by 8.3 per cent in 2019 with revenues of $135.9 billion, accounting for 51 per cent of the country's exports (NASSCOM 2019).

Limited data exist on the composition of the Indian BPO industry, specifically the breakdown between call centre, back office and professional service work. In the mid-2000s, call centres accounted for 60–65 per cent of BPO work, with 35–40 per cent for back office and service work (Taylor and Bain 2005). BPO activities service a range of industries, including insurance, retail, banking, healthcare, telecommunications and hospitality. About 80 per cent of call centres in India in the mid-2000s were subcontractors rather than captive units of lead firms (Batt et al. 2005).

The 'Global Call Centre Report', which surveyed 2500 call centres employing nearly half a million workers in 17 countries including India between 2003 and 2006, found an average worldwide annual employee turnover in call centres of about 20 per cent compared with 40 per cent in India (Holman et al. 2007). Such a finding can be either positive or negative, depending on the motivation prompting the worker to leave the firm. In the case of India, much of the turnover has been reported to be the result of poaching (Batt et al. 2005) or better career prospects (James and Vira 2012). Consequently, it was thus less likely to be a reflection of discontent among workers, despite some negative attributes of the work with respect to training, monitoring, employee relations and internal firm mobility, but also night work (Ramesh 2004; Mirchandani 2004). In India, working hours have been adapted to meet the needs of north American and European customers, which requires workers to work nights because of the time differences. Indian call centres compensate for this feature by offering free transportation and meals to staff, but it nonetheless impairs workers' social and family life. For women, it often means that they quit call centre work upon marriage (Tara and Ilavarasan 2009).

Workers in call centres and back office service work are well educated, with the majority having tertiary education. Nonetheless, when they begin working in a call centre they receive training of approximately 13 weeks for those working in a firm serving an international market and seven weeks for those serving the domestic market (Batt et al. 2005). Training mostly focuses on accent neutralization and cultural awareness or the learning of firm-specific knowledge such as company workflow or product awareness. For call centre work in particular, there appears to be scant opportunities for improving one's technical skills. Given that many of these workers are at the formative stages of their career, the lack of technical skill building and the low-skilled nature of call centre work risks eroding the technical skills learned during their tertiary education. As Ramesh (2004, p. 496) argues, 'most of these youngsters are in fact burning out their formative years as "cyber coolies" … BPO work does not provide any scope for skill upgradation.'

Other researchers have contested this viewpoint, arguing instead that the relative working conditions, particularly pay, in call centre and other BPO work are good but that, more importantly, work experience in this industry improves career prospects. James and Vira (2012) document how Indian call centre workers were able to achieve significant pay increases by jumping to other call centre firms. They note that analyses of working conditions in the industry have been too narrow, focusing on experiences within specific call centres and not considering intra-firm mobility and the cross-industrial career paths of call centre workers. In their two-year survey of call centre workers in India's National Capital Region, they found that 'good starting salary' attracted workers to call centres. Though earnings are low by international standards, by Indian standards they are high, amounting to 'twice the earnings of an entry-level high school teacher, accountant or entry-level marketing professional with a graduate degree' (James and Vira 2012, p. 854). Moreover, while there was limited upward mobility within their place of work, they were able to improve their working conditions by moving laterally between firms and negotiating pay increases. The authors find annual rates of employee turnover of over 50 per cent and overall average pay increases resulting from one job switch of 33 per cent, with an additional 20 per cent pay increase for those who switched twice during the study period. The authors also found, however, that 12 per cent of workers who switched firms took a pay decrease, favouring daytime call centre work for the domestic market over night-time international work, despite earnings being lower in domestic firms (James and Vira 2012).

13.3 CROWDSOURCING VIA DIGITAL LABOUR PLATFORMS

Like BPO, labour platforms are viewed positively by the Indian government as an important source of jobs and foreign exchange, with the government investing in digital infrastructure and training programmes to support their expansion.

Digital labour platforms differ in many ways from BPO work, though there are also similarities. The profile of the workers – young, well educated and often with IT backgrounds – is similar, as is the workers' contribution being made through a global services supply chain. On online labour platforms, workers located in a country provide a range of services to the client(s) through platforms which are based in different locations. Depending on the platform, these services can be 'micro-tasks', named as such because they take a relatively short amount of time and consist of having workers perform data annotation, data entry, data cleaning, content moderation, copywriting or audio transcription. Workers see the task posted and, as long as they have the right qualifications (usually a minimum threshold for their rating and experience, though they may also be required to pass unpaid qualification tests), can access the job, complete it and submit it. Either the client or the platform sets the price as a piece-rate and there is no negotiation.

The other type are 'freelance' platforms, whereby individual workers offer their services as programmers, graphic designers, statisticians, translators and other professional services occupations. Freelancing platforms such as Upwork, Freelancer.com, Workana, Jovoto and others match the services of freelancers, taking a commission on the workers' earnings. Most of these platforms are designed so that workers set up individual profiles and indicate their expertise and their rate, with the final price for their work set via a bargaining process with the client.

While there are some activities that cannot be easily transferred to labour platforms – particularly those activities that are performed in captive BPO units because of their proprietary nature – many other services can potentially be outsourced to an online platform as opposed to a subcontracted BPO enterprise. These include IT services such as website design and programming, translation, audio transcription, data entry, data cleaning, data science and data analytics. In other words, both micro-task platforms and freelance platforms are potential recipients of BPO work. In addition, platforms also perform work that did not exist previously such as data annotation and categorization for the training of artificial intelligence systems. And while it is true that lead firms could move BPO functions across different subcontractors in the Global South, the ease and speed with which this can be done via platforms is much greater.

There are also important distinctions with how workers are treated on labour platforms, with important consequences for their working conditions. Most notably, labour platforms have, for the most part, classified their workers as independent contractors; thus the workers are not privy to the protections and benefits that an employee working in a BPO firm would receive, even if the work they are doing is for a client located abroad. This poses problems in ensuring adequate social and labour protection for workers but also in applying local regulations.

The other principal distinction concerns ease of movement between platforms. While it would seem that platform workers would have ease of movement across platforms and therefore greater opportunity for improving their individual career prospects, the reality is quite different. Platforms are designed in such a way that it is difficult for workers successfully to move between platforms as workers must develop their work histories, reputation with their clients and ratings if they are to be eligible for jobs. As documented in Chapter 12, the continued receipt of work is contingent on positive ratings. Thus, while it might be easy to set up an account on a platform, depending on the location or country, setting up a successful account that leads to work requests being received requires important investments on the part of the worker (Aleksynska et al. 2018). For these reasons, many researchers have argued that the concentration among platforms is indicative of a monopsony (Kingsley et al. 2015), with low labour supply elasticities that correspond with monopoly power (Dube et al. 2018).

13.4 INDIAN PLATFORM WORKERS: MAIN CHARACTERISTICS

In 2017, the ILO conducted a global survey of 2350 micro-task workers on five major micro-task platforms: Amazon Mechanical Turk (AMT); CrowdFlower (renamed Figure Eight);[4] Clickworker; Microworkers; and Prolific. The Indian sample comprised 343 workers (15 per cent of the sample), with a large proportion of workers drawn from AMT (73 per cent). The survey captured information on workers' socio-demographic characteristics, their employment history, the tasks they performed and their working conditions, including their income, working hours and social security contributions. The survey was listed as a 'task' on these platforms and the tasks could be completed on a first-come, first-served basis (for more details, see Berg et al. 2018).

Figure 13.1 shows that micro-task workers were geographically dispersed throughout India, with about one-third located in Tamil Nadu (more than 100 workers) followed by Maharashtra, Karnataka, Telengana and Kerala (21 to 50 workers from each state), which have technology hubs and a high proportion of workers with technical expertise. The results of the demographic survey indicated that a high proportion of micro-task workers resided in cities (77 per cent), were predominantly male (80 per cent), young (on average 31.5 years) and highly educated (88 per cent had a university degree). Twenty five per cent of the workers reported that they had been performing tasks on micro-task platforms for more than five years.

When asked about their motivation for engaging in crowdwork, 26 per cent stated it was due to their preference to work from home which could be partly due to the long commuting that workers might have to do, whereas 25 per cent stated it was to complement their pay from other jobs. The other reasons for undertaking crowdwork included better pay compared to other available jobs (9 per cent) and a lack of other employment opportunities (6 per cent). For some micro-task workers, it was also to earn money while attending an educational institution (7 per cent) or as a form of leisure as they enjoy doing such work (17 per cent).

About 11 per cent of workers could only work from home, with strong gender differences: 19 per cent of women gave this reason compared to 9 per cent of men. Women's motivation to stay at home and perform crowdwork, despite being well educated, is largely due to the gender roles and expectation that women should take care of children and perform household tasks. Crowdwork provides them with the opportunity to be engaged in some form of work and to earn some income. As one women respondent explained, 'I am a housewife and [there is a] lot of work to be done inside the home like cooking, maintaining children. During leisure time I want to do some work with earnings. So I preferred crowd source which has no investment' (Respondent on AMT, India).

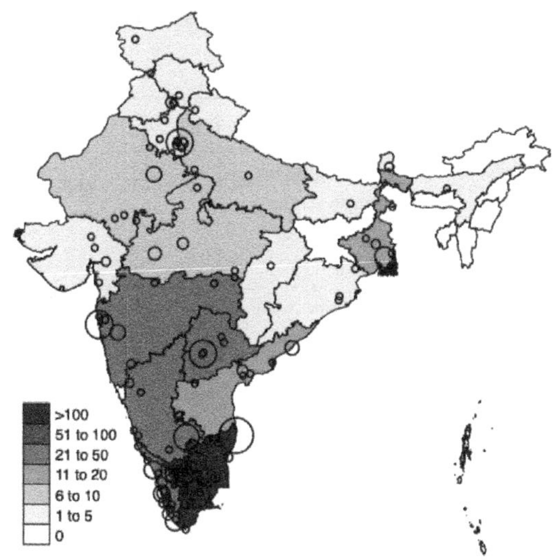

Note: Circles represent the numbers of crowdworkers per locality.
Source: ILO global survey of crowdworkers (2017).

Figure 13.1 Geographical distribution of Indian crowdworkers

On average, in a typical week, micro-task workers spent 32 hours working on platforms, of which 25 hours were paid work and seven hours were unpaid work, comprising activities such as looking for tasks, earning qualifications to become eligible for work and communicating with requesters. Overall, 40 per cent of workers reported that they worked seven days per week and 25 per cent worked regularly for six days per week (Table 13.1). A large proportion of workers worked during the night (10 pm to 5 am; 57 per cent) and during the evening (6 pm to 10 pm; 67 per cent) mainly as a result of time zone differences which meant that tasks posted in north America during the day were available to Indian workers in the evening or at night. As one worker explained 'Most crowdworking sites are from US and giving more jobs only on their day times, so people like me from other countries not able to work on day time. We have to lose our social commitments in order to work in night time' (Respondent on AMT, India). The need to adapt to the temporal distribution of jobs on platforms (O'Neill 2018) is similar to what has been observed in Indian BPO call centres wherein workers work night shifts in order to serve US clients in real time (Vira and James 2012). Night work affects the health and social life of both groups of workers. While BPO call centres offered better pay and the comradery of colleagues, platform work offered the comfort of performing work from one's own home.

Table 13.1 Percentage of workers reporting that they work long or unsocial hours, by sex

		Total	**Men**	**Women**
Working more than 10 hours a day	1 to 10 days per month	44.5	44.0	45.7
	>10 days per month	17.7	18.7	13.3
Days per week doing crowdwork	6 days per week	24.9	27.6	13.3
	7 days per week	40.0	39.7	41.7
Times of day	Evening	66.6	66.9	65.0
	Night	57.1	58.0	53.3

Source: ILO global survey of crowdworkers (2017).

A salient feature of digital platforms is that, once a worker has selected the tasks to be performed, the tasks and work process may be overseen by an algorithm rather than a human. The supervision and control of workers' tasks through algorithms can lead to unfair rejection and have implications for future earnings and the ratings of workers. Due to the lack of communication channels on these micro-task platforms, workers are often not given proper feedback and they miss the opportunity to learn from their mistakes and improve their future performance. Non-responsiveness might also arise as the requester themselves might not know why the work was rejected due to the lack of transparency found in the algorithms (Pasquale 2015). Almost eight out of ten Indian workers in the ILO survey had seen their tasks rejected, many voicing concerns about unfair rejections.

> Some providers ask us to transcribe many details from image, sometimes up to 50 data images. If we make one mistake in one image, they reject our entire hit. There are some work providers who provides several hits. If we make a mistake in one hit, they will reject all the hits. (Respondent on AMT, India)

The tendency to supervise, control and monitor work is also prevalent in call centre jobs. Monitoring is done through controls such as remote listening, call recording or surveillance cameras (D'Cruz and Noronha 2010). Yet with call centre work, because there is a human manager on the premises, there is a greater possibility to raise and resolve concerns. With micro-task platforms such voice is limited, if it exists at all, leading to frustration among workers.

13.5 EARNINGS ON DIGITAL LABOUR PLATFORMS

Table 13.2 gives information on the earnings of Indian workers on digital labour platforms, for both paid and unpaid work, disaggregated by sex. There is a wide degree of dispersion in earnings: average hourly earnings for paid working time total US$3.20 compared with median earnings of just US$2.00. While some higher earners drive up average earnings, about half of workers earn less than US$2.00 per hour. When we account for unpaid working time, the picture worsens. When analysing platform work, it is important to consider unpaid working time – activities such as searching for tasks or taking unpaid qualification tests – as this is a necessary part of work on platforms. As discussed previously, 22 per cent of working time (seven out of 32 hours), or nearly 15 minutes of every hour, were spent on unpaid tasks. Accounting for this time leads to a fall in average earnings to just US$2.30 per hour and a drop in median earnings to US$1.50 per hour.

Differences in earnings by sex are not pronounced, with similar average and median earnings for paid and unpaid work combined (average earnings of US$2.30 for both men and women and median earnings of US$1.50 for men and US$1.60 for women). Looking only at paid work, women performed slightly better although the share of women in the sample, and most likely in online platform work overall, is much smaller.

Overall, the average hourly earnings of Indian micro-task workers for both paid and unpaid work are slightly higher than that of the average earnings of workers in other developing countries (US$2.03) (Rani and Furrer 2021), but lower than the average earnings of workers in developed countries (US$4.30). For example, American workers on AMT earn 2.5 times the average earnings of Indian workers (Berg et al. 2018); such disparities have also been observed by other researchers on freelance platforms (Beerepoot and Lambregts 2015; Galperin and Greppi 2017). However, in spite of these disparities, most studies conclude that overall earnings on micro-task platforms are low (Hara et al. 2018; Bergvall-Kareborn and Howcroft 2014), and typically less than the legally mandated minimum wage. Platforms classify their workers as independent contractors, removing themselves from any legal and social responsibility, including minimum wages. Nonetheless, a sizable proportion of Indian workers are dependent on platforms to meet their household expenditures: income from crowdwork comprises 78 per cent of household income for those dependent on it on a full-time basis (58 per cent).

We now turn to a comparison of the earnings of online workers with those in the offline labour market. A number of basic demographic characteristics such as age, gender, place of residence and educational level are similar between the two samples (online micro-task workers and offline workers performing similar tasks in BPO centres and other companies), making the comparison meaningful.[5]

Offline workers who performed similar tasks in the local labour market earn 1.5 times more than online micro-task workers (US$4.40 and US$2.60, respectively), with no pronounced differences between men and women. Moreover, there is a greater concentration of online workers with low earnings whereas the distribution in earnings among offline workers with similar tasks is comparatively less skewed for both men and women (Figure 13.2). In addition, the 'typical' (median) micro-task worker earns much less than the platform average (US$1.50) which means that half of workers earn less than US$1.50 per hour for the time they spend working (Table 13.2).

Table 13.2 Hourly earnings, paid and unpaid work, by sex (in US$)

	Paid work			**Paid and unpaid work**		
	Median	**Mean**	**N**	**Median**	**Mean**	**N**
All workers	2.00	3.20	286	1.50	2.30	286
Sex						
Male	1.90	3.10	233	1.50	2.30	233
Female	2.50	3.30	53	1.60	2.30	53

Source: ILO global survey of crowdworkers (2017).
Note: Data trimmed at 1 and 99 per cent.

In addition, we have performed regression analysis to compare the average hourly earnings of online micro-task workers and offline workers controlling for individual characteristics and the nature of tasks,[6] all else being equal, for all workers and separately for men and women. The analysis shows that, for the overall sample, online micro-task workers earn 64 per cent less than offline workers with male online micro-task workers earning 62 per cent less and female workers earning 69 per cent less than offline workers performing similar tasks (ILO 2021).

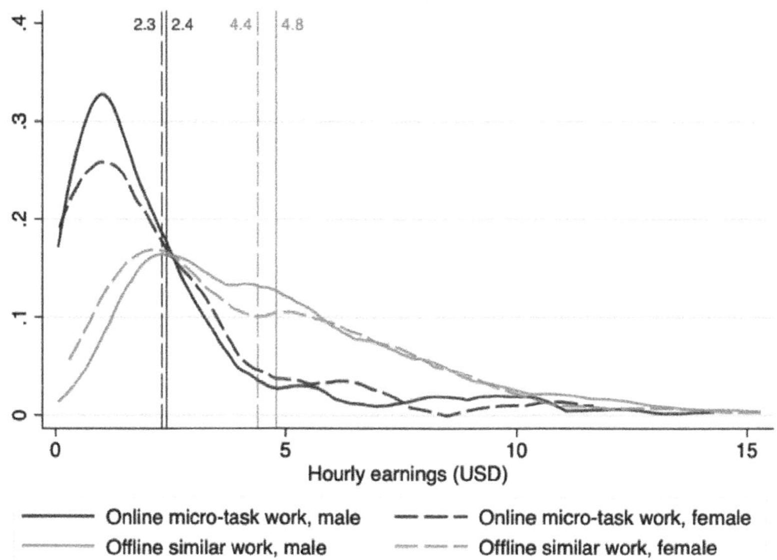

Note: Data trimmed at 1 and 99 per cent; vertical dashed lines show mean values.
Source: ILO global survey of crowdworkers (2017) and National Sample Survey Organisation (2018).

Figure 13.2 Distribution of hourly pay: Online and offline work by sex

Apart from low remuneration among micro-task workers, a major difference between BPO outsourcing work and platform work is that there is space for negotiation with regard to earnings and other benefits in the BPO model as there is a clear-cut employer–employee relationship. This is absent in micro-task platforms as tasks are posted with a predetermined price by the client or platform, leaving it to the discretion of workers to perform the task and with weak communication channels between client and micro-task workers.

13.6 CAREER PROSPECTS OF INDIAN PLATFORM WORKERS

An important finding from the literature on call centres is the ability of workers to improve their earnings by moving to other firms within the industry or to use their improved communication skills from call centres to obtain better jobs in other industries of the economy (James and Vira 2012). In addition, because earnings in the BPO call centre industry are high compared with local wages, many workers have been able to save money and use it to further their studies thus improving their long-term career prospects (Vira and James 2012).

With platforms, it is less evident that this is possible. Although we do not have information on the post-platform career trajectories of workers, evidence gleaned from the survey data is not promising. In particular, there are limited returns to tenure. There are increases in earnings during the beginning stages of crowdwork but, after two years of experience, there are no further increases (Rani and Furrer 2019). Moreover, while one would have expected efficiency on the part of workers in finding jobs to improve with experience, highly experienced workers still spend 19 minutes per hour doing unpaid tasks compared with 22 minutes per hour for new entrants, indicating limited improvement.

The lack of correlation between tenure and earnings is also reflective of the monopsony-type market of labour platforms (Kingsley et al. 2015; Dube et al. 2018). The excess supply of labour on platforms due to their global reach and ease of entry means that there is little need for businesses using the platforms to price their task higher, as there is a ready supply of labour regardless of the price. An analysis of tasks on AMT found that about 25 per cent of tasks were valued at $0.01, 70 per cent offered $0.05 or less and 90 per cent paid less than $0.10 (Ipeirotis 2010). As one Indian AMT worker remarked, ‘I have been working on Amazon MTurk for the last six years, but the amount of jobs and the payment hasn’t improved a bit even though I have got a 98.4 per cent approval rating.’

Another obstacle that Indian workers face, alongside other workers from the Global South, is that they are often blocked from performing certain tasks, contributing to lower earnings vis-à-vis American or European crowdworkers. On both micro-task and freelance platforms, there is an option geographically to select which workers can perform the jobs which can negatively affect earnings (Berg et al. 2018).

For micro-task workers, it is also questionable whether the tasks impart any skills that would benefit them if they searched for work elsewhere, in addition to making it difficult to prove that they have been doing work. While call centre work imparted few technical skills, it did improve workers' oral communication skills which not only made them more desirable because of their improved English but allowed them to perform better in job interviews (Vira and James 2012). Micro-task work does not confer this advantage.

The nature of micro-tasks does not seem to improve future employability. Other than time management and organizational skills, there are few learning opportunities for these highly educated workers. Many of the respondents were worried about their limited prospects of career advancement, as they were often not able to utilize their skills. There was also insecurity among workers about how this work was perceived by others and how they should reflect it in their CVs as they feared that it might not be valued as 'real' work (Berg et al. 2018). This is contrary to what was observed in call centres which not only provided mobility but, due to high incomes in comparison with the local labour market, was also a source of prestige among these workers.

13.7 CONCLUSION

India has long been a beneficiary of outsourcing, with the BPO industry providing an important source of foreign exchange, jobs and technology transfer over the past few decades. In the past decade, however, some outsourcing has shifted from BPO to directly outsourcing to individual workers via digital labour platforms, but there are also new tasks such as data annotation and content moderation, among others, which have been outsourced through these platforms.

In this chapter, we have discussed the working conditions of crowdworkers on digital labour platforms, comparing these with BPO workers using secondary empirical evidence. We found that the working conditions of crowdworkers are notably worse when measured in terms of earnings and career prospects but also with regard to the lack of social dialogue between workers and management. Other working conditions not addressed in this study, but likely to be of concern, include the social isolation common to working from home. Even more problematic is the tendency of platforms to classify workers as independent contractors, meaning that workers are not privy to the rights and benefits associated with an employment relationship such as paid annual and sick leave, or employer contributions into health insurance or retirement which workers in BPO firms commonly receive.

Platform work is conducted at home to faceless clients located throughout the world with management by an algorithm. It is difficult for workers to demonstrate to potential job recruiters that they have been engaged in this work and that they have learned marketable skills. While call centre work imparted few technical skills, it has been shown to improve workers' career prospects as their enhanced oral communication skills were highly marketable, in addition to improving workers' performance in job interviews (Vira and James 2012). Moreover, as work on platforms is broken into small and easily manageable micro-tasks, there is a tendency towards deskilling. Other than time management and organizational skills, there are few learning opportunities.

The other larger source of concern is the macroeconomic benefits of platform work and whether this is a beneficial development outcome for a highly educated workforce. Platform work is disconnected from the local economy and has fewer multiplier effects. Given the important investments undertaken by the Indian government and private citizens in IT education, it is necessary critically to assess whether such platforms should be promoted or whether these resources would not be better served by being invested in local economic opportunities. These may confer greater social and economic advantages both for the individual worker and society at large.

NOTES

1. See http://ilabour.oii.ox.ac.uk/online-labour-index/.
2. www.nasscom.in.
3. www.hindustantimes.com/business-news/bpo-industry-challenged-outsourcing-giant-india-losing-to-china-small-countries/story-13T7SBBiHb8ZujWCHe8B7N.html.
4. Now acquired by Appen.
5. The ILO global survey of crowdworkers conducted in 2017 asked respondents to describe the tasks they performed on the platforms. A taxonomy of 10 task categories was developed based on the services or projects offered by the platforms, or based on the skills required to perform the tasks. These task categories, except for surveys and experiments, and content access, were then matched to the closest International Standard of Industry Classification (ISIC). Two activities – office administrative and office support (ISIC 63) and other business support activities (ISIC 82) – were found to be similar to micro-tasks. The sample size of workers performing these tasks in the offline labour market was 1,570 (1,243 males and 327 females) and micro-task platforms was 252 (202 males and 50 females), and they were spread in major cities throughout the country. For details on methodology and analysis, please refer to Appendix 4B.2 in ILO (2021).
6. The comparison of average earnings is between the tasks performed by online micro-task workers (except those performing tasks related to surveys and experiments and content access) and similar tasks in the offline labour market. The results are available upon request from the authors.

REFERENCES

Aleksynska, M., A. Bastrakova and N. Kharchenko (2018), *Work on digital labour platforms in Ukraine: issues and policy perspectives*, Geneva: ILO.

Basant, R. and U. Rani (2004), 'Labour market deepening in India's IT: an exploratory analysis', *Economic and Political Weekly*, **39** (50), 5317–5326.

Batt, R, V. Doellgast, H. Kwon, M. Nopany, P. Nopany and A. da Costa (2005), 'The Indian call centre industry: National Benchmarking Report Strategy, HR practices and performance', CAHRS Working Paper Series 05-07, Ithaca, NY: Cornell University.

Beerepoot, N. and B. Lambregts (2015), 'Competition in online job marketplaces: towards a global labour market for outsourcing services?', *Global Networks*, **15** (2), 236–255.

Berg, J., M. Furrer, E. Harmon, U. Rani and M. Six Silberman (2018), *Digital labour platforms and the future of work: towards decent work in the online world*, Geneva: ILO.

Bergvall-Kareborn, B. and D. Howcroft (2014), 'Amazon Mechanical Turk and the commodification of labour', *New Technology, Work and Employment*, **29** (3), 213–223.

D'Cruz, P. and E. Noronha (2007), 'Technical call centres: beyond "electronic sweat-shops" and "assembly lines in the head"', *Global Business Review*, **8** (1), 53–67.

D'Cruz, P. and E. Noronha (2010), 'The exit coping response to workplace bullying: the contribution of inclusivist and exclusivist HRM strategies', *Employee Relations*, **32** (2), 102–120.

Dube, A., J. Jacobs, S. Naidu and S. Suri (2018), 'Monopsony in online labour markets', NBER Working Paper No. 24416, Cambridge, MA: National Bureau of Economic Research.

Galperin, H. and C. Greppi (2017), 'Geographical discrimination in the gig economy', accessed 17 November 2020 at http://dx.doi.org/10.2139/ssrn.2922874.

Hara, K., A. Adams, K. Milland, S. Savage, C. Callison-Burch and J. P. Bigham (2018), 'A data-driven analysis of workers' earnings on Amazon Mechanical Turk', in CHI'18: Proceedings of the 2018 CHI Conference on Human Factors in Computing Systems, Paper 449, New York: Association for Computing Machinery, accessed 17 November 2020 at https://doi.org/10.1145/3173574.3174023.

Hirschheim, R. and J. Dibbern (2009), 'Outsourcing in a global economy: traditional information technology outsourcing, offshore outsourcing, and business process outsourcing', in R. Hirschheim, A. Heinzl and J. Dibbern (eds), *Information systems outsourcing: enduring themes, global challenges, and process opportunities*, Berlin: Springer, pp. 3–21.

Holman, D., R. Batt and U. Holtgrewe (2007), *The global call center report: international perspectives on management and employment*, Ithaca, NY: Authors.

ILO (2021), *World employment and social outlook: the role of digital labour platforms in transforming the world of work*, Geneva: ILO.

Ipeirotis, P. G. (2010), 'Analyzing the Amazon Mechanical Turk marketplace', *XRDS: Crossroads, the ACM Magazine for Students*, **17** (2), 16–21.

James, A. and B. Vira (2012), 'Labour geographies of India's new service economy', *Journal of Economic Geography*, **12** (4), 841–875.

Kingsley, S. C., M. L. Gray and S. Suri (2015), 'Accounting for market frictions and power asymmetries in online labour markets', *Policy and Internet*, **7** (4), 383–400.

Messenger, J. C. and N. Ghosheh (2010), *Offshoring and working conditions in remote work*, Basingstoke: Palgrave Macmillan.

Mirchandani, K. (2004), 'Practices of global capital: gaps, cracks and ironies in transnational call centres in India', *Global Networks*, **4** (4), 355–373.

NASSCOM (National Association of Software and Service Companies) (2019), *Industry performance: 2018–19 and what lies ahead*, New Delhi: NASSCOM.

National Sample Survey Office (2018), *India periodic labor force survey, 2017–2018*, Delphi: Ministry of Statistics and Programme Implementation.

O'Neill, J. (2018), 'From crowdwork to Ola Auto: can platform economies improve livelihoods in emerging markets?', in H. Galperin and A. Alarcon (eds), *The future of work in the Global South*, Ottawa: International Development Research Centre, pp. 28–31.

Pasquale, F. (2015), *The black box society: the secret algorithms that control money and information*, Cambridge, MA: Harvard University Press.

Ramesh, B. P. (2004), '"Cyber coolies" in BPO: insecurities and vulnerabilities of non-standard work', *Economic and Political Weekly*, **39** (5), 492–497.

Rani, U. and M. Furrer (2019), 'On-demand digital economy: can experience ensure work and income security for microtask workers?', *Jahrbücher für Nationalökonomie und Statistik*, **239** (3), 565–597.

Rani, U. and M. Furrer (2021), 'Digital labour platforms and new forms of flexible work in developing countries: algorithmic management of work and workers', *Competition and Change*, **25** (2), 212–236.

Tara, S. and V. Ilavarasan (2009), '"I would not have been working here!": parental support to unmarried daughters as call center agents in India', *Gender, Technology and Development* **13** (3), 385–406.

Taylor, P. and P. Bain (2005), '"India calling to the far away towns": the call centre labour process and globalization', *Work, Employment and Society*, **19** (2), 261–282.

Vira, B. and A. I. James (2012), 'Building cross-sector careers in India's new service economy? Tracking former call centre agents in the National Capital Region', *Development and Change*, **43** (2), 449–479.

14. The geographic and linguistic variety of online labour markets: The cases of Russia and Ukraine

Mariya Aleksynska, Andrey Shevchuk and Denis Strebkov

14.1 INTRODUCTION

The turn of the millennium marked the emergence of a new type of labour market in which workers – often referred to as freelancers – digitally deliver services to clients through dedicated online labour platforms.[1] One important feature of these labour markets is that they potentially do not have geographical boundaries. In contrast to place-based digital platforms, such as taxi or delivery apps, online labour platforms match workers from any location to global pools of work opportunities without requiring workers' physical mobility (Horton 2010; Malone and Laubacher 1998; Graham et al. 2017). As such, they help to overcome distances and constraints of local demand through 'virtual migration' (Aneesh 2006). Given this, the development of online labour markets has been heralded as leading to the 'flat world' (Friedman 2007) and making spatial location increasingly irrelevant to economic activity.[2]

The current view that online labour platforms lead to the emergence of a 'global gig economy' and a 'planetary labour market' (Graham and Anwar 2019) is driven by researchers' almost exclusive focus on the functioning of English-language platforms (Kässi and Lehdonvirta 2018). As such, this view may be overly simplistic. Indeed, growing evidence shows the existence of Chinese online platforms which are comparable to their western counterparts in terms of the number of registered users (Kuek et al. 2015; To and Lai 2015). Online platforms also operate in other languages such as Spanish (Galperin and Greppi 2019) and Arabic (Kuek et al. 2015, p. 17). Ignoring these non-English-language platforms may not only lead to an underestimation of the size of online labour markets but also to a distortion of their real geography and to a misleading presentation of these countries as sheer suppliers of labour to the Global North. This may have significant policy implications for understanding the emergence of new economic centres and skill divisions across countries and in terms of developing governance tools for online labour markets.

Russia and Ukraine offer other examples of what can be called the multi-scalarity or multidimensionality, as opposed to the 'flatness', of online labour markets. The argument developed here is that this manifests itself in online platforms having geographic and linguistic variety as well as a resulting geographic and linguistic sorting of workers and clients. Both Russia and Ukraine are considered to be among the global leaders of online platform work, featuring among 'top ten' suppliers of workers to English-speaking platforms (Graham et al. 2017; ILO 2019). As such, many online workers in these countries are indeed part of the global online labour market. At the same time both countries also have several dozen Russian-language online labour platforms. These platforms create a regional online labour market. They match Russian-speaking workers and clients located anywhere but, as will be shown, primarily in the former Soviet Union (USSR) space. The number of workers on these platforms rivals the number of Russian and Ukrainian workers on English-speaking platforms (Shevchuk and Strebkov 2015, 2021; Aleksynska 2021). Moreover, in Ukraine, there are also Ukrainian-language online labour platforms, serving almost exclusively its domestic market. Thus, workers from Russia and Ukraine work not only on a global online market, largely mediated by the English language, but also on a regional online market which is exclusively mediated by a different common language (Russian), as well as on respective domestic (national) online markets.

The objective of this chapter is three-fold. First, it discusses possible reasons for the co-existence of domestic (national), regional and global online labour markets, and why this is neither exceptional nor temporary or transitory. The focus of the chapter is mainly on high- and medium-skilled occupations such as programming, design, writing, translating, engineering, marketing or consulting, even though low-skilled and routine microwork, such as image recognition or transcribing (see, for example, Berg et al. 2018), may follow similar patterns. Second, using Russia and Ukraine as examples, the chapter demonstrates the evolution of the three types of online labour markets: domestic; regional; and global. Third, using survey data from both countries, it examines worker characteristics on each of the markets, as well as the differences in outcomes, including as regards earnings, working hours and work satisfaction. Attention is also paid to informality and the skill composition aspects involved in the three types of market. The chapter concludes by discussing the policy implications surrounding the multidimensionality of online labour platforms.

14.2 RATIONALE FOR AND IMPLICATIONS OF THE VARIETY IN ONLINE LABOUR MARKETS

There are several reasons for the coexistence of a variety of online labour markets. If the global dimension has been highlighted in the pioneering studies of English-speaking online labour markets, the rationale for the existence of domestic and regional markets is less obvious. At least two sets of reasons stand behind their emergence.

The first set of reasons includes the linguistic, cultural and geographic proximity of workers and clients. Indeed, the role of common language as a factor that eases communication has long been established in the economics literature: it significantly enhances international trade, foreign direct investment and international migration (Melitz and Toubal 2014). In online labour markets, a shared language similarly creates a 'linguistic labour pool', allowing the matching of workers and clients speaking this language regardless of location. A common language, coupled with a similar culture, also helps to enhance trust (Guiso et al. 2009). In turn trust has long been viewed as key to economic transactions in the face of imperfect information and incomplete contracts (Algan and Cahuc 2014). Trust favours cooperative behaviour and facilitates mutually advantageous exchanges. Greater trust in cross-border exchanges increases trade and investment between countries, yet trust itself may be affected by conflicts and any recent history of traumatic experiences (Guiso et al. 2009; Algan and Cahuc 2014; Alesina and La Ferrara 2002). Given this, domestic and regional online labour markets have strong potential to be places that enhance trust and help reduce information barriers, although they may be vulnerable to trust erosion in the case of political conflicts.

In addition, sharing a common culture may also matter for workers in some particular professions such as designers, for whom it is important to share the same tastes, traditions and cultural codes as their clients. Thus, regional markets, while allowing national borders to be transcended, are places where culture-specific traits, including etiquette, taste and tradition, but also familiarity with similar technical norms and standards, can yield additional economic benefits. Geographic proximity, in its turn, also allows workers to stay within the same, or proximate, time zone.[3]

Evidence shows that, with regard to online labour markets, clients indeed often have a preference to work with spatially and culturally close contractors (Gefen and Carmel 2008; Ghani et al. 2014; Hong and Pavlou 2017). Some clients may prefer domestic workers even in situations in which all potential candidates are native speakers and share a similar culture (Galperin and Greppi 2019).

To respond to the linguistic, cultural and geographic preferences of clients, some platforms allow clients to choose the language and location of workers that they want to work with, locking out other potential contractors from executing tasks (Berg et al. 2018). While such practices may instil discrimination, they also stimulate the development of regional and local online markets.

The second set of reasons for geographical multidimensionality in online labour markets relates to the issue of workers' skills. On the one hand, there is an important issue of skill transferability, as not all workers may be able to execute all available tasks. Some skills may find their application in national labour markets only, such as legal practice or engineering, which require knowledge of national norms and technical standards. Some may be perfectly transferable, for example certain information technology (IT) skills, while others may be only partially transferable, for example writing skills in a regional language such as Russian. Moreover, technological developments carry with them increasingly higher demand for digitally delivered services in domestic and regional markets, leading to the proliferation of domestic and regional online platforms that value non-transferable or poorly transferable skills. On the other hand, there is also the issue of skills recognition. The economic literature on the transferability of migrants' skills (for example, Chiswick and Miller 2009) has long argued that, even with the most perfectly transferable skills, there is also the issue of their recognition. Clients from different cultures may not be familiar with the real value of certain diplomas, credentials and foreign experiences. They may also have certain stereotypes about the quality of skills, especially when they cannot verify them. 'Virtual migrants' may thus face similar dilemmas to physical migrants. As a result, even the most proficient experts, with the most transferable skills, may *prefer* to operate within a regional digital market where their skills may be recognized better than in international markets, and hence better remunerated.

Given these reasons, there is scope not only for international but also for regional and national online labour markets. Moreover, such regional and national markets may be quite 'sticky': because they fulfil important functions, such as easing communication and valuing common culture and skills, some workers and clients may consciously target them as their primary and long-term preference. Others may have no other choice than to stick to these markets, either because of language barriers or because of the poor transferability of their skills. Furthermore, it may be wrong to consider that all workers systematically test and try national and regional markets before moving to international ones; rather, some may stay on these markets permanently. Rating systems and reputation-building on some platforms, which are not transferable to others (De Stefano 2016), may further enroot this stickiness.

This variety in online labour markets has several implications. The first is that each market may attract a different type of worker. In other words, what can be observed is a certain sorting of workers across the range of online labour markets. The second implication is that each market may provide different outcomes for workers. This chapter examines both issues using the example of Russian and Ukrainian online freelancers. Lastly, the 'stickiness' of different markets means that different regulatory responses might be appropriate for different online markets.

14.3 THE CASES OF RUSSIA AND UKRAINE

The cases of Russia and Ukraine offer interesting examples of how a multitude of online labour markets emerges and develops. Both countries share a common history, most recently as part of the same country, the former USSR. Even though the share of Ukrainians speaking the Russian language has been declining since the country's independence in 1991, 28 per cent of Ukrainians still mainly use Russian in their daily lives and an additional 25 per cent are equally at ease in Ukrainian and Russian (KIIS 2019). There are continuing economic and cultural ties between the two countries, despite the deep political and military conflict that has been going on since 2014.

Both countries also share a similar and intertwined history of the development of online platform work. This mode of work emerged in the early 2000s and rapidly took off thanks to the development of better and more widespread access to digital technologies, developments in online labour platforms themselves and the growth of entrepreneurship culture more generally (Shevchuk and Strebkov 2015). The Russian-based platform FL.ru (formerly Free-lance.ru) and the Ukrainian-based platform Freelancehunt.com were both launched in 2005, becoming the leading platforms in each country, respectively.

The advent of the global economic recession of 2008–2009 fostered the expansion of online work in both countries. Businesses started to rely increasingly on outsourcing as a new business model and more workers started to consider new modes of employment such as online freelancing. Further, 2013–2014 marked deep political changes in Ukraine that had immediate economic repercussions: a rise in unemployment; the freezing of salaries; and a significant decline in economic growth. In Russia, a new economic recession in 2014 manifested itself notably in a dramatic currency depreciation which meant that working for clients from the United States (US) and Europe became particularly attractive to Russians. These issues also contributed to an increased interest in online platform work in both countries.

By 2020 there were several dozen online labour platforms operating in the Russian language, including six platforms reporting over 1 million users each (Shevchuk and Strebkov 2021). At least half a million registered workers from Ukraine were found on the six largest platforms through which Ukrainians access digital work, including on domestic and international markets, representing roughly 3 per cent of the employed population (Aleksynska et al. 2018).

Online work in both countries immediately featured international, regional and national dimensions. The regional dimension of digital work has been largely mediated by the Russian language. It is also driven by Russian-language platforms located primarily in Russia as well as in Ukraine. For example, the first successful Russian-language digital labour platform Weblancer.net, founded in 2003, was based in Ukraine. Around 2018–2019, about 12 per cent of Russian-language freelancers on the largest Russian platform FL.ru were from Ukraine (Shevchuk et al. 2021). Reciprocally, in the same period, over 30 per cent of freelancers operating on the Ukraine-based but Russian-language platform Freelancehunt.com were from Russia (Aleksynska 2021).

The conflict between Russia and Ukraine, which started in 2014, has had important repercussions in terms of shaping regional and national online labour markets. Using the publicly available personal profiles of all clients and workers on FL.ru, Shevchuk et al. (2021) show that the conflict has resulted in a sizeable outflow of Ukrainian freelancers and clients from this platform. Even so, more Ukrainians remain on this platform than have left it, lending evidence to the 'stickiness' hypothesis regarding this regional market.

These multiple dimensions of online labour markets can be visualized in Figure 14.1 which depicts three geographical zones: Russia; Ukraine; and the rest of the world. Each zone has its own online platforms through which domestic clients and workers can be matched (grey circles, grey flows). Some platforms operate in a language that is common between countries (white circles). Regardless of the platforms' location such platforms can serve both domestic markets in which the language is spoken (grey flows) but also the regional market (white flows). Moreover, clients and workers may ignore the platforms' location: they may be located in the same country while the platform is elsewhere (dotted flows). Finally, there are platforms located in a different zone and operating in a different language such as English (depicted in black). They allow the matching of workers and clients located anywhere and hence they feature a global dimension (black flows), a regional dimension (white flows) and a domestic dimension (grey flows).

The size of the circles and arrows in Figure 14.1 is not representative of the real size of the platforms and workflows between them. Nevertheless, some of these flows can be quantified (see also Table 14.1). Based on Russian and Ukrainian surveys of platform workers, described in Section 14.4, 31 per cent of Ukrainian freelancers operate exclusively in the domestic Ukrainian market (the total of the grey flows in Figure 14.1 within Ukraine). Thirteen per cent operate in the regional market of whom the vast majority (11 per cent) work for Russian clients (some of the white flows). The remaining 56 per cent work for clients located in the rest of the world, primarily in English-speaking countries (black flow). In the case of Russia, if the white circle can be considered as one of the six largest platforms, FL.ru, 64 per cent of Russian freelancers operate exclusively in the domestic Russian market (grey flow). An additional 22 per cent work for clients from the former countries of the USSR, of whom the majority (15 per cent) work with clients from Ukraine (white flow), while 21 per cent find clients in other countries around the world (see also Shevchuk et al. (2021) for a more detailed picture of workflows in the regional Russian-language market based on web-scraped data).

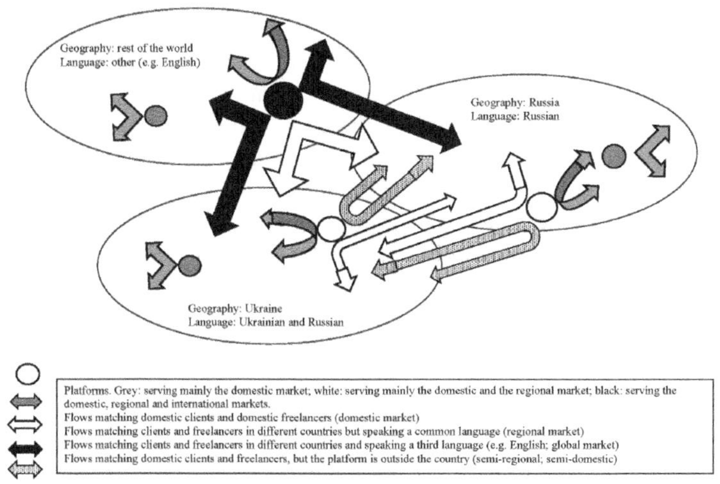

Figure 14.1 *The multidimensionality of digital online labour markets. Example of English, Russian and Ukrainian linguistic spaces*

While Russia and Ukraine are the focus of this chapter it is nonetheless important to mention that Russian is also widely spoken in several other countries. It is the second official language in Belarus, Kazakhstan and Kyrgyzstan, allowing workers from these countries also to join the Russian-language online labour market. Indeed, the presence of workers from other former USSR countries has been growing steadily since 2005 on FL.ru, reaching 8.7 per cent in 2018 (Shevchuk et al. 2021). The share of such workers on the largest Russian-language Ukrainian platform Freelancehunt.com was similarly around 7 per cent in 2019, with workers coming mainly from Belarus and Kazakhstan (Aleksynska 2021). The domestic digital markets of these countries also seem to be developing in many ways as a result of local-market versions of the Ukrainian and Russian platforms. For example, the Ukrainian platform Kabanchik.ua has created a Belarusian version, Kabanchik.by. These regional flows are omitted from Figure 14.1 for simplicity, but their importance can be expected to grow in the future.

14.4 DO WORKERS' CHARACTERISTICS AND OUTCOMES DIFFER DEPENDING ON THE TYPE OF MARKET?

To examine the differences in worker characteristics and outcomes across different online markets, this section draws on quantitative data from two surveys, conducted separately in Russia and Ukraine.

The Russian survey was hosted on FL.ru which has about 1.6 million users. FL.ru is a typical general purpose platform largely for high- and medium-skilled projects in categories such as programming, websites, texts, translating, multimedia, engineering, marketing, legal, consulting, etc. The data used here come from the fourth wave of the study, carried out from December 2018 to January 2019. The details of the methodology and the questionnaire are provided in Shevchuk and Strebkov (2021). The analytical sample includes only active freelancers located in Russia who perform their work through online platforms and who completed at least two work projects in 2018. The sample encompasses 1,255 respondents: 680 men and 575 women. Based on the question 'From which countries and regions did you have clients during 2018?' three groups of Russian freelancers were defined: (1) those operating exclusively in the domestic Russian market; (2) those doing work for clients from former USSR countries (except Russia); and (3) those working for clients in other countries around the world. The last two categories have a slight overlap since the most active freelancers (7 per cent) use a wide range of job-seeking channels and have clients from different countries.

The Ukrainian survey is a representative survey conducted among Ukrainian-based online freelancers operating on any online labour platform. This survey was conducted in December 2017, targeting respondents aged 18 and older who identified themselves as individuals performing work through at least one internet platform for pay in the 12 months preceding the survey (for details, see Aleksynska et al. 2018). The analytical sample presented here includes 588 respondents: 312 men and 276 women. The Ukrainian questionnaire contained the following questions: 'What is the country of origin of your main clients?' and 'How much of your work is performed for clients within Ukraine? And outside Ukraine?' Combining the answers to these questions, it is possible to divide all workers into those working mainly on the domestic, the regional (former USSR) and the international labour markets.

Table 14.1 demonstrates how the structure of freelancers differs depending on the market in which they are operating. This descriptive evidence suggests that in both countries men are more likely to work on international labour markets than women.

Workers with IT specializations, who arguably have the most transferable skills, are more likely to be oriented towards both regional and international markets. Designers and graphic artists also tend to orient towards external markets and – consistent with the culture hypothesis – Ukrainian designers in particular have an especially high propensity to work on the regional market while text writers in both Ukraine and Russia are also strongly oriented towards regional markets. This is where their (common) language skills seem to be valued the most. Possibly this is also because such workers do not have other transferable skills which would allow them to switch to another market. In contrast, business services, marketing and engineering specialists have a higher chance to stay on domestic markets.

The Russian sample shows that experienced freelancers (those with longer tenure as a freelancer) are more likely to orient to external markets. The more experienced and skilled freelancers from Russia may also pick out complex and lucrative projects from the regional Russian-language market and are competitive on the global market. However, the relationship between tenure and markets served is not the case in the Ukrainian sample. If anything, the distribution of tenure by markets in the Ukrainian sample may reflect an avoidance by novice freelancers of the regional Russian-language market, as well as the 'stickiness' of the regional market for freelancers with the longest tenure who might have started working on that market when it was at its peak.

There are also several implications for workers in different markets. The first of these is in terms of earnings. Previous research has shown that, although platform freelancers in Russia and Ukraine earn incomes that are slightly above the average wages in their countries, earnings differ significantly in terms of their socio-demographic and professional characteristics (Aleksynska et al. 2018; Shevchuk and Strebkov 2021). There is also a clear difference in earnings depending on the type of online labour market: platform workers have higher earnings on external markets as opposed to domestic ones (Table 14.1). For Russian freelancers, the global market offers the best likelihood of higher earnings. Ukrainian freelancers have a greater chance to earn higher earnings on both regional and international markets. Meanwhile, lower earnings are more likely to be observed for workers in both countries on the domestic markets.

Table 14.1 *Structure of online workers by type of market: local, regional or international*

		Type of market for Russian freelancers				Type of market for Ukrainian freelancers			
			Domestic	Regional	Global		Domestic	Regional	Global
		Total	Only Russia	Former USSR (excl. Russia)	Other countries	Total	Only Ukraine	Former USSR (excl. Ukraine)	Other countries
Sex	Female	*46*	48	41	40	*47*	50	49	44
	Male	*54*	52	59	60	*53*	50	51	56
Field of specialization	IT/programming/website creation	*26*	22	31	34	*20*	13	28	23
	Design/graphics	*35*	33	38	39	*9*	9	13	8
	Photo/audio/video	*15*	15	18	14	*10*	9	1	12
	Text writing	*26*	27	33	18	*34*	34	42	32
	Translation	*6*	5	6	12	*6*	4	5	7
	Business services, engineering	*30*	35	18	25	*21*	29	10	20
Freelance tenure	Less than one year	*19*	22	11	12	*57*	60	42	58
	1–4 years	*43*	41	50	43	*37*	34	46	36
	Over 5 years	*38*	36	38	45	*7*	7	12	6
Monthly earnings	1st quartile	*26*	28	26	18	*31*	38	24	27
	2nd quartile	*19*	20	15	14	*23*	27	15	23
	3rd quartile	*26*	24	28	27	*24*	20	34	23
	4th quartile	*29*	27	31	42	*23*	15	26	27

		Type of market for Russian freelancers				Type of market for Ukrainian freelancers			
			Domestic	Regional	Global		Domestic	Regional	Global
		Total	Only Russia	Former USSR (excl. Russia)	Other countries	Total	Only Ukraine	Former USSR (excl. Ukraine)	Other countries
Working night hours	Regularly (Russia) or typically (Ukraine)	*17*	14	24	24	*30*	28	32	34
Work satisfaction	Rather dissatisfied than satisfied	*19*	20	16	14	*6*	6	9	6
	Neither satisfied nor dissatisfied	*34*	35	35	29	*38*	45	38	34
	Satisfied	*47*	45	49	57	*56*	48	53	61
Number and share of respondents		*1,255 (100%)*	*798 (63.6 %)*	*278 (22.2 %)*	*266 (21.2 %)*	*588 (100 %)*	*182 (31.0 %)*	*78 (13.3 %)*	*328 (55.8 %)*

Note: Earnings: in the Russian sample, the earnings question was 'What was your personal average monthly post-tax income in 2018 including freelance and all other paid activities?' with eight categories; in the Ukrainian sample, the earnings question was open-ended: 'In a typical week, how much did you earn from platform work?' The answers have been converted into a monthly US dollar format. Here, the answers from both surveys are grouped into four quartiles, with a similar number of respondents in each. Earnings in the Russian sample 1st quartile: USD 400 or less; 2nd quartile: USD 401–600; 3rd quartile: USD 601–1000; 4th quartile: USD 1001 and above. In the Ukrainian sample, 1st quartile: USD 99 or less; 2nd quartile: USD 100–199; 3rd quartile: USD 200–399; 4th quartile: USD 400 and above.
Source: Own computations based on Russian and Ukrainian survey data.

These outcomes, however, have to be set against the high toll of informality of online work. In the Ukrainian sample, only 24 per cent of the respondents report being registered with the public authorities, hence paying social security contributions and income tax on their online activities. This share is actually higher among workers operating on the domestic market, compared to those operating on regional or international markets.[4] In other words, despite attractive earnings, online workers remain largely outside the scope of social protection and also risk encountering sanctions for tax avoidance; moreover, it is those with higher earnings that are least likely to be reporting them.

The second implication is in terms of working hours. Previous studies have revealed that freelancers in Russia and Ukraine routinely work non-standard hours and much more so than regular employees (Aleksynska et al. 2018; Shevchuk et al. 2019; Shevchuk et al. 2021). Part of the reason is that the majority of online freelancers combine online work with other offline activities. But it is also possible that the type of market served influences work schedules. In the Russian survey, respondents were asked: 'How often do you have to work late at night, that is from 9 pm to 6 am?' Seventeen per cent of freelancers reported working at night almost every day, but this share was lower for those who worked only on the local market (14 per cent) and higher for those who had regional or international clients (24 per cent for both). In the Ukrainian survey, respondents were asked: 'Please indicate during which of the following times of day you typically complete platform work (mornings, afternoons, evenings or nights).' Several answers were possible. Whereas 30 per cent of all freelancers typically worked at night, from 10 pm to 5 am, this share rose to 34 per cent of those working for international clients but dipped to 28 per cent of those working for local clients.

Freelancers in both countries report the highest satisfaction from working for international clients. In Russia the most dissatisfied freelancers are found on the domestic market while in Ukraine on the regional one. If it is particularly likely that higher earnings contribute to the highest work satisfaction on international markets, the effect of the earnings factor on regional markets is more nuanced and probably mitigated by other concerns.

14.5 CONCLUSIONS

The development of online labour markets offers an opportunity to understand better how the processes of digitalization and globalization are creating new and complex economic geographies by fundamentally altering the way people work and interact. This chapter adds to a growing literature that challenges the 'flat world' view of online labour markets. Using the examples of Russia and Ukraine, it contributes to the literature in at least two ways.

First, it theorizes that there is a strong rationale for the existence of geographic and linguistic variety when it comes to online labour markets. This rationale is linked, on the one hand, to the socio-economic benefits of operating in the context of a common language, similar culture and proximate geography. On the other hand, this rationale is rooted in the issues of skill transferability and skills recognition, suggesting that skills can be valued differently in different markets.

Second, the chapter demonstrates that this rationale is valid in the case of Russian and Ukrainian online freelancers. In addition to local and international markets, freelancers from these countries can operate in a regional online labour market, mediated by the Russian language where it is held in common. The chapter shows that the distribution of workers' characteristics differs across these three types of market. Moreover, there are key differences in workers' outcomes in terms of earnings, working hours and work satisfaction. Each market features a diversity of situations where various advantages and disadvantages may coexist. For example, if work on international markets pays more, it comes with a toll of a higher incidence of night work and is only possible for workers in some specific occupations.

The existence of an array of online labour markets has numerous policy implications. The first of these is in terms of governance. To date, much debate has centred on the international regulation of online labour markets. For example, in 2019, the Global Commission on the Future of Work of the International Labour Organization called for the 'development of an international governance system for digital labour platforms that sets and requires platforms (and their clients) to respect certain minimum rights and protections' (ILO 2019). According to Berg et al. (2019), such an international governance system could not only put in place minimum standards, but also develop the infrastructure necessary for facilitating payments to social security systems and establish a representative board to adjudicate disputes between platforms, clients and workers. This chapter has shown that, in addition to such an international governance system, there is scope for the domestic regulation of local online labour markets, in line with their general local labour market regulation. Regional regulation, in contrast, may not be politically feasible in the current post-USSR context. Yet, in its absence, compliance with international regulation may become problematic as the latter risks disregarding the regional specifics entailed by online labour markets.

A serious challenge to any regulation of work on online platforms relates to the high degree of informality of this work. In addition to the low rate of registration with public authorities, only about 12–15 per cent of freelancers in the regional Russian-language market conclude legal written contracts with their clients (Shevchuk and Strebkov 2018, 2021). Also, at least one-third of

Ukrainian freelancers work directly with their clients, bypassing the platform after the initial contact (Aleksynska et al. 2018).

The second implication of the presence of variety in online labour markets is in terms of skills policies. The rapid development of online labour markets has led some governments to conceive digital strategies to take advantage of the developmental benefits of their workers' access to online platforms (Graham et al. 2017). Currently the Trade Chamber of Ukraine is also developing a Digital Agenda for Ukraine which includes, as one of its proposals, an increase in state-supported training for professional IT specialists. This comes in the context of the share of technical graduates having already risen from 21 per cent to 27 per cent of all graduates between 2015 and 2017 (AVentures Capital et al. 2019). It is natural to expect that many of these specialists will work on global markets, whether through online labour markets or other opportunities. However, the concern is that such governmental efforts, coupled with certain neglect of other specializations, can lead to future skill imbalances and the overproduction of specialists with specific skills. It also raises broader questions as to whether local university systems are serving the interests of international employers rather than local economies and whether such a skills skew is not void of risks in the future. The high degree of informality of online work also means that public investments in these programmes result in private benefits that are not shared on a wider basis in society. Importantly, the existence of domestic and regional online markets on which different skills are differently valued also means that it may be more optimal in the long run for a country to produce a broad range of specialists to diversify their economic opportunities, including through online labour platforms.

NOTES

1. Support from the Basic Research Programme of the National Research University Higher School of Economics for the second and the third authors is gratefully acknowledged.
2. In this chapter, the term 'online labour market' is used to describe the array of online labour platforms that mediate digitally delivered services.
3. Even if for some workers it may be advantageous to be in different time zones, such as for the bookkeeping of US firms in India.
4. There is no equivalent question in the Russian survey.

REFERENCES

Aleksynska, M. (2021), 'Digital work in eastern Europe: overview of trends, outcomes, and policy responses', ILO Working Paper, 32, Geneva: ILO.

Aleksynska, M., A. Bastrakova and N. Kharchenko (2018), *Work on digital labour platforms in Ukraine: issues and policy perspectives*, Geneva: ILO.

Alesina, A. and E. La Ferrara (2002), 'Who trusts others?', *Journal of Public Economics*, **85** (2), 207–234.

Algan, Y. and P. Cahuc (2014), 'Trust, well-being and growth: new evidence and policy implications', in P. Aghion and S. Durlauf (eds), *Handbook of economic growth*, Amsterdam: North-Holland, pp. 49–120.

Aneesh, A. (2006), *Virtual migration: the programming of globalization*, Durham, NC: Duke University Press.

AVentures Capital, Aventis Capital and Capital Times (2019), 'Software development report in Ukraine, Poland, Belarus and Romania', accessed 24 November 2020 at https://software-development-cee-report.com/.

Berg, J., M. Furrer, E. Harmon, U. Rani and M. S. Silberman (2018), *Digital labour platforms and the future of work: towards decent work in the online world*, Geneva: ILO.

Berg, J., M. Cherry and U. Rani (2019), 'Digital labour platforms: a need for international regulation?', *Revista de Economía Laboral – Spanish Journal of Labour Economics*, **16**, 104–128.

Chiswick, B. and P. Miller (2009), 'The international transferability of immigrants' human capital skills', *Economics of Education Review*, **28** (2), 162–169.

De Stefano, V. (2016), 'The rise of the "just-in-time workforce": on-demand work, crowdwork and labour protection in the "gig-economy"', ILO Conditions of Work and Employment Series 71, Geneva: ILO.

Friedman, T. L. (2007), *The world is flat 3.0: a brief history of the twenty-first century*, New York: Picador.

Galperin, H. and C. Greppi (2019), 'Geographic discrimination in the gig economy', in M. Graham (ed.), *Digital economies at global margins*, Cambridge, MA: MIT Press, pp. 295–318.

Gefen, D. and E. Carmel (2008), 'Is the world really flat? A look at offshoring at an online programming marketplace', *MIS Quarterly*, **32** (2), 367–384.

Ghani, E., W. Kerr and C. Stanton (2014), 'Diasporas and outsourcing: evidence from oDesk and India', *Management Science*, **60** (7), 1677–1697.

Graham, M. and M. A. Anwar (2019), 'The global gig economy: towards a planetary labour market?', *First Monday*, **24** (4), accessed 24 November 2020 at https://doi.org/10.5210/fm.v24i4.9913.

Graham, M., I. Hjorth and V. Lehdonvirta (2017), 'Digital labour and development: impacts of global digital labour platforms and the gig economy on worker livelihoods', *Transfer*, **23** (2), 135–162.

Guiso, L., P. Sapienza and L. Zingales (2009), 'Cultural biases in economic exchange?', *Quarterly Journal of Economics*, **124** (3), 1095–1131.

Hong, Y. and P. Pavlou (2017), 'On buyer selection of service providers in online outsourcing platforms for IT services', *Information Systems Research*, **28** (3), 547–562.

Horton, J. (2010), 'Online labor markets', in A. Saberi (ed.), *Internet and network economics. 6th International Workshop, WINE 2010, Stanford, CA, USA, December 13–17, 2010, Proceedings*, Berlin: Springer-Verlag, pp. 515–522.

ILO (2019), *Work for a brighter future: Global Commission on the Future of Work*, Geneva: ILO.

Kässi, O. and V. Lehdonvirta (2018), 'Online labour index: measuring the online gig economy for policy and research', *Technological Forecasting and Social Change*, **137**, 241–248.

KIIS (2019), 'National opinion poll', accessed 24 November 2020 at www.kiis.com.ua/?lang=ukr&cat=reports&id=832&page=1.

Kuek, S., C. Paradi-Guilford, T. Fayomi, C. Imaizumi, P. Ipeirotis, P. Pina and M. Singh (2015), *The global opportunity in online outsourcing*, Washington, DC: World Bank Group.

Malone, T. and R. Laubacher (1998), 'The dawn of the e-lance economy', *Harvard Business Review*, **76** (5), 144–152.

Melitz, J. and F. Toubal (2014), 'Native language, spoken language, translation and trade', *Journal of International Economics*, **93** (2), 351–363.

Shevchuk, A. and D. Strebkov (2015), 'The rise of freelance contracting on the Russian-language internet', *Small Enterprise Research*, **22** (2–3), 146–158.

Shevchuk, A. and D. Strebkov (2018), 'Safeguards against opportunism in freelance contracting on the internet', *British Journal of Industrial Relations*, **56** (2), 342–369.

Shevchuk, A. and D. Strebkov (2021), 'Freelance platform work in the Russian Federation, 2009–2019', ILO Working Paper.

Shevchuk, A., D. Strebkov and S. Davis (2019), 'The autonomy paradox: how night work undermines subjective well-being of internet-based freelancers', *ILR Review*, **72** (1), 75–100.

Shevchuk, A., D. Strebkov and A. Tyulyupo (2021), 'The geography of the digital freelance economy in Russia and beyond', in M. Will-Zocholl and C. Roth-Ebner (eds), *Topologies of digital work: how digitisation and virtualisation shape working spaces and places*, Basingstoke: Palgrave Macmillan.

To, W.-M. and L. Lai (2015), 'Crowdsourcing in China: opportunities and concerns', *IT Professional*, **17** (3), 53–59.

PART IV

Case studies across the globe: Location-based labour platforms

15. Aliada and Alia: Contrasting for-profit and non-profit platforms for domestic work in Mexico and the United States

Andrea Santiago Páramo and Carlos Piñeyro Nelson

15.1 INTRODUCTION

The emergence of the platform economy has given rise to location-based platforms for domestic work in both Mexico and the United States (US). This chapter discusses two different approaches to using online platforms for domestic workers and the new hiring schemes that such platforms have induced. The first approach is Mexico's Aliada: this is an online for-profit service that functions as an intermediary between domestic workers and employers. The non-profit online platform Alia encompasses the second approach: it has been created by the National Domestic Workers Alliance (NDWA) in the US for the benefit of domestic cleaners. The overlap between work and digital technologies in these two cases opens up a new field of debate, highlighting challenges as well as opportunities to enhance social justice. Technology is a tool that produces new realities and social relationships and can transform attitudes, perspectives and practices. Platforms can therefore either reproduce precarity under new and 'modern' systems or else they can play a role in tackling inequality.[1]

Comparing the Mexican and US cases helps us to see the different effects that for-profit and non-profit platforms can have on domestic workers and to examine the cultural assumptions behind the design of each type. In the Mexican case, Aliada allows us to identify some of the problems that workers face in the platform economy when rights are not recognized and the responsibilities of the platforms are denied. While the Mexican case illustrates new forms of precarious living conditions brought about by working for location-based platforms for domestic work, the US case shows the potential of a non-profit digital platform to alleviate precarity and the effects created by the lack of recognition of workers' rights. In turn, Alia helps us see how technology is being used towards improving workers' living conditions in a social context where the state has not guaranteed basic rights to domestic workers.

This chapter discusses the extent to which online platforms either reproduce precarity or potentially help improve labour standards among domestic workers. Thus, by analysing Aliada, we see how its promises of freedom and a better future for domestic workers have not happened in reality. Quite the opposite: this for-profit platform reinforces inequality by denying workers' rights. Alia, on the other hand, has the potential to enable benefits for domestic workers in the US, yet it is too soon to value its success or failure. However, its implementation opens a real possibility for this workforce to gain de facto benefits.

15.2 WORKING CONDITIONS, THE MEANING OF DOMESTIC WORK AND ORGANIZING DOMESTIC WORKERS

Table 15.1 provides a comparative overview of the main demographics of domestic workers in Mexico and the US. One out of every ten working women in Mexico are domestic workers and, of these, one in three are the primary economic providers for their families. There are 2.2 million paid domestic workers in Mexico, 98 per cent of them being informal workers. This informality of the industry means that workers are excluded from social security programmes. Job insecurity, a lack of rights and a lack of recognition of this type of work as a dignified form of employment have been constant problems for this group of workers. The work is also highly feminized, being performed almost exclusively by women (95 per cent), with low levels of schooling (43 per cent have completed only elementary school) and low economic resources (63 per cent are unable to cover their basic necessities), while many are of indigenous origin (28 per cent) (Bensusán 2019, p. 11).

Between the 1930s and the 1970s, African-American domestic workers were the largest racial group in the US workforce; however, a demographic change took place in which large numbers of African-American women started to access better-paid jobs, such as nurses and clerical workers, transitioning out of domestic work. The number of African-American domestic workers considerably declined: in 1950, 42 per cent of African-American women in employment were domestic employees – a percentage that had fallen to 20 per cent by 1970 and to just 6 per cent by 1980 (Nadasen 2015, p. 152). The need to satisfy the demand for domestic workers triggered a massive influx of immigrant women of colour, mainly from Latin America, the Caribbean and south-east Asia. This trend continues today: 78 per cent of domestic workers were born abroad, 60 per cent of them in Latin America (Burnham and Theodore 2012,

p. 41). It is estimated that there are about 2.2 million domestic workers in the US today – coincidentally a similar estimate to the one for Mexico.

Table 15.1 *Domestic workers in Mexico and the United States*

	Mexico	United States
Domestic workers	2.2 million	2.2 million
Women	95%	95%
Occupation	Housekeepers, nannies, domestic cleaners, chauffeurs	Housekeepers, domestic cleaners, nannies, caregivers
Ethnic minorities/ indigenous	28%	54%
Immigrants/internal migration	29%	46%
Undocumented	n/a	36%
Less than 12 years of education	85%	32%

Source: CONAPRED (2015); Bensusán (2019); Burnham and Theodore (2012, pp. 11–12); Theodore et al. (2019); Xantomila and Román (2017).

Domestic cleaners in the US tend to be the least visible group of domestic workers partly because most are undocumented women of colour from the Global South. These factors influence them being hired frequently off the books and without written contracts (Theodore et al. 2019). The inherent characteristics of the job also make it more difficult for domestic cleaners to organize and demand benefits; cleaners are likely to have more than two clients per week, working for periods of four to six hours per job (Hondagneu-Sotelo 2007). In addition, they often clean when the client and their families are not in the house. The lack of interaction with clients makes it even harder for workers to negotiate better working conditions (Cruz and Abrantes 2014; Hondagneu-Sotelo 2007).

Domestic workers in Mexico and the US face similar problems: the occupation still lacks a positive perception for most people; domestic workers are one of the most exploited and informal workforces; and most domestic workers face social discrimination due to their ethnicity (Mexico) or immigrant status (US), or a mix of both. Domestic work continues to be seen as an unskilled form of labour, apparently requiring little knowledge to carry out the tasks that those performing them do. This belief is reinforced by the lack of professional training many such workers have when hired and because 'they mirror work that has traditionally been carried out by women in their own homes' (Oelz and Rani 2015, p. 11). While domestic workers perform tasks of care and housework that are primarily assigned to women due to social norms, performing these tasks in other homes means they take on new meanings and forms of social undervaluation due to the worker's own ethnic origin, social class and, in some cases, citizenship; transforming house chores into 'dirty work' or 'pitiful work' and delegating them to the most marginalized or vulnerable in the social pyramid (Devetter 2013). Those that perform this work are continually exposed to multiple forms of discrimination – for example, the use of derogatory names and the separation of food or utensils from those of the families that they work for. Generally, outside the realm of public scrutiny and lacking state intervention to regulate work dynamics, paid domestic work has been structured around informal agreements between subjects with a great disparity of power and benefits, both symbolic (prestige, status, acknowledgement and respect) and socio-economic (income, assets, education and care).

The organizing of domestic workers in Mexico and the US has followed different struggles. Domestic workers have different resources around which to mobilize for their rights. In Mexico, most domestic workers are in informal employment arrangements. Even so, domestic workers' rights are included in the Federal Labour Law (*Ley Federal del Trabajo*, chapter 13) and the Law on Social Security (*Ley del Seguro Social*) which establishes mandatory enrolment in social security programmes. Yet a diversity of barriers (among them the power imbalance, poor information about domestic workers' rights, historic discrimination and bureaucratic processes) blocks workers from accessing their rights. Domestic workers in the US, in contrast, are not included in the two most essential pieces of labour legislation: the National Labor Relations Act (which regulates the right to form a trade union and bargain collectively); and the Fair Standards Labor Act (which regulates the minimum wage, overtime and child labour).[2] In both countries an overwhelming majority of domestic workers, legally protected or not, are unable to access their fundamental rights.

Domestic workers in Mexico and the US share several features as well as sharp differences. Both workforces are comprised heavily of women, with a strong presence of ethnic minorities; they also perform many of the same sub-occupations. They do have strong differences in terms of their educational level and undocumented status and institutional settings, however. Domestic workers on both sides of the border have therefore adjusted their organizational strategies based on their national backgrounds.

Organizing domestic workers by trade unions in Mexico and the US has been historically difficult due to the dispersion of work locations and the stigma associated with this work that causes many domestic workers to reject identifying as such or embracing a collective identity. Lack of time and resources for organizing, widespread ignorance about their rights and the sheer variety of working profiles and hiring schemes (live-in; live-out with one or more employers; and now location-based platforms) also hinder organizing capacities within this group of workers. Nevertheless, the first National Union of Domestic Workers (*Sindicato Nacional de Trabajadores y Trabajadoras del Hogar*) was created in 2015 by a group of already organized domestic workers, advocating rights for domestic workers in different public spaces in Mexico. In addition to the National Union, there are other local domestic worker organizations that, for decades, have been fighting for domestic workers' rights in their states, including *Centro de Apoyo y Capacitación para Empleadas del Hogar* in Mexico City, *Red de Mujeres Empleadas del Hogar* in Guerrero and *Tzome Ixuc* in the state of Chiapas. Of the total number of domestic workers in Mexico, however, less than 1 per cent organize collectively to demand their rights.

Turning to the US, the immigration status of most domestic workers also plays against organizing this workforce. The NDWA is the latest national domestic worker organization to emerge in the US. It was created in 2007 and is currently powered by 62 affiliate organizations in 19 states. Each affiliate organization is independent; NDWA plays the role of an umbrella organization at the national level. That is, it focuses on capacity-building for domestic workers at local level but also coalesces its human and economic resources and power to lead national and local campaigns and programmes for the passing of legislation granting rights to domestic workers (Piñeyro Nelson 2018). Additionally, the NDWA is trying to secure de facto benefits for domestic workers, particularly cleaners, through an online platform it has developed called Alia, i.e. an alternative way for cleaners to access benefits not granted by the state.

15.3 THE PROMISES OF ALIADA AND ALIA AND THEIR OPERATION

The rise of the platform economy (offering short-term, temporary positions filled by independent contractors) has had an impact on domestic workers in Mexico and in the US. Location-based domestic work platforms first emerged in Mexico in 2013 within the framework of a growing digital economy, in a country in which 57 per cent of the working population has informal employment (Carmona 2019) and for whom the welfare state has never been a reality. In the case of the US, Care.com was the first location-based platform focused on domestic work, launched in 2006.[3] Handy, the other major platform in the area of domestic labour, was created in 2012.[4]

While some platforms do offer some benefits, most do not due to the lack of regulation and they are thus a prime source of low-wage labour (Berins Collier et al. 2017). In both countries domestic workers are generally excluded under these new hiring models from basic rights and social protections since legally they are not considered to be subordinated workers but independent ones (Moore 2018, p. 1228). Aliada (in Mexico) and Care.com or Handy.com (in the US) serve as intermediaries between domestic service workers and clients.

15.3.1 Aliada: A Mexican On-Demand Platform for Domestic Workers

Aliada was created by two young Mexican entrepreneurs and began operation in 2014, presenting an alternative for creating new and better opportunities within domestic service through technology. The values they boast to their users include efficiency, security, speed and flexibility. The platform works roughly by matching 'cleaning service providers' with 'users'. Those soliciting services enter their address; the specificities of their home (number of rooms, bathrooms); the type of services required (cleaning, laundry, ironing, cooking); the number of hours (minimum three, maximum eight); and the time and date for the job. Afterwards, an employee profile appears to the user, showing their evaluations (ratings), a description of their skills and the guarantee that their personal information, address and background have been verified by the platform. The platform sets an initial hourly charge rate. Some workers can set the price themselves over time although others are excluded from this benefit and the criteria for who is eligible to do so are not clear. Some interviewees commented that this might depend on the number of services performed and the evaluations received.

Aliada is presented as 'a link or intermediary and should never be considered as a company offering cleaning services or a cleaning agency'.[5] In this way, the platform avoids the obligation of offering employment benefits to employees such as paid leave, social security and access to health services. Nevertheless, the platform intervenes in the establishment of service standards, in hiring protocols which encompass psychological and background checks, in knowledge assessments and when a user reports a problem with the employee. There is, therefore, a relationship of subordination that, legally, would determine the obligation of these platforms to guarantee workers' rights. Within this framework, however, employees are still considered to be independent workers and therefore lack any kind of social protection.

The 2019 reforms to the Federal Labour Law in Mexico do not extend to domestic workers using online platforms. In this sense, employees hired using these systems are under similar conditions to digital delivery workers from companies such as Uber or Rappi. Many domestic workers, however, use Aliada to make private arrangements, thus gaining more clients alongside those obtained through their own personal networks. Platform work provides this and flexible schedules, both of which are perceived by domestic workers as positive. Aliada charges a service fee for domestic work services made through the platform; however, no labour benefits are offered and, as service demand varies according to season or the requirements of clients, no income stability is guaranteed. Some workers, therefore, prefer combining different hiring schemes to gain greater stability or put in place a backup.

Comprehensive occupational training is not offered by Aliada although it does require employees to pass an online test demonstrating basic knowledge related to cleaning services. Variations in training were reported by domestic workers. Some described receiving basic training in cleaning or talks about expected performance and behaviour with clients which emphasized desired attributes such as responsibility and honesty. For some domestic workers, entering Aliada was the first time they learned how to use a smartphone and download an app. In many cases, relatives or friends of the workers helped them navigate this system.

15.3.2 Alia and the Fight for de facto Rights for Domestic Cleaners

Alia is a non-profit online platform providing benefits to domestic cleaners. The project was started by NDWA Labs, the 'innovation arm' of the alliance, and is funded by different groups like Google and the Open Society Foundation. Alia's website was officially launched in December 2018 (Binns 2019).

The platform functions by allowing clients to make contributions to the Alia account of a particular cleaner of between \$5 and \$40 per job, which functions as a kind of 'tip' on top of the agreed rate. Workers can access benefits such as paid time off (via a prepaid Visa Incentive card worth \$120 for each day off); disability insurance; accident insurance (comprehensive coverage starting at around \$15 per month); and critical illness insurance. The scheme also offers life insurance (\$5,000 policy at just \$2 per month), although workers who want to acquire this benefit must become members of NDWA and pay \$2 per month. This is because NDWA's life insurance coverage works as a group policy allowing NDWA to gather only workers' general information (name, postal code, gender, date of birth and the name of the worker's beneficiary). They do not require a social security number or have to disclose immigration status. Each of the cleaners sets their own time off and insurance options.

Cleaners join the platform by creating an account with a personal email, phone number and local postal code[6] and find their client(s) by entering their phone number. Each client also has to create an Alia account, select a cleaner and the amount of the contribution to be made each time the worker cleans the house, and enter credit card information.

Alia can operate as non-taxable income for the employee due to the structure of the site. Clients buy credits each time they deposit money in a cleaner's account: \$1 equals one credit. In other words, clients buy credits from Alia and these credits form 'a non-taxable income to cleaners'. To put it simply, employers are giving extra money for the cleaner's wellbeing which they can build up to pay for their benefits. Buying Alia credits may also help employers being taxed by such an operation: as stated on the platform's website: 'Alia does not report nor share any information about contributions outside the Alia program, including reporting to the IRS, because Alia makes no payments directly to cleaners.'[7] Each client should consult with a tax lawyer or accountant, however, if they need to include the value of the benefits they receive through Alia in their tax returns. Credits are deducted when a cleaner signs up for accident insurance or life insurance.[8]

Alia also offers other important features. For example, neither cleaners nor clients pay fees for their Alia accounts. Additionally, Alia's credits do not expire (the account may be closed if it is inactive for more than 18 months) and, if the worker passes away, the executor of the estate is able to access the credits in the account. In the case of deportation, the cleaner only needs to provide a mailing address where the Visa Incentive card with the total number of outstanding credits can be sent.

One negative, however, is that cleaners cannot currently make contributions to their Alia account which would allow them to save more money to use for any of the eventualities described above (sickness days, insurance, etc.); only clients can buy credits in the scheme.[9] Also, for the time being, Alia is only focusing on independent cleaners, but the NDWA is planning to include cleaning businesses that want to access Alia on behalf of their workers and to extend Alia to cover nannies and caregivers (Binns 2019).

Domestic workers in the US are still excluded from basic labour protections despite a number of local laws approved in recent years. While it is important to continue getting legislation passed for domestic workers, creating enforcement mechanisms for these laws remains a challenge. Alia, meanwhile, aims to offer benefits for domestic workers immediately; if successful, it will be a significant achievement. When only 4.9 per cent of domestic cleaners and 6.3 per cent of nannies receive employer-provided health insurance, and less than 3 per cent of nannies and maids are covered by a pension plan (Sherholz 2013), having an option like Alia could be a game-changer for domestic workers as well as for other informal workers.

15.4 PLATFORM REALITIES: PAY AND WORKING CONDITIONS

For-profit platforms have created new values and narratives to understand work, new ways of organizing and managing workers and new types of relationship between clients and service providers. They promise greater freedom, flexibility and autonomy and better working conditions. In spite of the derogatory names, better pay – at least if counted on a per hour basis – can be seen as an improvement but domestic workers' rights are still denied and inequality is not being fought. Instead, the platforms have facilitated new forms of exploitation and new ways of silencing critics and workers' demands.

At the same time, we have also seen the emergence of non-profit platforms like Alia to help domestic workers improve their working conditions despite the lack of legal recognition in the US.

15.4.1 Aliada: Disconnecting Workers' Rights

The self-management narrative incorporated in online platforms like Aliada gives the illusion that the subject is the key to change and that there exists a possibility of choice and control over certain variables such as income, schedules and work location. The reality is that such platforms operate on the basis of a profound power imbalance between users and service providers in terms of the capacity to decide the conditions of exchange. For example, the amount of time that each employee works, which in turn determines their income, is established by the particular requirements of each employer. Based on the demand, the workers choose to accept one, two or more services each day, or to work on Sundays to compensate for days when no work is available. While no service can exceed eight hours, there is no limit to the number of services that may be offered.

Additionally, because the cost of the service is calculated by the hour and not by the day,[10] and the cost as fixed by the platform is higher than the national average generally paid to domestic workers,[11] users rarely contract services for the full eight hours to reduce costs, even if more time is needed for deep cleaning. The work therefore tends to be more intense given the time compression: the cleaning must be done in fewer hours and without rest (and always with an eye on the evaluation). Workers are also incentivized to search for other services or work during the day so that, where they are only contracted for three hours, which that platform establishes as the minimum, they are able to count on additional income to raise their monthly salary. Because of this, the system of online platforms can be more strenuous than other previous hiring models.

Certainly not all cases are the same; for some workers the online platform offers a good alternative due to the flexibility of choosing jobs by the day or the hour because they see it as a complementary job, or it allows them to do other activities. For domestic workers who have children, working under this flexible scheme can balance paid work with caregiving tasks with their families. There are others that choose recurring services for the same clients, which allows a certain regularity that, nevertheless, is not completely guaranteed because users can always cancel services. For those that mostly depend on the platform for their principal source of income, the organization of work around the demand for services entails constant uncertainty with regards to income, the distances they will need to cover throughout the week and the planning of their week. At the same time, cancelling a cleaning job entails a penalty for the workers, if the cancellation is made within 24 hours of the upcoming job, except where the platform decides the cancellation was justified.

Another example of the power imbalance within the platform is the access to information that each party has about the other and with regards to safety measures. The profile of each employee, including their personal information, is shown on the platform so that users can decide who to hire. The client is able to see comments left by other users and have the certainty that the person they hire has passed toxicological and psychometric tests, and even a criminal background check. The suspicion that employees are not trustworthy and could steal from the user is an existing stigma exploited by the platforms in order to market the advantages of their services: the 'safest way to clean your house'. Cleaners, on the other hand, do not know the background of users and are unable to consult their references because there are no public evaluations for them, even in cases of harassment or abuse. The platform controls the totality of the information and thus cleaners are impeded from having the same experience as users: the ability to make decisions based on the experiences of other people.

The framework that cleaners do have for their 'decision' is constructed based on the specific demands of the user. This framework is also constructed and materialized with a system that grants very unequal benefits to each party and with the reinforcement of certain longstanding cultural prejudices and stereotypes in Mexico – that these activities should be carried out by women, who are seen as untrustworthy until proven otherwise. Stigmas about domestic workers being potential thieves continue to be reproduced within this apparently 'inclusive' platform. One of the founders of Aliada said the following in a radio interview:

> At the beginning … we thought that there was going to be a daily call [from clients] saying that their homes were being robbed. [However] after 140 000 services … [Aliada's employees] have passed [security measures] and responded perfectly; to start providing services, all the 'Allies' must pass the same filters as Uber drivers with 'Blacktrust' [the name of a certification exam] which includes psychometric, toxicological and quality tests.[12]

Therefore we are facing a system that, far from bending the arc of domestic service towards a more just society, in fact maintains the same structures of privilege where the costs are, once again, incurred by workers.

With the platforms, new difficulties arise in respect of the organizing of domestic workers. The first is that, given that communications and the administration of the work is primarily done via the app and on an individual's account, employees are prevented by the platform from meeting each other and keeping in contact. Some of our interviewees who have been working for some years and have seen the development of the platform commented that, in the beginning, there were WhatsApp groups administered by the platform to notify employees of possible jobs. These groups began to be utilized by workers as a space to air collective grievances but these were later deactivated by the site. Fear of being deactivated can have a negative impact on collective organizing.

At the same time, organizing potential is further eroded when there is no clarity of who the employer is. How and to whom should demands be made when the platform reiterates that they are just the intermediary and the workers actually work for themselves? As one worker mentioned: 'We are employees that work on our own account and supposedly Aliada works for us. This is how they want us to see it. Aliada works for us because we pay them to look for jobs' (Interview, 2018). At the same time, when the work found on online platforms is perceived as temporary, secondary and complementary, or is constantly changing, the incentives for organizing are few given that employees have little spare time. In interviews, most employees stated that they do not have free time as the little time they do have is spent on domestic labour in their own homes.

Finally, the platform also reflects ideal types of worker. The worker on an online platform is portrayed as a 'professional' and, being so, is distanced from the image that has historically captured them as 'backwards', given their place of origin or ethnic background. For some workers, this has represented a positive change and they have perceived more respectful treatment and greater recognition of their work. Given that many workers have suffered discrimination, this is one of the aspects that they value as the most positive. Nevertheless, for some workers, being a cleaning 'professional' entails presenting a different appearance or origin. This is causing the production of new categories that are creating distance between those cleaners that see themselves as 'professional' and 'modern' and those that do not fit into these categories. Being from Aliada meant for one of the interviewees not being a 'standard' domestic worker:

> There are many of us that are well-dressed, well-groomed, well-made up, well-combed women. You say that's not a domestic worker. [In Aliada] we are more of that type, like another level of domestic worker. Because those classic ones from a little town, there are hardly any … I don't know, in Aliada there is a certain level of prestige with their workers, to put it that way. (Interview, 2018)

However, despite the differences in appearance and social background, all of them are exposed to situations of job insecurity.

15.4.2 Alia: The Potential of a Non-Profit Platform for Domestic Workers

It is simply too soon to analyse the impact Alia has had on domestic cleaners: it was launched quite recently and, during 2020, the Covid-19 pandemic slowed its use. Nonetheless, Alia can potentially develop into a significant app granting rights to domestic workers in the US. The key to this is its own expansion. If NDWA can get Alia to cover nannies, caregivers and live-in domestic workers, the app will have a bigger pool to develop many more accounts which means more workers will get some of the benefits denied by the state. NDWA is also trying to enable workers to add money into their Alia accounts. If this also happens it will allow domestic workers to strengthen this on-demand app.

Alia does have the potential to overcome some of the structural inequalities that domestic workers, cleaners in particular, face in the US and elsewhere. Securing workers' basic benefits while also protecting domestic workers' immigrant status is a huge development, more so when xenophobic discourses come to the fore. In addition, for Alia to work, workers need to convince their employers to enrol in this app (as mentioned before, Alia is not mandatory). These interactions promote employee–employer connections, which is a way to secure workers' benefits outside of the state's parameters. If Alia becomes a successful on-demand app boosted by employers and employees, it will show local and national authorities how 'small', but significant, changes can have an impact on domestic workers' lives and pressure the state apparatus to rise to the occasion. In addition, Alia shows how, unlike Aliada whose primary purpose is to exploit workers to the benefit both of clients and the developers of the app, workers can use technology to improve their working conditions.

Nevertheless, Alia does have important issues. The main one is that workers getting benefits is dependent on employers. While there are employers who value domestic workers' labour, such as Hand in Hand, the network of employers of domestic workers in the US, most do not. Overall, employers want to get the best deal for the lowest price and many do not care to think about what is best for the employee and what is a fair labour exchange, especially because of the social understanding of domestic work as cheap and dirty labour (Federici 2012; Romero 2002). That is why apps like Aliada or Handy have such success. Hence, until the social conceptions of paid domestic work change, or the law changes to enforce domestic workers' rights, these types of app will rely on employers' goodwill and, as history shows, that is not the best situation. Therefore we expect Alia to face difficulties getting employers on board, yet if it can add the features mentioned above (to cover nannies, caregivers and live-in domestic workers and to enable workers to add money into their Alia accounts), it is more likely to grow and get more employers involved.

Despite the challenges, Alia is a promising platform that can offer access to benefits for a workforce continually denied them. In this sense, Alia could be a model to follow for other domestic worker organizations and unions. Moreover, Alia could lead the way for other informal workers to achieve de facto rights and push for federal and local governments to support this initiative both economically, by adding money to workers' accounts, and symbolically, by promoting it through official channels. However, that this non-profit app is up and running does allow domestic workers to consider alternatives to abusive for-profit platforms. That in itself is already a win.

15.5 CONCLUSION

Both in Mexico and in the US, domestic workers have lacked strong social protection from the state. On-demand platforms have arisen in a context in which the rights of this group of workers have not been guaranteed and in which many domestic workers still face discrimination and undervaluation regarding their work. The mass nature of the internet and the development of software facilitating the creation of for-profit labour platforms and non-profit platforms to support workers are giving rise to new social and economic phenomena within the history of domestic service work. Like any revolution, the digital one represents new challenges and opportunities for workers as well as for society as a whole.

The Mexican case shows that location-based domestic work platforms like Aliada can involve new forms of job insecurity and precarity under a tricky narrative that hides the mechanisms which produce an imbalance of power between users and workers. The case of Alia shows that, in the face of platforms that seek to eliminate the role of the employer in order to avoid giving social protection to workers, there are initiatives from domestic workers themselves that, via the use of technology, allow them potentially to access social benefits; given the laws that currently prevent them from accessing basic rights, technology thus allows them a margin of action in which to improve their working conditions.

Online platforms cannot be taken in themselves as antagonists of social justice nor as liberators of traditional forms of power. In no case can we opt for a focus of technological determinism that would put technology as the driver of all changes whether positive or negative. Platforms, like all technological tools, are drawn from the socio-cultural assumptions that shape them and can be oriented in different ways according to social, economic and political context. Platforms have the potential either to reproduce precarity among domestic workers or to help improve their conditions of work.

The guidance of technology and the development of the platform economy to transform working conditions for the better is thus a political and moral decision. We must be attentive to new forms of inequality and the asymmetry of positions regarding the services market that location-based platforms have sharpened. Without doubt we must create effective mechanisms to take advantage of technology in favour of domestic workers, and unions must consider new organizing strategies given the changes to models of work in the platform economy.

Finally, collective organizing is still a major challenge for this industry. Ways to make organizing more effective will have to emerge, among other things from a consideration of the following issues:

1. The integration of workers on digital platforms into the demands of unions and domestic worker organizations.
2. Diversifying the organizing strategies of domestic worker organizations by having greater regard for the use of technology.
3. The strategic use of digital connections in order to establish spaces where worker networks can be created in spite of the fragmentation and dispersion of workplaces.

NOTES

1. These findings are, on the one hand, part of an eight-month ethnography and participant observation exercise carried out in 2017 at the New York office of the NDWA. A total of 39 in-depth interviews were conducted with domestic workers, NDWA staffers, public officials and members of staff of other domestic worker organizations. Regarding Aliada, the findings are the result of eight in-depth interviews with domestic workers that work in Aliada as well as digital ethnography carried out in 2018 and 2019.
2. Home care assistants were recently incorporated into Fair Labor Standards Act minimum wage standards. Nevertheless, nannies, domestic cleaners and live-in domestic workers continue to be excluded from this legislation.
3. www.care.com/company-overview.
4. www.handy.com/about.
5. Aliada's terms and conditions can be found here: https://Aliada.mx/terminos.
6. A worker deciding to take paid time off will need to provide a mailing address in order to receive a non-reloadable Visa Incentive card: https://cleaner.myalia.org/terms-and-conditions.
7. Alia's website does not give a clear answer as to why this form of Alia's credits is taxable and why others are not. 'We simply cannot advise you on your taxes' is stated on the website. They do share a link for more information as to the explanations of the Internal Revenue Service on the matter. The main issue is whether or not workers categorize themselves as childcare providers. For more, see www.irs.gov/businesses/small-businesses-self-employed/what-is-taxable-and-nontaxable-income.
8. www.myalia.org/how-it-works.

9. It is not clear why domestic cleaners are not entitled to buy credits. We suppose that it may be due to the undocumented status of most cleaners which complicates their chances of opening bank accounts.
10. Live-in and live-out traditional hiring schemes in Mexico paid workers on a per day basis – another way in which Aliada has changed the market.
11. In 2015, 70 per cent of domestic workers gained a maximum of two minimum wages per day (Cebollada 2016, p. 15), amounting approximately to $7; with Aliada they can receive $9 for services of only three hours.
12. This interview was done on Martha Debayle's radio programme on 24 January 2017.

REFERENCES

Bensusán, G. (2019), *Perfil del trabajo doméstico remunerado en México*, México: ILO.

Berins Collier, R., V. B. Dubal and C. Carter (2017), 'Labor platforms and gig work: the failure to regulate', *IRLE Working Paper 106-17*, accessed 4 December 2020 at http://irle.berkeley.edu/files/2017/Labor-Platforms-and-Gig-Work.pdf.

Binns, C. (2019), 'Housecleaning with benefits', *Stanford Innovation Review*, accessed 28 November 2020 at https://ssir.org/articles/entry/housecleaning_with_benefits#.

Burnham, L. and N. Theodore (2012), *Home economics: the invisible and unregulated world of domestic work*, New York: National Domestic Workers Alliance.

Carmona, Y. (2019), *Panorama de la situación de trabajadores y trabajadoras en empleo informal en la Ciudad de México: Recomendaciones para una ciudad más inclusiva*, Manchester: Women in Informal Employment Globalizing and Organizing.

Cebollada, M. (2016), *Las personas trabajadoras del hogar remuneradas en México: perfil sociodemográfico y laboral*, México, DF: Consejo Nacional para Prevenir la Discriminación.

CONAPRED (Consejo Nacional para la Prevenir la Discriminación) (2015), 'Condiciones laborales de las trabajadoras domésticas. Estudio cuantitativo con trabajadoras domésticas y empleadoras', accessed on 4 December 2020 at www.conapred.org.mx/userfiles/files/TH_completo_FINAL_INACCSS.pdf.

Cruz, S. A. and M. Abrantes (2014), 'Service interaction and dignity in cleaning work: how important is the organizational context?', *Employee Relations*, **36** (3), 294–311.

Devetter, F. X. (2013), '¿Por qué externalizar las tareas domésticas? Análisis de las lógicas desigualitarias que estructuran la demanda en Francia', *Revista de Estudios Sociales*, **45**, 80–95.

Federici, S. (2012), *Revolution at point zero: housework, reproduction, and feminist struggle*, Oakland, CA: PM Press.

Hondagneu-Sotelo, P. (2007), *Domestica: immigrant workers cleaning and caring in the shadows of affluence*, Berkeley, CA: University of California Press.

Moore, L. (2018), 'Transformative labor organizing in precarious times', *Critical Sociology*, **44** (7–8), 1225–1234.

Nadasen, P. (2015), *Household workers unite: the untold story of African American women who built a movement*, Boston, MA: Beacon Press.

Oelz, M. and U. Rani (2015), 'Domestic work, wages, and gender equality: lessons from developing countries', Working Paper 5/2015, Geneva: ILO.

Piñeyro Nelson, C. (2018), 'Organización, emociones y resistencia de las trabajadoras del hogar latinas y caribeñas en le Ciudad de Nueva York, Estados Unidos', *Revista Latinoamericana de Antropología del Trabajo*, **2** (3), 2–24.

Romero, M. (2002), *Maid in the USA*, 10th edition, New York: Routledge.

Sherholz, H. (2013), 'Low wages and scant benefits leave many in-home workers unable to make ends meet', *Economic Policy Institute*, 26 November, accessed 28 November 2020 at www.epi.org/publication/in-home-workers/.

Theodore, N., B. Gutelius and L. Burnham (2019), 'Workplace health and safety hazards faced by informally employed domestic workers in the United States', *Workplace Health and Safety*, **67** (1), 9–17.

Xantomila, J. and J. A. Román (2017), 'Las trabajadoras del hogar, 5% de la población ocupada del país', *La Jornada*, 30 March, accessed 4 December 2020 at www.jornada.unam.mx/2017/03/30/sociedad/039n1soc.

16. The role of worker collectives among app-based food delivery couriers in France, Germany and Norway: All the same or different?

Kristin Jesnes, Denis Neumann, Vera Trappmann and Pauline de Becdelièvre

16.1 INTRODUCTION

Exploitative labour practices in the platform economy are widely documented (Chesta et al. 2019; Kirk 2020; Dinegro Martínez 2019; Polkowska 2019). Platform workers have nevertheless challenged the poor working conditions and low pay offered by many location-based platform companies and, in the platform food delivery sector, couriers in several countries have started to mobilize and protest (Tassinari and Maccarrone 2020; Cant 2019). Such couriers are mostly self-employed or have insecure employment contracts and they do not share a physical workplace with fellow couriers. These traits, together with the characteristics of this workforce – many are young or migrants, or both (Vandaele et al. 2019) – have made it hard for trade unions to organize in this industry.

The couriers often look to organize themselves, however. They have done this in innovative ways by relying on online communities, such as Slack or WhatsApp, to build solidarity and mobilize their fellow couriers. These online communities also compensate, to some degree, for the lack of a shared workplace (Tassinari and Maccarrone 2020; Woodcock and Graham 2019). Protests are thus often bottom-up initiatives, led by self-organized worker collectives and/or grassroots unions (see Vandaele 2018, p. 20),[1] while there are also cases of cooperation between worker collectives and unions.

The existing literature lacks an exploration of this interplay between new and 'old' collective actors in the struggle to ensure that the interests of workers in the platform economy are effectively represented. This chapter fills this gap by examining and contributing to a better understanding of the couriers' road to mobilization, as well as of the role of new worker collectives and how they engage with unions in France, Germany and Norway. Based on Kelly's theory of how individuals become collective actors (Kelly 1998), the argument is made that there are many similarities in the processes of mobilization in France, Germany and Norway, with a prominent role for worker collectives. We look in particular at instances of where these collectives have sought to work with unions and access union-organizing expertise. The alliances that emerge differ between the country cases due to a diversity in action repertoire and context. This indicates a path-dependent development in line with the institutional context of industrial relations systems, although the German case demonstrates that there is some leeway for the actors.

16.2 LOCATION-BASED PLATFORM FOOD DELIVERY IN FRANCE, GERMANY AND NORWAY

The global platform food delivery market was valued at $96 billion in 2019 (Imarc Group 2019), and the Covid-19 pandemic has made food delivery via location-based platforms even more widespread. As is the case with other companies organized similarly, the logic of 'winner takes all' is important in this market (Valenduc and Vendramin 2016). In some cities, several companies are still competing to establish a dominant market position – and cost, at the expense of working conditions and pay, is a typical arena of competition – while in other cities one company has achieved monopoly (Dealroom 2017).

At the time of our case study, there were three main food delivery companies in France, namely: Uber Eats (since 2015); Foodora (from 2015 to 2018); and Deliveroo (since 2015). Foodora ceased its operations in France in 2018 due to competitive pressure as well as legal cases. Both Uber Eats and Deliveroo use self-employed contractors as couriers and pay on a per delivery basis, in contrast to Foodora which used to employ its couriers.

In Germany, Foodora and Deliveroo were the biggest players in the platform food delivery market before Takeaway.com bought Delivery Hero's German businesses, consisting of Pizza.de, Lieferheld and Foodora, in April 2019 and formed Lieferando. Deliveroo left the market in August 2019. Deliveroo had offered two employment options up to the beginning of 2018 – self-employed contractors and marginal part-time employment[2] – although from the beginning of 2018 couriers were only engaged as self-employed contractors. Foodora employed couriers on marginal part-time employment contracts while Lieferando operates today on the basis of offering couriers contracts of employment.

In Norway, Foodora was the main operating company at the time of the study despite the presence of a few local market exceptions. It started operating in Norway in 2015. The company hires its couriers on marginal part-time employment contracts (of 10 hours per week) which guarantees them certain collective rights, including the right to negotiate collective agreements and to strike. The couriers obtained a collective agreement in September 2019. After this study was completed, the Finnish company Wolt emerged as a competitor to Foodora's dominant position. Wolt engages contractors; and Foodora has started to follow suit after Wolt's expansion.

16.3 METHODS AND DATA

Our research is primarily based upon interviews with 64 interviewees in the three countries.[3] Interviewees were identified from among the key actors involved in protest actions and were either unionized couriers or couriers who supported such protests. To establish contact, a multi-tiered recruitment approach was used: union representatives facilitated field access and we contacted couriers through official channels or addressed them at events or outside hotspot restaurants. The subsequent strategy was based on the snowball principle. We also interviewed union officials. In total, 15 couriers working for Deliveroo and 10 union officials were interviewed for the French case study during the period 2017–2019; interviews were conducted with a total of 17 couriers and three union officials across both Foodora and Deliveroo in Germany during 2018–2019; and 16 couriers and three union officials were interviewed concerning Foodora's operations in Norway during 2018–2019. Participants were interviewed both individually and in groups using a semi-structured interview schedule.[4]

Based on the interviews, a comparative analysis was conducted encompassing inductive and deductive approaches. Inductively, the mobilization process and its inherent dynamics for each country are reconstructed via developing the core themes and comparing these to evaluate if, and how far, they had also occurred in the other countries. Deductively, categories were created according to Kelly's mobilization theory and union strategies, while all material was then recoded. Both approaches reveal commonalities and differences across the countries.

16.4 FOUR SIMILARITIES IN THE MOBILIZATION PROCESS

Kelly's framework on mobilization (1998, 2018) includes five elements. First, workers must identify issues of injustice. Second, blame for these injustices must be placed externally, for example at the door of the management or the company. Third, there is a process of social identification whereby these injustices come to be perceived as collective issues. Fourth, some workers must claim leadership of the mobilization process. Fifth, and finally, the group must feel efficacy; that is, that they are able to achieve something together. This section focuses on the first four elements of Kelly's framework as they highlight several similarities between the case studies.

16.4.1 Issues of Injustice: Dissatisfaction with Low Pay and Working Conditions

The first key element in mobilization is that individuals are able to identify shared issues of injustice. Our data reveal that, from the onset, couriers were dissatisfied with the low pay and poor working conditions, reflecting the latent instability of the cash nexus in a capital–labour relationship between the platform and the worker (Joyce 2020). This was regardless of the existence of different contractual arrangements and pay systems.

In France, couriers were paid on a piece basis for each delivery. This led to variations in payment per delivery and, hence, one of the main demands of the couriers was an hourly wage. One courier in France stated: 'We want to be paid for each hour we worked. Being paid just per ride is not profitable.'

In Germany, couriers expressed discontent at huge wage arrears and the generally low wage which was often below the statutory minimum regardless of the contractual arrangement. Some couriers had a minimum of 10 hours per month in their contracts, but workers are, according to German labour law, eligible to demand a minimum number of hours per week. In general, however, it was not possible to realize such a demand.

Couriers in Norway had a minimum of 10 hours per week in their contracts, with the possibility of working more hours. They also had contracts of employment and were protected by the Norwegian Working Environment Act which covers most work-related issues and allows couriers to negotiate collectively. Nevertheless, the hourly wage of the couriers (about €11.50) and the additional bonus per delivery (about €1.50 during weekdays and €2 at weekends) had not increased between 2015 and 2019, and so they were also dissatisfied with the levels of pay.

On top of these wage-based claims, couriers in all three countries, regardless of their contractual status, had to pay for their own work equipment such as the bike and bicycle repair as well as the costs of maintaining a smartphone or phone subscription. This effectively rendered the payment even lower, as illustrated in the following quote from a Norwegian courier:

> Compensation for the use of own equipment is quite central because, as it is now, they [the company] in a way transfer the costs of running a business over to the employees ... We use our own bikes, our own phones, our own clothes, all of our own equipment, and we fix the bikes ourselves ... So, if you had a realistic calcu-

lation of the costs of doing the job, which the individual employee pays, it would reduce the real salary quite sharply.

While low pay was thus the main issue that united couriers in all three countries, poor working conditions was another. In France and Germany, couriers related their experiences of getting disconnected from the app without being given any reason. Other work-related issues of concern included a lack of training and career opportunities; a lack of spaces in which to meet other couriers; and distances that were too lengthy, which led to both a loss of pay and physical tiredness. In France, couriers are considered as self-employed contractors and are therefore not protected by the labour law, nor do they have any social protection. Accidents are common among couriers, some of whom therefore sought to obtain social protection by organizing, although others did not want to do so as they considered themselves independent.

16.4.2 Blaming the Food Delivery Platform

According to mobilization theory, injustice must not only be perceived but needs to be attributed. In the case of the couriers, blame was indeed laid at the door of the company and its management, as expressed by this Norwegian courier:

> I do not think that the leadership in Foodora will do anything [to improve our working conditions] if they do not have to. That is my impression. They are not looking to optimize our everyday life in any way, nor our safety. If so, they would have made some other choices.

In addition to low pay and poor working conditions, the way the companies had set up the labour process was in itself a 'trigger of mobilization' (Tassinari and Maccarrone 2020, p. 49; see also Vandaele et al. 2019 and Veen et al. 2019). Although the companies did provide a basic number of meeting places for couriers, the couriers found or created their own online digital communities, such as through WhatsApp, where they discussed grievances and blamed the company for them. Regarding this, a German courier for Foodora expressed the shift from individual problem perception to collective reflection during a social media group discussion:

> In this WhatsApp group, we talked openly about what really bothered us. That is when I realised for the first time that I was not alone with my problems. Instead, others got even more upset about them. So, there was a first collective awareness of the problems: that this was the general consensus and that it wasn't really okay to deal with employees the way Foodora did. After the problems started to accumulate and sometimes serious things happened, such as salaries not being paid on time or

> wrongly paid regarding holidays or sick days, these were things that threatened the existence of the company. Then [we] decided to meet and talk about it personally.

This reflects earlier findings that differential subversive spaces may emerge, both virtual (via social media networks) and physical (in respect of meeting points for couriers in the city), where the control mechanisms imposed by the app's algorithms cannot discipline workers (Briziarelli 2018). Indeed, frequent changes in the algorithm concerning deliveries, working hours and payment gave the couriers a feeling of arbitrariness which was a breeding ground for resentment. The 'worker-subject' as a 'quantified self' (Ivanova et al. 2018) is required to perform and compete, but the moment performance is no longer reflected in payment, the company risks workers no longer following this imperative.

> The drivers were divided into two parts, the good ones, so to speak, and the bad ones. The good ones, i.e. the 50 per cent better ones [those riders who deliver the fastest], were allowed to take the shifts first; and what was left, the rest were allowed to take. And that creates such fierce competition between people and it also causes a lot of frustration … With Foodora in particular, it's totally crazy. People have to act accordingly and if that doesn't work, you get a warning and after three warnings you're out.

To be more precise: management of the workforce by an algorithm highlights that a small change made on a computer in an office somewhere can have far-reaching and negative consequences for workers within a locality. Consequently, it is doomed to create constant antagonism. Once a worker has identified it, 'algorithmic despotism' can fuel a desire for workers to have a greater say in the workplace (Griesbach et al. 2019). Arriving at such a stage enhances the action of laying the blame for poor working conditions and low pay on the company, as this quote by a German courier illustrates: 'The displeasure was significant and multiplying at that time because any innovation introduced by Foodora has always been to the detriment of the couriers, always to complicate our job situation … On the whole, more problems came up than were solved.'

Therefore, the insecurity that the couriers felt did not only stem from their contractual situation. It also consisted of a persistent uncertainty about what change might come next in how they were managed, even from one shift to the next. The platform company itself was held to blame for such changes that were perceived as entirely arbitrary and certainly remote from the control and input of the worker.

16.4.3 Social Identification Processes and the Establishment of Worker Collectives

The couriers established a 'we', a collective identity, quite soon after platform companies entered the market. This was the case despite differences between the couriers themselves: they were students or migrants, and sometimes both; and they had different levels of dependency upon the job as a courier. Furthermore, it is impossible to ignore the spatial challenges of couriers in meeting up and getting to discuss work-related issues. The wide use of the pronoun 'we' in interviews is a strong indicator of the couriers' social identification as a group, as expressed in the following quote from a Norwegian courier: 'Why did I join the union? *We* want to have better salary and working conditions. *We* have to be together and be like a team to succeed.'

In all three cases, the 'we' developed quite quickly from conversations in social media forums into the development of worker collectives: CLAP (*Collectif des livreurs autonomes de Paris*; the Independent Couriers Collective of Paris) in France; *Liefern am Limit* (Delivering to the Limit) in Germany; and *Riders Club Norway* and then the local union in Norway. In all three cases, a similar process of mobilization could be observed: the couriers communicated through social media platforms; 'labelled' themselves as a group; and established a collective, creating a Facebook page and designing a logo for it to encourage feelings of identity and recognition. This formalizing of belonging and identity had a consolidatory effect on the collective itself and led, in all three cases, to strong identification as a worker collective.

In France, a Facebook group was the starting point for CLAP in Paris in early 2017. The group rapidly gained additional members[5] and, in March 2017, during a demonstration, the formation of CLAP was announced by the leaders of the initiative. CLAP eventually organized several demonstrations and strikes.

In Germany, a feeling of solidarity arose through discussions between Deliveroo riders in a Cologne-based WhatsApp group named *Arsch Huh* Riders[6] which established a strong mutual support network among the couriers, extending to self-organized repair shops. Core members of the WhatsApp group then initiated the media campaign *Liefern am Limit*, which strengthened the couriers' feeling of belonging to a group. In the following period *Liefern am Limit* developed into a mouthpiece for courier interests not only in Cologne but across Germany. For a short period of time, in Berlin, Foodora couriers created an online media campaign called 'Deliverunion' where, with the help of an independent grassroots trade union, *Freie Arbeiterinnen- und Arbeiter-Union* (FAU, Free Workers' Union), they also tried to organize couriers. However, after a certain period of media attention, the initiative dried up without having achieved significant results for the workers (Trappmann et al. 2020).

In Norway, some of the most enthusiastic couriers first established a club on Facebook in 2017 – Riders Club Norway – where they organized social activities, biked together and started talking about work-related issues. These discussions eventually developed into a feeling of having to do something about their poor working conditions. The couriers benefited from contracts of employment and thus could organize and negotiate collectively. The question was how: by self-organizing or through a union, as the following quote illustrates:

> The group had a basis in Riders Club Norway where we talked specifically on what we should do about our working conditions and whether we should organize ourselves. There was a debate about whether we should do it internally in the union or whether we should create our own organization … There was some disagreement, but then we concluded that if we want to achieve anything, we must do it within the trade union system.

The process of social identification was then relocated to the local *Norsk Transportarbeiderforbund* (NTF, Transport Workers Union) branch in Oslo and in other cities where new members were recruited.

16.4.4 Leadership: Key People in the Worker Collectives

In all three cases, there were key individuals in the respective collectives who were central to the process of organizing couriers. All three organizations – CLAP, *Liefern am Limit* and Riders Club Norway in combination with the local union – had media-friendly spokespersons who enjoyed a high level of recognition among other couriers. These key individuals worked more hours than the average courier and were enthusiastic cyclists, often promoting riding for environmental reasons. The presence of these key individuals was important to the mobilization process in all three countries.

Thus, in France, it was no more than two key individuals who started CLAP, with the aim of raising demands on behalf of all the couriers; and who quickly engaged in social media groups to make themselves better known.

In Germany, mobilization depended on couriers being active and achieving strong publicity via concerted media campaigns. In some cases, those involved even gained public honours in recognition of their commitment to workers' rights.

In the Norwegian case, a couple of couriers from Riders Club Norway enthusiastically convinced others inside and outside the collective to join NTF from the moment they decided upon this strategy and throughout the events that followed. Some eventually became union representatives and they have been highly active both in terms of strike activities and in the media.

16.5 VARIATION IN THE RELATIONSHIPS BETWEEN WORKER COLLECTIVES AND UNIONS

In all three cases, the worker collectives worked together at some stage with unions. This was important as regards the processes of mobilization and indeed its outcomes in practice. However, the degree of formalization of the collaboration varied between the countries due to the wider institutional context.

In France, the couriers did not want to involve the unions other than in receiving different kinds of support. Some of the couriers were sceptical of unions who, in general, suffer from a poor public image in France. The fieldwork was also conducted at a time when the 'yellow vests' protest movement (*les gilets jaunes*) demonstrated in the streets, and some of the couriers were also engaged in those protests. Hence, the unions did not get the chance to play an active role in the couriers' mobilization process.

This arm's length strategy went both ways: CLAP was, at first, formally excluded from the *Confédération Générale du Travail* (CGT, General Confederation of Labour), which did not necessarily want to spend resources on seemingly 'unorganizable' food delivery workers. CGT mainly represents workers and, as the couriers were considered self-employed contractors, it was complicated for the union confederation to organize them.

Even so, CLAP did receive assistance from individual union representatives who supported their case by helping them to articulate their demands, structure themselves and identify the right means of action. All this was important to the success of the mobilization process. This help from CGT was not official at the beginning and the internal processes of information and validation about including the couriers within the union took time. Eventually, CGT assistance became official and included both financial support (travel and equipment, for example a computer) and material aid (room loans, occasional help with communications and drafting leaflets), although CLAP remained independent throughout. CGT also supported the couriers by lobbying the government and supporting CLAP's demonstrations and events.

The mobilization process therefore reflected the weak industrial relations system in France. In the beginning, union confederations did not want to represent these workers who were not salaried, but then CGT (as well as other union confederations or unions) tried more proactively to get involved in mobilizing activity among the couriers – including also outside Paris.

In Germany, *Liefern am Limit* approached *Nahrung Genuss Gaststätten* (NGG, Food, Beverage and Catering Union), which has an office in the centre of Cologne, asking for help with organizing couriers and providing institutional backup, a potential strike fund, legal advice and more practical support in terms of training, funding for banners and travel. It was thus a smaller union that was happy to open its doors to the collective and set the example for progressive new forms of collaboration. NGG already had some experience with organizing workers who were young, self-employed and working in businesses with a low threshold to entry in terms of qualifications and training. The union also had a track record of being active in companies like McDonalds and it had been instrumental in establishing a works council at Foodora the year before, in 2017. In contrast, bigger unions, such as ver.di, were reluctant to organize the couriers.

After six months of unofficial support, NGG offered *Liefern am Limit* to become an official union-organizing project. *Liefern am Limit* wanted a strong partner because, as one courier commented: 'Tipping manure in front of a head office is one thing; but negotiations happen on paper.' Thus, the campaign shifted from being one that was supported to one that was, after a period, institutionalized within the union. To maintain a close link between *Liefern am Limit* and the union, one of the leaders of the collective was hired as a union official who, together with other core members, began to organize the workforce within the union. This had many positive effects since the union gained inside knowledge on top of the relationship of trust it enjoyed with the courier representative.[7] Ex-couriers and now union members experienced at organizing couriers were also sent out from Cologne to spread their knowledge to other locations in Germany, with *Liefern am Limit* and Cologne becoming known as the stronghold of the couriers' fightback in Germany.

Once collaboration was established, however, the form of organization followed the traditional path of German industrial relations, concentrating on securing the representation of workers' interests via works councils. With the help of NGG, *Liefern am Limit* pushed towards establishing works councils at Foodora in five other German cities. The union helped with the process of setting up works councils, providing training and support in the process of negotiating their establishment.

In Norway, NTF first got in touch with the couriers through its local branch in Oslo which ordered a mountain of food through Foodora, making sure that several couriers had to show up at the union's office. The union then talked to the couriers about what unions do and handed out recruitment leaflets, and this convinced some of the couriers to become members. After that, there was a quiet period with no new members until Riders Club Norway decided to approach the union.

The couriers decided to join NTF after having several discussions in Riders Club Norway on working conditions and pay. When meeting the couriers, the union lined up many of its most senior officers: an important sign to the couriers that the union was serious about organizing and supporting them.[8] After this meeting in 2017, several more couriers signed up with the union and formed a local branch. After organizing around 100 of the 600 couriers working in Oslo, the branch made a demand in February 2019 for a collective agreement with Foodora, stepping up its campaign in May with a strike notice. The couriers eventually went on strike in August 2019 after negotiations at the national mediator's office.[9] By that time NTF had merged with *Fellesforbundet* (United Federation of Trade Unions), which supported the strikers with strike benefits and various campaign materials in addition to the backing of senior officials. After a five-week strike, during which the union more than doubled its membership, the parties agreed on a collective agreement. Similar to Germany, the main leaders during the negotiations and the strike were then hired by the union to recruit new members in other platform companies.

In Norway, in line with the traditions of collective bargaining, there had been little evidence of protest activities among couriers prior to their five-week strike. Behind the scenes, however, the couriers, within a union framework, had been working on a two-year process of developing a collective agreement, organizing the couriers and negotiating with the company. This process could be said to be quite typical of the road to a collective agreement in Norway, reflecting its industrial relations regime, but it was ultimately successful in delivering the benefits of a collective agreement to the couriers.

16.6 VARIATION IN EFFICACY

Small successes pave the way for further mobilization, according to Kelly's theory on mobilization. This was the case in all three countries, but the level of efficacy perceived by groups of couriers varied in line with the institutional context.

In France, there were two notable success stories. First, in 2016, Deliveroo changed couriers' contracts from a per hour basis to a per delivery one. In the beginning, this new piece-based contract was for all couriers. After demonstrations took place, the company decided that the per delivery contract would be applied only to new couriers. This gave the couriers involved in the demonstrations that feeling of collective efficacy. Second, there was the question of the introduction of supplementary health insurance by Deliveroo in August 2017. Pressure from CLAP-organized couriers, who demonstrated in support of the benefit, was instrumental in getting the platform to introduce this form of insurance. According to Deliveroo, delivery workers benefiting from the scheme receive up to €30 per day for a maximum of 15 days of absence on top of social security pay. This was introduced across the different countries in which Deliveroo operated.[10]

In Germany, *Liefern am Limit* had a significant media profile which motivated many to continue and resulted in new active couriers being recruited. A series of demonstrations followed, accompanied by a sense of efficacy that demands could be met after putting the companies under substantial pressure. With the union's support, *Liefern am Limit* couriers were able to establish a works council at Deliveroo. However, Deliveroo then switched to using self-employment contracts, instead of hiring marginal part-time workers, which led to the dissolution of the works council as the self-employed do not have the right to organize one. Nevertheless, the initial success led to a domino effect, encouraging further action and resulting in the establishment of works councils within Lieferando in Cologne, Hamburg, Münster, Nürnberg, Frankfurt and Stuttgart.

Small success stories were also important in organizing couriers in Norway. First, the couriers, with the help of the union, gained more hours per week in their contracts. The couriers achieved this by the use of §14-4-a (1) of the Norwegian Working Environment Act which states that workers can demand extra hours in their contract if they work more than the set number of hours over a period of time. With support from the union, some of the couriers used this regulation to obtain extra hours and some even gained full-time contracts. This was important in organizing terms as it showed that collective efforts matter when it comes to individuals' contracts:

> Interviewee: More people joined as we got some victories. For example, when we demanded a position corresponding to actual working hours.
> Interviewer: Did that get you more members?
> Interviewee: Yes, because then we had a concrete result to refer to. We could say 'look here: within this system we can use the law to fight and improve your rights.'

Media attention to the strike in Foodora and support both from politicians (it took place in the middle of an election campaign) and the public are other elements that strengthened couriers' feelings of success. During the strike, about 150 new couriers decided to organize in the union, partly due to the visibility of the strike (the couriers set up a base in a central square in Oslo and ran an active social media campaign with the hashtag *#rosastreiken* ('pink strike')), and partly through the couriers' active attempts to organize new couriers. All this eventually led to the couriers being successful in establishing a collective agreement. This agreement included a wage increase, compensation for equipment (for example bikes and clothing), extra pay in the winter and an early retirement pension.[11]

16.7 CONCLUSION

In this chapter we have examined the mobilization of couriers and their interplay with unions in three different institutional contexts – France, Germany and Norway – contributing to a better understanding of the issue. The chapter shows that couriers can indeed be mobilized: couriers in all three countries, initially self-organized, later engaged with unions and with no little success.

With the aid of Kelly's theory on how individuals become collective actors, we found many similarities in the process of mobilization. Couriers shared the same grievances; they placed the blame for these on companies and on management; they quickly formed worker collectives; and they took charge of the mobilization process. Another similarity is how the collectives mobilized others in a way that created a shared social identity, critical to workers coming together in support of aims. The worker collectives led creative campaigns in social media and on the streets with support from, or in collaboration with, unions. This was done in a way that complemented union activities.

Therefore, our cases are good examples of the new forms of cooperation emerging in the field of industrial relations. This can potentially have positive effects on working conditions. As described by Hyman and Gumbrell-McCormick (2017, p. 13): 'synergies between the organisational capacity of the "old" and the imaginative spontaneity of the "new", drawing on the strengths of each, is an important means to build effective resistance to the re-commodification of labour.'

The cases diverge in terms of efficacy, however. This is where institutional context plays an important role. Although the couriers had similar grievances and blamed the companies for them, the possibility of these being addressed varied due to the different employment relationships among the couriers and the disparities in employment relations systems. As a result, the action repertoire followed the 'typical' pattern of industrial relations in each of the countries concerned. By adding this institutional context to Kelly's mobilization theory, we are able to provide a better understanding of differences in engagement with unions and in the outcomes of mobilization processes.

In France, due to couriers' status as self-employed contractors, they were not in a position to communicate with the platform company at all, as one courier told us: 'We cannot negotiate with the platforms … For them we are not representative.' Consequently, couriers mobilized in the streets and also addressed their claims for better protection to the government. In Germany, the couriers from *Liefern am Limit* mobilized independently at first but quickly joined a union and followed the traditional German path of establishing works councils. The hope of more independent action with FAU did not turn out successfully. In Norway, the couriers joined a union and followed the traditional pathway there of dialogue and collective agreement.

These differences in action repertoire and strategies reflect the industrial relations systems in place at national level. This can also be read as a pertinent comment on the debate as to whether national- or sectoral-level industrial relations define new industries: the national level seems indeed also to be relevant to new industries.

NOTES

1. Examples of these new worker collectives include the Asociacion de Personal de Plataforms in Argentina; KoeriersCollectief/Collectif des coursier-e-s in Belgium; Foodstersunited in Canada; Riders Union and its offshoot Deliverance Milano in Italy; #niunrepartidormenos in Mexico; Riders Union in the Netherlands (although later on also a part of the main union confederation); and Riders x Derechos in Spain.
2. Deliveroo directly announced, one day after the election of the works council election committee in December 2017, that the employee model would be replaced in the future by recruiting couriers on self-employment contracts.
3. Empirical fieldwork was carried out independently and followed a field logic in each case with slightly different interview schedules in each respective language. The empirical data were sufficiently coherent to sustain comparative analysis.
4. The semi-structured interview schedules used in the three case studies included open questions about the perceived origin and course of the protests, about the (economic) significance of the job in the life of the couriers and about their motivations to engage collectively; as well as questions regarding their contractual situation, their everyday working life and their attitudes towards unions. This was complemented by sociodemographic and, in some cases, biographical background information.
5. A Facebook group with about 4,500 members in August 2020.
6. A Cologne saying that can be loosely translated as 'Backsides up, Riders'.
7. As a result of the merger, the union integrated the most capable leader who had single-handedly designed the social media campaign. In particular, the way of campaigning and brand-building (he was also a graphic designer) was indispensable and led to him being retained as project secretary for a longer period. He also acted as a hub for communication with ordinary workers. These were two crucial skills that the union could not do without if they wanted to remain involved in the courier struggle.
8. NTF, later Fellesforbundet, is a member of the Norwegian Confederation of Trade Unions (Landsorganisasjonen i Norge – LO). In 2018, LO launched LO-self-employed, a strategy towards the self-employed, indicating that the confederation is more open to including self-employed and workers with non-standard employment relationships as members.
9. According to Ilsøe and Jesnes (2020), the following strike notices were sent out by Fellesforbundet: 86 couriers on 28 June, 16 couriers on 28 June, 40 couriers on 6 September, 53 couriers on 20 September, 25 couriers on 2 October (in Trondheim) and 31 couriers on 2 October. The strike ended on 27 September and the strike notices after that date were therefore not taken into account.
10. www.20minutes.fr/societe/2629099-20191016-deliveroo-met-place-couverture-maladie-livreurs-france.
11. *Tariffavtale for distribusjon og budtjenester i FOODORA 2019–2020, Foodora/Fellesforbundet* (Collective agreement for distribution and courier services in Foodora, 2019–2020). The agreement is a company agreement and covers couriers on employee contracts in all cities in Norway. It was renegotiated in November 2020.

REFERENCES

Briziarelli, M. (2018), 'Spatial politics in the digital realm: the logistics/precarity dialectics and Deliveroo's tertiary space struggles', *Cultural Studies*, **33** (5), 823–840.

Cant, C. (2019), *Riding for Deliveroo: resistance in the new economy*, Cambridge: Polity Press.

Chesta, R., L. Zamponi and C. Caciagli (2019), 'Labour activism and social movement unionism in the gig economy: food delivery workers' struggles in Italy', *Partezipacione e Conflitto*, **12** (3), 819–844.

Dealroom (2017), 'Food delivery tech: battle for the European consumer', *Dealroom.co* blog, March, accessed 12 January 2021 at https://blog.dealroom.co/wp-content/uploads/2017/10/Food-Tech-Prez-FINAL.pdf.

Dinegro Martínez, A. (2019), 'App capitalism', *NACLA Report on the Americas*, **51** (3), 236–241.

Griesbach, K., A. Reich, L. Elliott-Negri and R. Milkman (2019), 'Algorithmic control in platform food delivery work', *Socius*, **5**, 1–15.

Hyman, R. and R. Gumbrell-McCormick (2017), 'Resisting labour market insecurity: old and new actors, rivals or allies?', *Journal of Industrial Relations*, **59** (4), 538–561.

Ilsøe, A. and K. Jesnes (2020), 'Collective agreements for platforms and workers: two cases from the Nordic countries', in K. Jesnes and S. M. N. Oppegaard (eds), *Platform work in the Nordic models: issues, cases and responses*, Copenhagen: Nordic Council of Ministers, pp. 44–55.

Imarc Group (2019), 'Online food delivery market: global industry trends, share, size, growth, Opportunity and forecast 2020–2025', accessed 12 January 2021 at www.imarcgroup.com/online-food-delivery-market.

Ivanova, M., J. Bronowicka, E. Kocher and A. Degner (2018), 'Foodora and Deliveroo: the app as a boss? Control and autonomy in application-based management – the case of food delivery riders', Working Paper No. 107, December, Düsseldorf: Hans-Böckler-Stiftung, accessed 12 January 2021 at www.boeckler.de/pdf/p_fofoe_WP_107_2018.pdf.

Joyce, S. (2020), 'Rediscovering the cash nexus, again: subsumption and the labour-capital relation in platform work', *Capital and Class*, **44** (4), 541–552.

Kelly, J. (1998), *Rethinking industrial relations: mobilisation, collectivism and long waves*, London: Routledge.

Kelly, J. (2018), 'Rethinking industrial relations revisited', *Economic and Industrial Democracy*, **39** (4), 701–709.

Kirk, E. (2020), 'Contesting "bogus self-employment" via legal mobilisation: the case of foster care workers', *Capital and Class*, **44** (4), 531–539.

Polkowska, D. (2019), 'Does the app contribute to the precarization of work? The case of Uber drivers in Poland', *Partezipacione e Conflitto*, **12** (3), 717–741.

Tassinari, A. and V. Maccarrone (2020), 'Riders on the storm: workplace solidarity among gig economy couriers in Italy and the UK', *Work, Employment and Society*, **34** (1), 35–54.

Trappmann, V., D. Neumann and M. Stuart (2020), 'New forms of labour unrest in platform work: insights from Germany', Presentation at the 2nd Crowdworking Symposium, 8.-9-10.2020, Paderborn.

Vandaele, K. (2018), 'Will trade unions survive in the platform economy? Emerging patterns of platform workers' collective voice and representation in Europe', Working Paper 2018.05, Brussels: ETUI.

Vandaele, K., A. Piasna and J. Drahokoupil (2019), '"Algorithm breakers" are not a different "species": attitudes towards trade unions of Deliveroo riders in Belgium', Working Paper 2019.06, Brussels: ETUI.

Valenduc, G. and P. Vendramin (2016), 'Work in the digital economy: sorting the old from the new', Working Paper 2016.13, Brussels: ETUI.

Veen, A., T. Barratt and C. Goods (2019), 'Platform-capital's "app-etite" for control: a labour process analysis of food-delivery work in Australia', *Work, Employment and Society*, **34** (3), 388–406.

Woodcock, J. and M. Graham (2019), *The gig economy: a critical introduction*, Cambridge: Polity Press.

17. The pitfalls and promises of successfully organizing Foodora couriers in Toronto

Raoul Gebert

17.1 INTRODUCTION

Workers in the platform economy, particularly those who offer delivery, tourism and transport services on platforms such as Foodora, Airbnb and Uber, do not have a clear status under Canadian labour law. As a result, unionization has largely escaped them so far.

The appearance of entirely new economic sectors, or the appearance of new groups of workers within already existing sectors, has always provided for disruptions or 'critical junctures' (Crouch 2005) that put existing institutional and legal frameworks to the test. Hence the spread of 'digitally organized, just-in-time work' (Huws et al. 2018) should thus be interpreted as a challenge to pre-existing institutions. At such a key moment, creative actors can become 'institutional entrepreneurs' (Fligstein 1997), however, employing their existing resources and networks in novel fashions and thereby leading the way towards new institutions, networks and alliances. This process may also be referred to as 'institutional experimentation' (Ferreras et al. 2020).

The institutional framework for unionization and collective bargaining in Canada is closely linked to the industrial era. In particular, labour relations legislation, based on the American Wagner Act of the 1930s, and court jurisprudence generally presuppose a stable employment relationship between employee and employer as well as a relatively fixed workplace. The unionization model is based on direct control of the organization of work and the face-to-face, onsite production or delivery of services. It does not lend itself easily to a decentralized organization of work, let alone when it is merely 'facilitated' by a digital platform. In that way, signing up platform economy workers becomes a test of trade union capacities for adaptation and experimentation which go beyond the renewal of traditional organizing capabilities or legal strategies.

There are, however, precedents for such institutional creativity among Canadian unions based on signing up new groups of workers in defiance of the traditional organization of work, enabling them to negotiate their working conditions collectively. Artists, for example, were the first to have to deal with a similar challenge to platform workers: powerful clients who set a pre-determined 'price' for the services to be provided alongside the existence of an informal but formidable structure of recruitment and discipline. At the same time, this position as a dependent service provider actually gives dual status since such artists are also self-employed under the law as they are (on paper) allowed to accept or refuse contracts at any time. For artists as for other professional groups, notably home childcare service providers in Quebec, access to collective bargaining, while preserving the legal status of self-employed workers, was achieved thanks to the creation of a new system of labour relations for each of these sectors specifically (the former in 1988; the latter in 2009) following decades of organized struggles and legal challenges (Bellemare and Briand 2012).

Workers in the platform economy have (unfortunately) not yet reached this stage. To make matters worse, such workers have not always been welcomed with open arms into the ranks of Canadian labour organizations. The difficult relationship between platform economy workers and traditional labour institutions stems from the fundamental insecurity created by technological disruption. For instance, when thinking of Uber drivers and their relations with Canadian unions, one immediately focuses on the manifest conflict with organized taxi drivers who are, in fact, represented in certain Canadian cities by a union that collectively negotiates wages and prices. Instead of looking to embrace new workers in the transportation sector, traditional unions have largely opposed the deregulation of the taxi sector altogether. This, of course, is not a peculiarly Canadian phenomenon (Haake 2017).

An application for certification from the United Food and Commercial Workers Union (UFCW) has been submitted to the Ontario authorities (OLRB 2020b) on behalf of Uber drivers in Toronto who the platform considers to be independent contractors. Indeed, Canadian labour relations research (Tucker 2017) has thus far focused on Uber and similar transportation services as a key indicator of the future of worker representation in the location-based platform economy. In contrast to Europe, where there are examples of unionization (Jesnes 2019), location-based food delivery services have largely flown under the radar not least since such workers are rarely unionized: there is just the one case in North America so far when, in January 2020, workers for the grocery delivery platform Instacart in a Chicago suburb voted to join UFCW (Kanu 2020). Consequently, scholarship on the phenomenon of food delivery unionization remains patchy.

In an attempt to remedy this situation, this chapter will provide an overview of the recent campaign by the Canadian Union of Postal Workers (CUPW) to organize the rapidly growing food delivery sector in Toronto under the banner 'Foodsters United', with bike couriers at Foodora as the starting point. The main takeaway from this case study is that resourceful, creative actors within the current union landscape can play a central role in opening up the pre-existing institutional framework to workers in the location-based platform economy. In particular, the insistence by the CUPW legal team on establishing 'dependent contractor' status (rather than a reliance on the more traditional status of 'salaried employee'), and the subsequent transformation of Foodsters United into a community union (albeit without legal bargaining certification), should be seen as elements of institutional experimentation. Such an approach went beyond mere 'resourcefulness' in organizing (Ganz 2000) or the creative interpretation of an existing legal framework. Nevertheless, even though CUPW and the Toronto bike couriers managed to combine well-resourced organizing, a superior communications strategy and some elements of institutional entrepreneurship – establishing 'dependent contractor' as an important precedent in Canadian labour jurisprudence (OLRB 2020a) – they still lost the overall struggle.

17.2 THE CANADIAN FOOD DELIVERY SECTOR AND THE STATE OF CANADIAN UNIONS

The bike delivery service Foodora, a subsidiary of the German multinational Delivery Hero, entered the Canadian market in July 2015. Paving the way for entry into Canada was the acquisition by Delivery Hero of Hurrier, a Toronto-based bike courier service. The multinational then marketed its food delivery services under its well-known, fuchsia-coloured brand Foodora. In October of the same year, Foodora then also started operations in Montreal, Canada's second-largest city. One year later, it began offering services in Vancouver before finally branching out into the nation's capital, Ottawa, and other Canadian cities during 2019.

After an aggressive marketing campaign and with the use of competitive pricing schemes, the company captured about 15 per cent of the Canadian market at the height of its success in this country (2019), a share similar to Uber Eats but still far behind the homegrown Skip the Dishes (originally a Canadian start-up out of Winnipeg but now a division of the United Kingdom multinational Just Eat) at about 29 per cent. The total Canadian market for platform-assisted food delivery was estimated in 2018 to reach C$1.5 billion (AIMS 2019).

Despite significant markups of 15 to 30 per cent on each delivered dish and a network of over 3000 restaurants, Foodora never broke even in its Canadian operations, leaving millions of unpaid bills in the wake of its departure from the country in May 2020. In a company news release, it stated 'Canada is a highly saturated market for online food delivery and has lately seen intensified competition. Foodora has unfortunately not been able to reach a strong leadership position and has been unable to reach a level of profitability in Canada that's sustainable enough to continue operations' (Foodora Canada 2020).

The Canadian labour movement, one of the most stable union microcosms in the world (the unionization rate in Canada has hovered around 30 per cent for the better part of the last two decades), decided in 2017 to open up to workers in the platform economy. With UFCW working on Uber drivers, it fell to CUPW to take up the efforts of the local Foodora couriers in Toronto to organize a union. This would turn out to be a landmark foray into a brave new world of labour jurisprudence, recognizing the dual status of food delivery couriers as independent service providers and as contract workers completely dependent on the company owning the platform via which they are offering their services. Unfortunately for the couriers, Delivery Hero's decision to pull the plug on its Canadian operations left the newly recognized bargaining unit without an employer to bargain with.

A concomitant legal case went into overtime on questions of compensation payments and unfair labour practices, but the positive precedent for Canadian platform economy workers that it set should be felt into the future.

Based on five semi-structured, qualitative interviews with union organizers, the next section of this chapter analyses the organizing process by which a rather traditional labour organization staged a massively successful unionization campaign on behalf of a dynamic, young workforce. We then examine the legal arguments presented to the Ontario Labour Relations Board, the province's labour tribunal, which eventually led to the recognition of 'dependent contractor' status for food delivery workers – an important precedent in Canadian jurisprudence. Finally, we discuss prospects for some additional union forays into the platform economy, pointing to a dynamic future for labour in this expanding sector of the modern, twenty-first-century service economy.

17.3 THE ORGANIZING DRIVE

The market for food delivered to the home is a rather recent addition to the repertory of mobile services provided in Canada and no Canadian labour union had ever attempted to organize food delivery platform workers. The way in which Foodora entered the market in Toronto proved to be the essential starting point for the Foodora organization drive. Hurrier had a tightly knit core group of bike couriers who had worked alongside each other for several years before being taken over by Delivery Heroes. It was this core group of workers that would become instrumental in creating a sense of unity and solidarity among the otherwise physically distanced and rapidly growing Foodora workforce. It is particularly remarkable that even the introduction of motor vehicle delivery would not break the organizational clout and narrative-shaping capacity of the original Hurrier workforce.

Being considered of dual status had not been a hindrance to collective bargaining as such, as Canadian unions had previously organized artists, home daycare workers and delivery drivers. Traditional bike couriers (for letters and parcels) had also been approached but no permanent organization had managed to gain a foothold. For food delivery couriers, it took a group of motivated workers with specific grievances to kickstart the campaign, but it was the relative novelty of the services being provided (and the related grievances), that made Foodora a Canadian first in bike courier organization, not the dual status itself.

When Foodora took over Hurrier, the main objective (besides entering the Canadian market) was to obtain Hurrier's logistics software. They then used the Hurrier app in other companies they operated around the world. As for the Hurrier workforce, they wanted to make as few changes as possible, being happy to rely on a positive identity and a highly motivated workforce in the face of the strong competitive pressures. Over time, they were hiring more and more people, including car drivers, and the relationship and identification with the company deteriorated. As one courier recalls:

> The transfer itself was ok [and proceeded] initially without the loss of the community vibe. [The money was even] better because of more advertising. I was able to get half an order to one full order per hour more. A positive difference. But then [the community vibe] slowly went away.

At this point, the main competitors in major Canadian urban centres were other multinational companies such as Uber Eats, Door Dash and Just Eat. CUPW had organized local unions for bike messengers in Toronto and Montreal during the late 2000s but these fell apart and lost certification even before the first collective agreement could be signed. This and the challenge presented by having a status as independent contractors were thus the main roadblocks to organizing food delivery workers. As a consequence, none of Foodora's competitors were unionized when the company took over Hurrier.

In terms of its operating practices, Foodora had established itself along a similar pattern to its European operations and was heavily relying on the efficiency of bikes over cars in congested urban centres, regardless of harsh Canadian winters. With their milder climates, Vancouver and Toronto are actually the only two Canadian cities where delivery on bikes could feasibly be considered a year-round endeavour. However, Foodora's reliance on bike couriers rather than drivers might ultimately have made it more vulnerable to an organizing drive. The strong collective identity of bike couriers would be put to the test with the mass hiring of new couriers, including drivers. One local organizer recalls:

> There is a scene [of bike messengers]. You can watch documentaries on it: it's the same in NYC, London, San Francisco and so forth. But that doesn't necessarily transfer over into organizing food couriers. At Hurrier, there was a lot of that [bike] culture, with many people coming from messenger services. It's not like that at Uber Eats. But it would be naïve to think that that was enough to organize the company. Especially because the Hurrier messengers were a small minority by then.

Above and beyond the delivery method, the key to managing the couriers as a workforce had been to sign them as independent contractors, thereby ensuring maximum flexibility for the company. The obvious downsides for the couriers were the instability of working hours, pay and benefits, as well as drastically uneven health and safety practices, little or no training opportunities and, finally, complete subordination to the algorithms of the company's platform.

In particular, the health and safety concerns of the delivery workers were enormous: Foodora paid the obligatory workers' compensation fees[1] but it never provided sick days, invalidity or health insurance benefits to its workers. Additionally, some high-profile cases of door incidents or other accidents (Ghebreslassie et al. 2018; Mojtehedzadeh 2019), coupled with Foodora's inaction to provide help to the couriers involved, pushed the pro-unionization narrative:

> Streetcar tracks are deadly for bikers. It happened to me in the middle of the night. I was frozen and in pain. I texted the dispatch to let them know I had been hurt. The first thing they texted back was: 'Can you complete the order?' [It made me feel:] 'I am just a number to you. You don't care about me.' It's a hard thing to feel.

Also, the question of providing and replacing broken or damaged bikes was never settled in the workers' favour. In the case of injury, workers could apply for compensation from the public accident insurance scheme but they could not get coverage for additional expenditures and replacement equipment. The calculation of working hours, in order to obtain a complete replacement of salary, was also problematic.

Foodora headquarters employees not taking safety issues seriously was of even more concern to women in the company:

> Other delivery companies often use a third-party piece of software to veil the telephone numbers of couriers and customers. But at Foodora, customers would save my number and contact me later. Much of it was sexual harassment. When I got

> orientation at Foodora, I asked them if they veiled the phone numbers and they said they didn't – and very casually so! The manager told me that 'their customers weren't like that'. [Protecting my safety] became a cost associated with my job. I had to use a commercial app for C$14.99/month to veil my number.

Pay and benefits were equally of concern. The going rate for a delivery was C$4.50 plus C$1 per kilometre between the restaurant and the client. Additionally, there was a C$5 bonus for every 20 minutes a courier had to wait at a restaurant for food to be prepared. Altogether, this meant reasonably good pay for the couriers, but change was in the air:

> We got a warning when they cut the pay in Montreal. They changed the pay structure there. They made it a lot worse. You no longer got kilometres. Only a small amount hourly plus C$2.50 an order. When we heard that, we were afraid they'd implement that in Toronto, where the cost of living is much higher. They had made that change everywhere in Canada but in Toronto.

To make matters worse, the app and the dispatchers became more and more stringent about measuring waiting time at the restaurant with the result that 19 minutes (and, by extension, 39 or 59 minutes) were not sufficient to get the C$5 wait-time bonus.

The fear of a top-down imposition of pay and benefits exemplifies the dual-status conundrum. On paper individual contractors were free to accept or reject the proposed payment scheme for their services but, in practice, these changes resulted in a top-down imposition of wages outside of any collective bargaining framework.

As a result of such grievances, the first grassroots organizing campaign among food delivery workers in Canada was kickstarted, symbolically enough, on 1 May 2019 based on the strong network among the ex-Hurrier delivery couriers. Foodora's strong reliance on bike couriers also represented the principal difference to other food delivery platforms, such as Uber Eats, adding to the explanation as to why it was the first to be targeted in Canada.

In early 2018, Foodsters United formed as a grassroots association of bike couriers within Foodora in order to spearhead a union drive there. After an initial, rather haphazard, attempt with a different national union (throughout 2018), Foodsters United voted unanimously to ask CUPW for help instead. The postal workers' union was selected because of its history of organizing bike couriers in Toronto and Montreal. The contrast in resources obtained from the national union was stark: an external organizer was sent to help Foodsters United host meetings and provide training for couriers as organizers and on how to have conversations with their co-workers. A local organizer recalls: 'There also was a big budget for swag. Gorgeous pink T-shirts and hundreds of posters. It was a huge way to recruit people and raise awareness. Couriers don't have a shared, physical workspace. But this way, the whole city became the workspace.'

17.4 THE CASE BEFORE THE ONTARIO LABOUR RELATIONS BOARD

Canadian labour law reserves collective bargaining for those with status as salaried employees. Challenging the notion of bike couriers being 'independent contractors' would therefore be at the heart of the unionization initiative. The Canadian labour movement had identified the pretence that platform workers were independent contractors as the primary obstacle to unionization. The Foodora case was thus part of an overarching legal strategy to pave the way for many different types of platform workers, namely by eliminating the presumption of independence and by reclassifying them as 'dependent contractors'. The aim was to level the institutional playing field, eventually paving the way for a specific labour relations framework for workers in the platform economy, allowing them to bargain collectively. Any type of platform work could have been chosen with which to challenge the notion of 'independent' contract work, but the couriers' strong sense of shared solidarity, especially in the wake of their employer's neglect of the obvious health and safety challenges, made them a particularly good target for an organization drive.

The exclusion of spatially mobile workers from standard bargaining rights became a widespread practice in Canada throughout the 1990s (Tufts 1998). In response, CUPW led a successful organizing campaign among rural and suburban mail carriers and, in 2003, Canada Post, which had been using an independent contractor model in order to save costs while ensuring door-to-door mail delivery in less densely populated areas, agreed to integrate its 8,000 rural and suburban carriers into a regular collective bargaining unit (CUPW 2017). Similarly, in 2017, delivery drivers for Canada Bread, one of Canada's largest industrial bakeries, earned the right to bargain collectively after proving they were employees not independent contractors (OLRB 2017). Classification of spatially mobile workers has also been at the heart of the recent battle surrounding Proposition 22 in California (Working Partnerships USA 2020).

Within these precedents, tribunals used the level of control over the workflow and economic dependency on a single employer as the main tests to determine the status of workers and thus their right to collective bargaining. Both these cases, however, relied on the tradition of 'salaried employee' status for their success.

CUPW filed for certification before the OLRB in July 2019, which promptly established an obligatory voting period even though over half the eligible couriers had signed a union card.[2] The Canadian labour relations framework resembles a decentralized model in which union accreditation generally has to be achieved in each individual workplace, with employer-wide or multi-employer collective agreements being rare. Determining the exact extent of the 'workplace' is a problem when it comes to platform economy workers and there were several arguments about whether car drivers and suburban Foodora workers in Mississauga, part of the Greater Toronto Area and some 25 km from Toronto, should have been included. Furthermore, any delay between the signing of membership cards and the certification vote gives employers the chance to 'flood the list' with additional employees being hired with the purpose only of casting a vote against unionization.

Shortly after voting began, the company launched a 'vote no' campaign to counter the unionization push. Its move to send emails and push notifications directly to couriers via its platform was the subject of an unfair labour practice complaint by CUPW. After the voting period had ended, the results were sealed pending the various disputes before the OLRB.

The hearings at the Board began in September 2019 and dragged on into 2020. The first issue was the question of how to classify Foodora workers: as employees, independent contractors or the hybrid 'dependent contractors' being sought by the union movement. The company claimed it had 'little or no control' over the couriers' work and suggested that they had a multitude of concurrent income sources (so-called 'double-apping'), thereby making them economically independent.

Another issue for the OLRB to consider was the list of workers entitled to vote in the certification procedure. Whereas the company claimed that almost 1,200 workers were eligible, as they had previously signed up to signal their availability for deliveries, CUPW maintained that only those couriers who had completed recent deliveries should be considered.

The issue of employee discipline became key to the determination of how much control Foodora actually wielded over its couriers. The union submitted in evidence an email sent to all couriers warning them they would be 'deprioritized' if they declined more than 15 per cent of the delivery requests sent their way. Additionally, even when the rate of declined deliveries was acceptable to Foodora, the platform closely monitored couriers' sign-in behaviour (the equivalent of 'clocking on') and performance scores based on customer reviews. Witnesses also reported that Foodora cancelled or shortened shifts, and unilaterally placed them on a break, on several occasions.

As its main line of defence, the company fielded its recruitment process (which did not rely on formal interviews or references and permitted couriers to work for Foodora's competitors) and bike couriers had to provide their own equipment in order to start making deliveries. Only the highly recognizable, fuchsia-branded delivery backpacks were mandatory equipment for the couriers to purchase: the price of C$50 was automatically deducted from their first paycheque.

CUPW was able to back up the evidence from its witnesses with more detail, including a 'notice of termination' sent to one courier and further documentation regarding the monitoring and follow-up of performance scores. In particular, the company was shown to accumulate a number of 'strikes', a measure against which couriers could be disciplined or prevented from working.

The Foodora couriers were joined by some 300 Uber drivers (backed by UFCW) in their quest to unionize platform workers in the Greater Toronto Area. Citing recognition as employees and better pay and working conditions, including protection against unjust dismissal, the drivers were met with the same categorical answer from their company: that they were independent contractors and thus not eligible to unionize. Delayed by the pandemic, the proceedings in this case finally started at the OLRB on 5 June 2020.

The unions' joint campaign ('Gig economy workers rising'), while not being a formal organization bundling resources or coordinating strategies, still managed to create a feeling of solidarity and also helped align the legal strategy between the two unions representing Uber and Foodora workers. One local organizer recalls:

> UFCW came to our rallies and OLRB hearings. We tried to coordinate with them, finding people to sign cards. Uber X and Foodora employed some of the same people. Foodora by then had plenty of car drivers. We wouldn't have been able to do it without car drivers. It helped us getting out of the bike messenger bubble.

There was also some sympathy action at Delivery Hero's headquarters in Berlin, staged by *Gewerkschaft Nahrung-Genuss-Gaststätten*, the German union organizing food delivery workers and arranged through the global union federation UNI of which CUPW is a member.

Closing arguments in the Foodora dispute were heard on 29 January 2020 and the decision on the employment status of the couriers was finally sent down on 25 February: the OLRB ruled that Foodora bike couriers did indeed have the status of 'dependent contractors'. Matthew Wilson, vice chair of the Board, said that couriers were 'a mere cog in the wheel that is powered by Foodora', before continuing, 'in a very real sense, the couriers work for Foodora, and not themselves' (OLRB 2020a).

The decisive arguments which won 'dependent contractor' status were the level of control, including discipline and workflow. Thereby, the Board rejected the argument that economic independence and concurrent work for competitors should be regarded as bellwethers of independent work. 'The focus is not on the frequency of exercising control. Rather, it is about the right and ability of the company to control how the work is performed' (OLRB 2020a).

While the decision has wide-ranging implications for the rights of other platform economy workers to organize, it was at that point only the first step for Foodora couriers in Toronto, with the precise list of eligible delivery workers yet to be resolved by the Board and the results of the unionization vote remaining sealed in the meantime.

Following the decision, couriers' health and safety concerns again came to a head in the midst of the health crisis of spring 2020. Among the more flagrant concerns, it was reported that, during the first wave of Covid-19 in Canada, the digital platform was now demanding couriers deliver medication to sick patients recovering from undeclared illnesses at home, although Foodora itself was unwilling to provide couriers with medical protective gear, sick leave or medical benefits (Mojtehedzadeh 2020).

17.5 FOODORA'S EXIT FROM CANADA AND THE SETTING OF PRECEDENT IN THE CANADIAN PLATFORM ECONOMY

Before a decision on the list of eligible couriers could be rendered, Foodora announced its departure from the Canadian food delivery market (Foodora Canada 2020), citing concerns over profitability and intense competition even as the home delivery market was booming due to the health crisis which shut down restaurants in Toronto for over two months.

Two days later, the company initiated bankruptcy proceedings, seeking protection from its creditors, citing debts of C$4.7 million. Among the creditors were outstanding payroll deductions on behalf of government agencies, such as the Canada Revenue Agency and the provincial Workers' Compensation Board, but also a total of C$243,000 in wages directly owed to headquarters employees. Couriers, whom the company still considered as independent contractors, were not listed as creditors. However, Foodora stated that the announced closure should constitute 'their two-week notice of termination', once again highlighting profound confusion over the exact status of couriers (notices of termination usually apply only to salaried employees).

While the rest of the certification battle had now become moot, CUPW immediately attempted to gain compensation payments for unfair labour practices and to make sure that any outstanding amounts were paid as part of insolvency proceedings. On 25 August 2020, a C$3.46 million deal was announced (CUPW 2020a) providing union members with some compensation for the departure from the Canadian market. In exchange, the union dropped its legal challenges concerning unfair labour practices. Having accomplished the main part of its legal strategy, namely proving that couriers were 'dependent workers' (albeit not salaried employees), CUPW could afford to drop that litigation.

The deal provided compensation for loss of employment based on the number of hours a courier would have been expected to work. Foodsters United helped the couriers calculate the amount they were owed. 'It was well received by many people. It rewarded people who had been doing the job full-time. Some people received up to C$6,000. It helped to show that there is a benefit to being in a union,' one local organizer recalled.

Despite the unfortunate circumstances, negotiating a deal with Foodora also propelled CUPW and Foodsters United into a new institutional role in place of a collective bargaining role: that of a structured support network and community-based network. For instance, a 'hardship fund' was created during the pandemic and the network continues to organize self-help workshops for bike couriers and delivery drivers. It thus aims to represent the concerns and address the needs of food delivery workers more generally, beyond seeking formal collective bargaining recognition at any one company. In turn, organizers hope that this encompassing, community-based approach will facilitate future unionization drives or set the stage for regional or provincial authorities to intervene, similar to the 'Justice for Janitors' campaigns elsewhere (Milkman 2006).

CUPW eventually announced that 88.8 per cent of Foodora couriers in Mississauga and Toronto had voted to join the union (CUPW 2020b). Given that the certification process was now moot, the OLRB had authorized the unsealing and counting of the votes. Further proof of the union's success was that all of the subgroups which Foodora and CUPW had argued about had voted in favour of certification. The common grievances among bike couriers and delivery drivers concerning payment schemes, control over work and lack of respect had carried the day. The organizing efforts thus turned out to be an impressive, yet bittersweet, success.

Whether or not more favourable labour legislation in Quebec and British Columbia will facilitate the organization of food couriers in Montreal and Vancouver, and whether the OLRB decision on 'dependent contractors' would set a legal precedent there might soon be put to the test. In the province of Ontario, however, they have effectively helped to move the goalposts for organizing platform economy workers. The fate of Uber drivers will be decided by the same vice-chair, dealing with some of the same preliminary issues (e.g. dual status and flooding the list), while the UFCW legal team will be able to rely heavily on the Foodora case law to make its request for certification.

The UFCW has also taken on the case of Uber drivers elsewhere in Canada and plans to file equally for union accreditation in British Columbia. A dispute involving Uber's arbitration clause (as an unfair bargaining practice) is also being heard by Canada's Supreme Court, with the UFCW having been joined to the case.

17.6 CONCLUSION

Whether the groundbreaking organizing drive of Foodora delivery workers in Toronto will prove to have a lasting effect on the representation of Canadian platform economy workers remains to be seen. Initial successes in organizing Walmart workers (ending with store closures and lengthy litigation) and Couche Tard convenience store employees in Quebec (ending with a loose framework agreement rather than a collectively bargained labour contract) have, unfortunately, not led to further organizing drives in the low-end retail sector. Creative institutional arrangements in the home daycare sector, in contrast, have emboldened and strengthened union organizations there.

On the one hand, the success at Foodora in Toronto might ultimately have stemmed from the presence of a favourable microcosm of highly motivated and tightly knit local organizers, particularly egregious health and safety concerns and the presence of an experienced legal team and a forceful communications strategy at an established, well-resourced national union in the midst of a quest for a new *raison d'être*. However, employers in the food delivery sector have now been forewarned that Canadian labour tribunals may very well reject their prosaic stories of independent contract workers providing flexible, mobile services; in that sense, they might well put other measures into place, both legal and organizational, with the aim of thwarting future organizing drives.

On the other hand, the precedent before the provincial labour board might pave the way for other Canadian platform economy workers currently in the process of organizing their union, thus feeding into the overarching litigation strategy of the Canadian labour movement. Additionally, it might lead national unions to take a closer look at investing serious resources in the food delivery sector as potential for union growth in Canada, in an era of ongoing deindustrialization and nearly complete public sector representation. According to some, such an approach is starting to take hold within CUPW: 'They knew that [platform work] is what [the employers] will try to do with other mail delivery workers. This is coming. [They] will have to fight it at some point.'

Labour scholars should view this case study as an example of institutional experimentation in the wake of the major disruption that platform work constitutes to traditional Canadian labour law, and in two specific ways. First, CUPW did not ask for 'salaried employee' status, in order to include platform economy workers within the existing labour relations framework, but helped to create a new legal status, that of 'dependent contractor'. This is reminiscent of what unions in the province of Quebec managed to achieve for home daycare workers, as well as unions representing Canadian artists across the country. Each of these two other cases actually resulted in a completely separate institutional framework for labour relations in those two sectors, something that might yet be a future solution for Canadian platform economy workers. Second, Foodora's departure led to the establishment of a CUPW-funded community union without legal bargaining rights. While that is also not a completely new approach, it is a relatively new development in Canada where the existing institutional location in which formal collective bargaining takes place has, in general, remained quite solid.

In summary, you can teach an old dog new tricks. When combined, the superior resources, communications and organizational capacities of traditional labour unions can be redeployed in dynamic, new sectors to gain collective bargaining rights for platform workers even when the institutional framework is not favourable. By picking up the case of Foodora bike couriers in Toronto, CUPW also managed to move the institutional goalposts surrounding the representation of workers in the platform economy, creatively skirting the existing organizational and legal framework and setting an important precedent under Canadian labour legislation through institutional experimentation.

NOTES

1. Workers' Compensation Boards are provincially regulated throughout Canada and provide insurance for workplace injuries and illnesses via programmes that are funded by employers requiring coverage. The Ontario Workers' Compensation Board charges an annual premium (or fee) and also applies a rebate to or surcharge on the employer following a comparison of the difference between the costs of expected and actual claims.
2. There is no automatic certification or 'card check' procedure in Ontario labour law, as opposed to the neighbouring province of Quebec and the federal Canadian Labour Code, under which the union could have benefited from instant certification.

REFERENCES

AIMS (Atlantic Institute for Market Studies) (2019), 'Food delivery apps user rate in Canada reached record level in May 2019, and likely expand after 2021', accessed 14 January 2021 at www.aims.ca/op-ed/food-delivery-apps-user-rate-in-canada-reached-record-level-in-may-2019-and-likely-expand-after-2021/.

Bellemare, G. and L. Briand (2012), 'La syndicalisation des services de garde au Québec: à pratiques innovatrices, des concepts nouveaux', *La Revue de l'IRES*, **75**, 117–141.

Crouch, C. (2005), *Capitalist diversity and change: recombinant governance and institutional entrepreneurs*, Oxford: Oxford University Press.

CUPW (Canadian Union of Postal Workers) (2017), 'Collective agreement for rural and suburban mail between Canada Post Corporation and the Canadian Union of Postal Workers', accessed 14 January 2021 at www.cupw.ca/en/collective-agreements/rural-and-suburban-mail-carriers.

CUPW (Canadian Union of Postal Workers) (2020a), 'CUPW and Foodora reach settlement for workers', *CUPW*, 25 August, accessed 14 January 2021 at www

.newswire.ca/news-releases/cupw-and-foodora-canada-reach-settlement-for -workers-829180991.html.

CUPW (Canadian Union of Postal Workers) (2020b), 'The results are in: 88.8% of Foodora couriers vote yes to union!', *CUPW Communiqué*, 16 June, accessed 14 January 2021 at www.cupw.ca/en/results-are-888-percent-foodora-couriers-vote -yes-union.

Ferreras, I., I. MacDonald, G. Murray and V. Pulignano (2020), 'L'expérimentation institutionnelle au travail, pour le meilleur (ou pour le pire)', *Transfer*, **26** (2), 119–125.

Fligstein, N. (1997), 'Social skill and institutional theory', *American Behavioral Scientist*, **40** (4), 397–405.

Foodora Canada (2020), 'Foodora Canada announces plans to close business while assuring support for employees', *Globe Newswire*, 27 April, accessed 14 January 2021 at www.globenewswire.com/news-release/2020/04/27/2022709/0/en/foodora -Canada-announces-plans-to-close-business-while-assuring-support-for-employees .html.

Ganz, M. (2000), 'Resources and resourcefulness: strategic capacity in the unionization of California agriculture, 1959–1966', *American Journal of Sociology*, **105** (4), 1003–1062.

Ghebreslassie, M., C. Taylor and A. Singh (2018), 'Ontario workplace safety board reviewing Uber Eats following marketplace investigation', *CBC (Canadian Broadcasting Corporation)*, 16 November, accessed 14 January 2021 at www.cbc .ca/news/canada/marketplace-food-delivery-apps-labour-issues-1.4895801.

Haake, G. (2017), 'Trade unions, digitalisation and the self-employed – inclusion or exclusion?', *Transfer*, **23** (1), 63–66.

Huws, U., N. H. Spencer and D. S. Syrdal (2018), 'Online, on call: the spread of digitally organized just-in-time working and its implications for standard employment models', *New Technology, Work and Employment*, **33** (2), 113–129.

Jesnes, K. (2019), 'Employment models of platform companies in Norway: a distinctive approach?', *Nordic Journal of Working Life Studies*, **9** (S6), 53–73.

Kanu, H. (2020), 'In-store Instacart shoppers in Chicago get OK for Union vote', *Bloomberg Law*, 8 May, accessed 14 January 2021 at https://news.bloomberglaw .com/daily-labor-report/in-store-instacart-shoppers-in-chicago-get-ok-for-union -vote.

Milkman, R. (2006), *LA story: immigrant workers and the future of the US labor movement*, New York: Russell Sage Foundation.

Mojtehedzadeh, S. (2019), 'Could bike couriers be unionized? Labour drive aims to set a precedent for the gig economy', *The Star*, 1 May, accessed 14 January 2021 at www.thestar.com/news/canada/2019/05/01/safety-risks-low-wages-in-spotlight-as -unionization-drive-rolls-out-for-bike-couriers.html.

Mojtehedzadeh, S. (2020), 'Foodora couriers now being asked to deliver medication – and they still have to buy their own protective equipment', *The Star*, 2 April, accessed 14 January 2021 at www.thestar.com/business/2020/04/02/foodora-couriers-worry -about-new-order-medication-sometimes-delivered-face-to-face.html?rf.

OLRB (Ontario Labour Relations Board) (2017), *International Brotherhood of Teamsters* v. *Canada Bread Company Limited*, 2017 CanLII 62172.

OLRB (Ontario Labour Relations Board) (2020a), *Canadian Union of Postal Workers* v. *Foodora Inc. d.b.a. Foodora*, 2020 CanLII 25122.

OLRB (Ontario Labour Relations Board) (2020b), *United Food and Commercial Workers International Union (UFCW Canada)* v. *Uber Canada Inc.*, 2020 CanLII 3649.

Tucker, E. (2017), 'Uber and the unmaking and remaking of taxi capitalisms: technology, law and resistance in historical perspective', Osgoode Hall Law School of York University, accessed 14 January 2021 at https://digitalcommons.osgoode.yorku.ca/cgi/viewcontent.cgi?article=3602&context=scholarly_works.

Tufts, S. (1998), 'Community unionism in Canada and labor's (re)organization of space', *Antipode*, **30** (3), 227–250.

Working Partnerships USA (2020), 'News coverage/gig workers rising', accessed 2 November 2020 at www.wpusa.org/news/campaigns/gig-workers-rising/.

18. Labour management and resistance among platform-based food delivery couriers in Beijing

Jack Linchuan Qiu, Ping Sun and Julie Chen

18.1 INTRODUCTION

According to official statistics, location-based food delivery platforms were already serving 113.6 million Chinese people by November 2015, a figure which had more than tripled, to 397.8 million, by March 2020 (CNNIC 2020, p. 41).[1] Food delivery platforms, indispensable to everyday life under Covid-19 lockdowns, have thus further extended the trend of 'big data' – in China as much as in the west – in which labour is organized through algorithms to constitute an essential public good subsequent to the private transactions in data, money, goods and services (Chen and Qiu 2019). How, then, do food delivery platforms in China structure labour and how do couriers resist top-down managerial manipulation? How can we make sense of the labour–capital–state relationships in ways that enrich our understanding of platform labour in China and beyond?

This case study is based on observations drawn from our ethnographic fieldwork, conducted in Beijing since August 2017, which include participant observation and interviews.[2] We frequented sites where couriers usually concentrate to rest such as on street corners, at takeout restaurants and courier 'stations'.[3] At these sites, we built rapport with workers in informal conversation before recruiting research participants or requesting interviews. Participant observation allowed us to see workers' interactions with food delivery platforms, their routines and acts of individual or collective resistance. This chapter also draws from two questionnaire surveys during May–August 2018 (N=1339) and May–August 2019 (N=771), which showed changing composition of full-time and part-time couriers in a one-year period. Although not all findings from surveys will be presented in this chapter, the shifting composition of the work force is significant for us to understand the labour–capital relationship in the platform economy in China.

First we provide a brief description of the rise of food delivery platforms in China, followed by delivering a critical overview of such platforms' labour control mechanisms through corporate policy, algorithms and technological design. We then look at labour market aspects of the organization of food delivery work before examining how couriers react to management by algorithms through parameters such as temporality, affective labour and gamification. Related to the control of working time, our surveys reveal the noteworthy trend of 'de-flexibilization', meaning that the workforce consists increasingly of full-time couriers working on fixed schedules, while the number of couriers working on flexible hours has in fact declined. We further demonstrate that couriers are not simply passive entities subject to a digital panopticon. Rather, workers have developed alternative ways of responding which can only be comprehended alongside corporate decisions and local state policies. Finally, we discuss how China's app-based food delivery services can be seen as an important case of digital utility that needs to be regulated as such.

18.2 THE RISE OF FOOD DELIVERY PLATFORMS

Similar to the emergence of western platform capitalism (Srnicek 2016), the growth of Chinese food delivery platforms benefited from post-2008 quantitative easing that helped inject massive investment into the financial and tech sectors. However, in China things are distinct in two respects. First, platform companies operate in a statist economy dominated by the Chinese Communist Party. The fast growth of the Chinese economy and the authorities' control over vast stocks of cash produce a very different dynamism between state and capital in which austerity is irrelevant while the party-state is both capable of offering massive support for tech start-ups and willing to do so. Second, Chinese labour supply has, for the first time in many decades, started to decline in overall quantity while employment has begun to shift from labour-intensive manufacturing to the service sector, especially among rural-to-urban migrants who supply most of the labour in food delivery.

Chinese food delivery platforms, similar to Deliveroo or Uber Eats, have grown through mergers and acquisitions. These are essentially financial operations geared towards rapid market expansion (Jia and Winseck 2018). Starting from 2011, China's food delivery sector has evolved through four stages: emergence, expansion, combination and competition. Up to 2015, the emergence phase witnessed initial investments from Baidu (China's dominant search engine); Alibaba (China's leading e-commerce company); Meituan (by then a group-buying platform comparable to Groupon); and American venture capital firms such as Sequoia Capital. From 2016 to 2019, more than 50 delivery platforms appeared in China but only a few have managed to survive due to competition and increased difficulty in financing (Forward the

Economist 2019). In 2017, the third largest platform, Baidu Delivery, was acquired by Ele.me and the market has since been dominated by a duopoly of Ele.me and Meituan. Following a short period of market stabilization, competition between them has since intensified and, after the arrival of Covid-19, inter-capitalist struggle has become more pronounced.

We cannot overstate the significance of the economic and social magnitude of food delivery platforms. Half of China's netizens used food delivery apps in 2019, which surpassed the popularity of ride-hailing services, while the total amount of location-based food delivery transactions now exceeds 600 billion yuan ($87 billion) (CNNIC 2020; State Information Centre 2020). Although there are no official employment statistics, it is reported that Meituan and Ele.me had more than 6 million couriers on their books in May 2020 (Tencent.com 2020).

18.3 LABOUR MARKET CHARACTERISTICS OF FOOD DELIVERY

The astonishing rate of market expansion, the relatively low entry barriers for workers and fierce competition between the platforms have resulted in a volatile labour market characterized by high labour mobility, informal employment and a proliferation of intermediary staffing agencies. According to our own survey data, food delivery couriers are predominantly young (79 per cent being under 30). They are low paid: full-time couriers in Beijing earned an average monthly income of 4048 yuan (approximately $589) in 2018, lower than the average monthly wage (4500 yuan) for urban workers in the catering industry (Beijing Statistics Bureau 2017). Furthermore, over 90 per cent of couriers are migrant workers from other parts of China, especially from rural regions, who face structural discrimination in seeking employment and in the provision of healthcare and social benefits while living in Beijing.

Different to the Chinese ride-hailing platforms expanding out of the pre-existing taxi industry (Chen 2018), location-based food delivery platforms have carved out a new, digitally mediated market in which the dominance of new employees is the most salient point. According to our 2018 survey, about 75 per cent of couriers have worked for less than one year and more than one-half have less than three months of work experience as couriers. Despite being new to the market, couriers are also quite fluid. With the exception of the newest couriers (with less than three months), it is common for them to hop from one delivery platform to another. For example, nearly 80 per cent of couriers with between three and 12 months of experience have changed platforms.

Throughout our fieldwork, we have observed a proliferation of informal types of employment in the food delivery labour market, similar to what has been recorded in the ride-hailing market (Chen 2018). However, because food delivery is less regulated than ride-hailing, the diversification of informal labour is more severe while the labour management structure is more hierarchical.

In making sense of the complicated employment relationships that have multiplied over the years, we use four main types to summarize the labour contracts on offer or lack thereof. These four types are also based on other structural elements such as the working app, working conditions and labour processes of food delivery work, each with its own employment relations and working conditions, as follows:

1. Platform-employed labour – workers directly hired by platforms and who have signed labour contracts with the platforms. This is the most formal and direct employment that would abide unequivocally by China's Labour Contracts Law.[4]
2. Outsourced labour – couriers recruited by intermediary firms such as subcontractors or staffing agencies. The latter employ couriers who, mostly, work on a full-time basis with fixed schedules. However, there are also informal labour agencies and brokers that do not sign contracts with workers. In any case, in this arrangement, there is no direct contractual relationship between the platforms and couriers. The managerial division of labour between intermediary firms and the platforms is not always clear, leaving space for the staffing agencies and the platform companies to evade accountability and responsibility. However, couriers' working arrangements mean that their working conditions are relatively stable, featuring less precarity than crowdsourced labour.
3. Crowdsourced labour – these are self-employed couriers working part time or on a casual basis. A small minority are recruited by staffing agencies to work full time but are denied a contract as outsourced labour since they are 'self-employed' which also makes them the least likely to be protected by the Chinese Labour Law. From a courier's perspective this mode of employment is both the most flexible and the most precarious while, for platforms and intermediary agencies, it offers the cheapest cost with the least accountability. In reality, platform algorithms often assign less desirable jobs to crowdsourced couriers because they are seen as more dispensable than platform-employed or outsourced labour.
4. Restaurant-employed labour – some large restaurants can afford to maintain their own couriers who may or may not be hired through intermediary staffing firms. The workers can work for the restaurants both as employees and self-employed. Where the restaurants hire couriers directly, there is an unambiguous employment relationship with the couriers and the restaurants bear clear management responsibility. However, the couriers who are hired through the staffing agencies sometimes would sign con-

tracts with them, but the restaurants, rather than the staffing agencies, would play the crucial role in labour management on a daily basis (e.g. scheduling shifts). This decouples management responsibility from the contractual employment relationship, producing ambiguity and elusiveness to the detriment of workers.

The discrepancy between each type of employment illustrates the varied levels of informality and labour rights, which, in reality, may also overlap if couriers choose, or are forced, to enter themselves into multiple types of employment relationships. For example, a courier can work full time with Meituan while at the same time providing crowdsourced, casual labour for Ele.me.

Over the years, the dominant mode of employment has shifted from platform-hired couriers, when the market was at the emergence stage, to a mix of platform-hired and crowdsourced couriers when the market was expanding. Nowadays, outsourced couriers are on the rise while platform-hired couriers are in decline as the platforms and intermediaries exploit legal loopholes to maximize their profits. Staffing agencies have thus come to play a major role in labour recruitment and management. It is noteworthy that most workers were simply not given the opportunity to become direct employees of the platforms. Temporary staffing agencies serve as the intermediary between couriers and platforms, whereas rapid market transformation has led both platform companies and workers to put greater reliance on the intermediaries. We found that, during 2018–2019, Meituan and Ele.me enlisted more than 30 staffing agencies to amass a pool of labour and to manage workers.

To reduce labour costs and control workers, food delivery platforms and intermediaries have formed convoluted and multilayered relations with one another. One platform may work with several outsourcing companies while each of the companies may further outsource to smaller staffing agencies, obfuscating the contractual structure and corporate liabilities. These smaller agencies are responsible for managing workers in certain areas of Beijing. Many of them have developed a top-down location-based management system which includes a hierarchy of management units with the director centre and *quyu* (areas) at the top, and in charge of a large urban district, to the *zhandian* (stations) and *xiaozu* (delivery groups) at the bottom, which take care of smaller neighbourhoods.[5] Usually each group has a leader and 10 to 30 workers. Some agencies use data from the platform to carry out real-time labour surveillance.

18.4 MANAGING FOOD DELIVERY COURIERS

To extract surplus value from the labour process, capitalists often rely on lengthening the working day and/or heightening the intensity of work, both of which may be carried out through the deployment of new technologies. As far as food delivery is concerned, labour–capital relations are manifested in (1) time control; (2) increasing demand for emotional labour; and (3) the transformation of the labour process into gameplay to encourage what Burawoy (1979) called 'making out' and further disguise exploitation.

18.4.1 Control of Time

Time is an important source of value in the on-demand economy, especially for location-based delivery platforms that prioritize catering to customers' immediate needs. Algorithms are thus deployed to manage couriers' working time. The platform companies constantly calculate and revise delivery times because they consider efficient delivery to be the weapon to attract customers, beat competitors and impress investors. For example, in its application for initial public filing, now the market leader Meituan boasted that it achieved an average of 30-minute delivery times (Meituan-Dianping 2018). The strive toward efficient delivery translates to an ever shortening time for couriers. According to courier Xiao Ji:[6] 'When I first worked for Baidu, the delivery time for each order was 45 minutes, but nowadays it has been reduced to 29 minutes! It's crazy. When you look at your mobile, it shows you have just a few minutes left, you have to run.'

Due to the platforms' overemphasis of speed and efficiency, workers are under great pressure to deliver on time in the midst of Beijing's notorious traffic jams that have not been ameliorated in recent years. Especially during rush hour, they have to compete against the limited delivery time set by the platforms. As the delivery app installed on their mobiles collects data about delivery times, it attempts to predict, manage and rearrange the labour process with increasing precision.

Furthermore, the design of the algorithms disregards workers' emotional responses. During interviews, couriers mentioned the pressure to race against 'platform time' as a major issue. They see themselves as having little bargaining power with the platforms. As the algorithm evolves, the platforms try to shorten delivery times in order to extract more surplus value. The use of cut-off times in this way has led to a spike in traffic accidents among delivery workers, as has been observed elsewhere such as in Bangkok (Arvidsson 2019).

Li Feng, another courier, was angry about the way Meituan calculates delivery distance and time:

> When estimating delivery time, it [the algorithm] predicts the time length based on linear distance. It is not the case when we deliver the food. Roads have many bends. And we also need to wait for traffic lights … Yesterday I delivered one order which was claimed as 5 kilometres by the system; however, I rode almost 7 kilometres. The system regards us as helicopters, but we are not.

Like most key players in the location-based platform economy, food delivery platforms such as Meituan and Ele.me employ a discourse in recruiting couriers which combines flexible schedules and good rewards, but in reality their ordering dispatch system is elusive and black-boxed while, as discussed earlier, the average wage for couriers is lower than that of workers in the catering industry.

A striking finding we noted from our survey data is de-flexibilization. That is, between 2018 and 2019, the proportion of part-time couriers declined from 60 per cent to 25 per cent while that of full timers increased from 40 per cent to 75 per cent. Full-time workers are required to comply with a compulsory eight hour day with little schedule flexibility. The rising number of full-time couriers in the place of couriers with flexible hours is a counter-intuitive trend because 'flexible hours' is deployed discursively and practically by digital platforms to attract casual workers while shirking employers' responsibilities (Rosenblat and Stark 2016). It is also recognized as the merit of working in the platform economy by some workers whose time is otherwise constrained by, for instance, their family obligations (Berg et al. 2018). We call this 'de-flexibilization' because the trend is for the labouring hours and the shifts to be increasingly rigid.

There are multiple contributing factors to de-flexibilization. First, as the market competition intensifies between Meituan and Ele.me, reliability becomes equally important as speed when it comes to on-time delivery. Flexible workers with a certain level of autonomy to decide when to work are considered as less predictable and less reliable than couriers who work full time and on fixed schedule. But flexible couriers still serve the purpose of vindicating the flexibility discourse in the platform economy, which is why they still exist. The platform algorithm nudges workers, sometimes quite forcefully, to shift from part-time to full-time work. We were told by one part-time courier:

> At the beginning, there were some orders. We log in [to the system] and deliver the food. In the second half of 2018, we received fewer orders; even during rush hours we did not have orders. They [the delivery platforms] do not assign you orders unless there are so many that the full-time workers cannot complete them.

Second, the proliferation of employment types and the management roles played by the staffing agencies mentioned earlier offer the platform companies a unique advantage to exploit a large pool of full-time employed and self-employed couriers. Platform companies in the United States are motivated to make the classification of platform workers as independent contractors into law – for instance, the passage of Proposition 22 in California in 2020, so that they can evade legal responsibility for the work force that relies on their platform to perform their work (Shapiro 2018). Food delivery platform companies in China, however, set in motion the trend of de-flexibilization to achieve the goal of retaining an indirect management control over scheduling and hours to maintain market competitiveness without shouldering employers' responsibility for full-time couriers.

Third, as digital platforms significantly reduce the barrier for market entry, the threat of labour oversupply may drive down wages, leading to economic slow-down in China, where there are few alternative job opportunities. This makes the full-time mode, with its relatively stable stream of income, more appealing for couriers despite insufficient labour protection. However, precisely due to the lack of access to institutional protection, full-time couriers found themselves struggling. To get a living wage, they work longer than eight hours *de facto*. One full timer reported to us: 'If the system does not assign you enough orders in rush hours, you have to wait for more [orders]. Otherwise, you cannot make ends meet.'

18.4.2 A World of Emotional Labour

As the platforms compete to attract and retain customers, delivery workers have to perform various kinds of emotional labour, in ways similar to flight attendants (Hochschild 1979, 2003), albeit under algorithmic control:

> We are told that there is a lot of dos and don'ts. For example, when delivering food, we are not allowed to enter the customer's room, receive any tips or ask for a good comment. We have to smile, knock at the door and hand over the change with both hands.

According to Hochschild (1979, 2003), emotional labour comprises the acts of expressing socially desired emotions during service transactions. In our case, it refers particularly to emotional regulation and the social performance of workers when dealing with customers. While couriers are optimizing themselves in order to maximize customer satisfaction in a customer-oriented culture, they must simply endure their own relative inferiority (van Doorn 2014).

Our interviewees revealed that they had consistently to make phone calls, set times and wait outside when a customer was not at home. Sometimes they have to apologize even if they are not responsible for an unexpected delay. The goal is 'make the customer satisfied and get a five-star comment', because customer ratings and comments are directly related to workers' salary and promotion through the ranks.

Many couriers perceive the superiority attached to customers to be 'unfair'. Figure 18.1 is a screenshot from an app which displays every step in the food delivery process after the courier has received the order. It shows the mobile phone number of the courier so that the customer can easily call him or her. Customers are entitled to 'cancel the order' or 'rush the food' (*cuidan*) at any time in the ordering process. Some workers complain that such a design disempowers them: 'They [the customers] can see everything, all the processes; but we do not know who they are. And when there is a problem, we cannot just cancel the delivery like they can.'

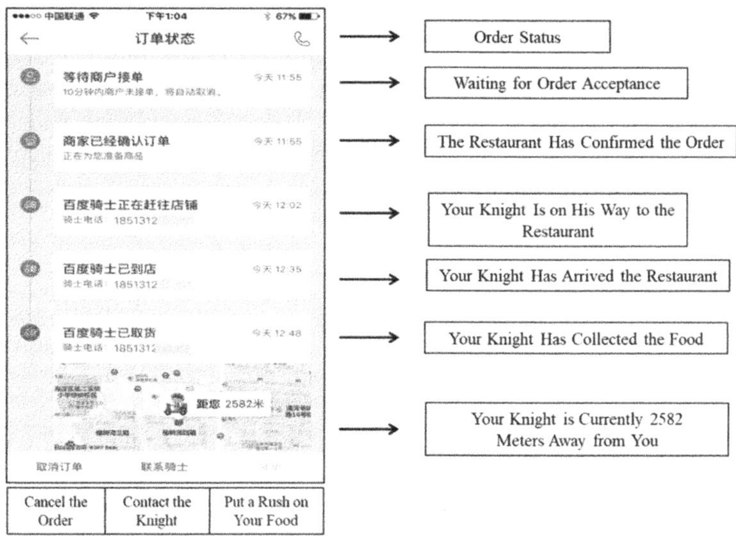

Figure 18.1 Order status in the app (visible to customers)

Algorithmic evaluation is legitimized on the basis of encoded assumptions of who matters and who does not. This echoes the findings of Hanser (2007) about 'distinction work' in which service work in Chinese department stores demonstrates the unequal power relationship between workers and shoppers with the former forced to acknowledge the superior class position of the latter. In the case of food delivery, 'customer superiority' is enforced through algorithmic design that disempowers couriers who are confronted not only with 'information asymmetries' (Rosenblat and Stark 2016) but also 'access asymmetries'. As such, the platforms have programmed and engineered workers to be servile labourers who have no choice but to accommodate the rising demands of emotional labour in respect of apologizing, communicating, coordinating, explaining and, most importantly, pleasing customers.

Table 18.1 The seven levels of knights in Baidu Delivery

Level of knight	Bonuses for each order	Accumulated points
Divine Knight (神骑士)	1.5 yuan (approx. $0.23)	6000
Sacred Knight (圣骑士)	1.2 yuan (approx. $0.18)	4100
Diamond Knight (钻石骑士)	1.0 yuan (approx. $0.15)	2800
Black Golden Knight (黑金骑士)	0.8 yuan (approx. $0.12)	1800
Golden Knight (黄金骑士)	0.5 yuan (approx. $0.076)	900
Silver Knight (白银骑士)	0.3 yuan (approx. $0.045)	400
Ordinary Knight (普通骑士)	0.1 yuan (approx. $0.015)	<400

18.4.3 Employees as Players

Food delivery platforms commonly adopt a game-like system and Chinese platforms are no exception (Mason 2018; Sun 2019). Couriers in China are referred to as *qishi* (knights) in most platform systems. The reference to 'knighthood' in Chinese implies masterful skills among workers riding their e-bikes (instead of horses); the emotional labour involved in expressing courtly professionalism; faithfulness; and a hint of gaming as the term frequently appears in electronic games.

To use Baidu Delivery as an example (although Ele.me and Meituan have a similar game-like hierarchy of couriers), its knights were organized into a hierarchy of seven levels whose specific names, from Ordinary Knight at the bottom to Divine Knight at the very top, are strongly reminiscent of video game terminologies. The overall game strategy of food delivery platforms is based on a 'scalable management technique' (van Doorn 2017, p. 903), as shown in Table 18.1. The hierarchy of knights reflected levels of labour based on, but not limited to, the number of their finished orders, travelling distance, length of working time and work performance reviews such as ratings and comments from customers.

The overall ranking system in Baidu Delivery was based on a points scheme under which the points required to maintain the level on a monthly basis were subtracted from the total number of accumulated points. If the accumulated points were insufficient to maintain the level of knighthood, the worker was demoted. A higher level of knighthood means that a worker must endure more pressure. 'I became a Black Golden Knight last month. I didn't expect that. See, if I want to keep being a Black Golden, I need another 832 points. This is a lot of work,' said Xiao Xu, whose mobile app gave him a reminder of his status: 'To maintain the same level, you need another 832 points.'

Table 18.2 Allowances for special order deliveries in Meituan

Special kinds of allowances	Requirements	Amount of allowance per order
Delivery distance	More than 3 kilometres	2 yuan (approx. $0.30)
Night shift	21:00–24:00	2 yuan (approx. $0.30)
	24:00–03:00	3 yuan (approx. $0.46)
Large order	Above 80 yuan (approx. $12.16)	2 yuan (approx. $0.30)
	Above 200 yuan (approx. $30.40)	5 yuan (approx. $0.76)
	Above 500 yuan (approx. $76)	10 yuan (approx. $1.50)
	Above 1000 yuan (approx. $152.04)	15 yuan (approx. $2.28)

Before summer 2018, worker income was a combination of base wage and bonuses. A shift system was the key to basic salary but this also implied limited flexibility for couriers because they could not log in and out as they pleased. For example, the base salary for couriers in Baidu Delivery was 3000 yuan ($457) and the bonus was calculated on the basis of knight level as well as the total orders delivered that month. From July 2018 onwards, many couriers in Baidu, Ele.me and Meituan have been transferred to an outsourced mode of employment as a result of which they have lost their 3000 yuan base salary, with their income thereafter being based on a piece rate. Each knight level corresponds to a different piece rate per delivery, each with its own level of privilege for bonuses, cancellation times and so on. For example, Ordinary Knights gets a bonus as low as US$0.015 per delivery and Divine Knights get a bonus of $0.23 per delivery. The accumulated points are key to moving up the levels in the 'game'. On Ele.me, as reported by an outsourced courier, the piece rate is 8 yuan (approx. $1.23) per delivery if one completes less than 1000 deliveries monthly. The rate rises to 9 yuan (approx. $1.38) per delivery if the completed monthly deliveries are between 1000 and 1300, and to 10 yuan ($1.53) per delivery if the completed monthly deliveries are over 1300.

To boost their business, platforms encourage couriers to deliver as many orders as possible in a limited time. According to Xiao Xu, 'They want us to work around the clock. In order to save time waiting for the lift, I kept climbing the stairs and my knees almost killed me.' The 'game', in this sense, is a very physical one.

In addition to ordinary deliveries, the platforms provide other incentives to boost income, including premia for long-distance deliveries, night shifts, bad weather, immediate deliveries and large orders (Table 18.2).

The hierarchical and 'game-style' system of evaluation of workers encourages competition between individual workers, different groups and sites since it acts as a basis for self-motivation and entrepreneurship. Such a system thus gains legitimacy not only through gaming workers' behavioural engagement (Rosenblat and Stark 2016) but also through engaging more workers into assisting its reproduction (Gillespie 2014). The platforms' increasing capacity to collect and tabulate social dynamics as information and worker evaluation standards exacerbates the poor working conditions of couriers by reducing them to being a 'mobile subject' (Platt et al. 2016 p. 2209).

The process of labour categorization is one way for the platforms to control and discipline workers. By promising the potential for salary increases, the system of categorization contributes to what Gillespie (2014) terms a 'calculated worker', based on worker management by algorithm (Rosenblat and Stark 2016).

Nevertheless, while the managerially imposed level of knighthood presumes the complete subordination of food delivery couriers, hampering their collective organization and ability to resist, the system can also be gamed both from within its schemes of calculated categorization as well as from outside them. For instance, the couriers may exploit the potential offered by other platforms (for example WeChat) and build alliances with other players (for example local state authorities) so as to 'short circuit' the 'circuits of labour' in the takeaway food sector (Qiu et al. 2014).

18.5 STRATEGIES OF RESISTANCE

Data and algorithms have become more than instruments in the discipline of labour by platforms. They have also triggered new tensions between capital and labour; between different platforms in a typical mode of inter-capitalist struggle (e.g. between Meituan and Ele.me); and between the companies and the Chinese authorities, especially local state governments. All these dynamisms complicate the picture, adding to the complex case of food delivery platforms in China being forced to respond to direct and indirect pressures of labour resistance. Despite the lack of resources, the power imbalance and information asymmetry, Chinese couriers do resist the logic of exploitation actively and collectively.

Food delivery is, of course, not unique in this respect: in other sectors such as ride-hailing, it has also been observed that the platforms face obstacles by way of labour resistance from worker-drivers who engage not only in traditional-style strikes but also in algorithm-based struggles with the purpose of gaming the system; for instance, when one platform provides higher 'bonuses', workers shift to that platform *en masse* (Chen and Qiu 2019). On other occasions, during work stoppages by ride-hailing drivers, those on strike used very low customer feedback scores to penalize strikebreakers (Wang 2016). This is a strategy that could be used during strikes by food delivery couriers although, so far, no such instances have been reported. Collective resistance, when undertaken on a large scale, would have unpredictable consequences in triggering government interventions: either crackdown on workers or pro-labour policy change, or both may occur at the same time.

In the food delivery sector, according to incomplete statistics contained in the *China Labour Bulletin*, couriers in Beijing, Shanghai, Tianjing, Chongqing and 19 other provinces have staged protests over pay cuts, low wages, wage arrears and unfair managerial practices by platforms. The total number of strikes has increased from nine in 2016 and 10 in 2017 to 57 in 2018 and 45 in 2019. While most of these strikes received scant coverage due to media censorship, the collective struggle will, structurally speaking, continue to intensify if the trend of labour 'de-flexibilization' and work intensification continues with more part timers forced to work full time and full timers being forced to work even longer.

With our surveys showing that 90 per cent of couriers are the main breadwinner in their family, food delivery labour is, in this sense, no longer 'flexible labour' at all in the contexts of social responsibility and social reproduction. Furthermore, the struggles of couriers are likely to persist when the survival needs of workers and their families clash with the capitalist and managerial logics of the platforms.

Besides the more dramatic moments of protests and strikes, couriers have developed ways of everyday resistance that bypass or fool the algorithms, sometimes even managing to subvert the system. Such strategic acts can arise from individual decisions but, more often, they are collective and pose alternative articulations to algorithmic control. One common tactic is to switch from one platform to another: couriers download multiple delivery apps and constantly switch between platforms to get more orders. This is workers' own 'planning' of their work schedules achieved as a direct result of a decision to game the system.

Another strategy was seen for instance during the 'bonus wars' among rival delivery platforms in spring 2014 and summer 2017. This is when the platforms were handing out considerable bonuses to expand market share and squeeze competitors into bankruptcy. Some workers formed alliances with restaurants, inputting fake orders online and instantly accepting them. After receiving the assigned orders, they would pretend to complete them without actually delivering anything, splitting the cash incentives with the restaurants. Because the platforms provide bonuses after reaching a certain number of orders, delivery workers in one *zhandian* would also help others to complete their orders so as to gain additional bonuses – again achieved through gaming the algorithms.

During rush hour, to ensure timely delivery without complaints, couriers sometimes transfer orders among themselves. Formally, they might use the *zhuandan* (order transfer) function in delivery apps to inform other couriers of an order; often, however, they just informally ask colleagues or friends to complete a delivery for them through WeChat groups, a virtual community through which workers may mitigate their vulnerability collectively. This is because the food delivery platforms constantly change their bonus policies, incentive information and digital management tactics. WeChat groups have become important sites for information sharing and circulation between dispersed couriers. They are also a critical place to build 'communities of practice' (Wenger 2000) as they allow couriers to share work experiences and strategies of how to game the system while informing each other of real-time traffic conditions.

18.6 CONCLUSION: FOOD DELIVERY AS DIGITAL UTILITY

In this chapter, we have examined China's location-based food delivery platforms, especially in Beijing, and the labour conditions therein. Using algorithms and real-time data, platform companies have gained tremendous managerial power both through technology design and the top-down manipulation of employment relationships, resulting in the stringent imposition of fast-paced capitalist logic upon couriers' working time structure, the requirement for emotional labour and the introduction of game-like systems of labour hierarchy.

However, Chinese delivery workers have found alternative ways to make sense of the algorithms and to develop new forms of resistance in creating strategies to work around or subvert platform control. Algorithmic labour management endeavours to embed inaccessible algorithms into the material condition of labour; in battling these, workers are seeking to overcome the logics of capital and the market, opening an algorithmic front for collective struggle while traditional forms of protest and strikes continue.

The struggle between platform domination and labour resistance has to be understood within the statist context of China in which local state authorities may become particularly important allies in labour rights advocacy in the platform economy. This is not to deny the increasing instances of authoritarian crackdown on labour activism. However, it would be wrong to assume the government is homogenously anti-labour, especially where the interests of local authorities are not captured by financial and/or high-tech capital.

The role of the state – especially city governments in China – is indeed highly significant. Several such governments, including the Beijing municipal authorities, intervened during the Covid-19 pandemic after realizing that food delivery platforms have become an essential life support mechanism for urban life under lockdown. Furthermore, the publication of a magazine article on 8 September 2020, entitled 'Delivery workers, trapped in the system', triggered a public outcry against platforms' excessive exploitation of food delivery couriers.[7] As a result, the local authorities in Beijing have exerted pressure on the platforms to ensure worker safety and welfare. Partly due to government pressure, and partly due to the criticisms voiced by public opinion, Meituan and Ele.me have begun to redesign their algorithms to offer more humane working conditions, for example by extending the delivery time. This shows that policy intervention, by the state and other public sector players, propelled by public pressure, can and indeed should make a difference with regard to platform labour being sustainable.

This is, however, where we need new concepts such as 'digital utility' (Chen and Qiu 2019) as a means of reorienting scholarly and policy agendas away from the technicalities of platforms to ultimate public interest outcomes. Couriers come from diverse backgrounds with high levels of labour mobility and the management system is complex and opaque, but this cannot conceal the growing centrality of location-based food delivery platforms to everyday life in urban China and, subsequently, their functioning as a digital utility. By 'digital utility' we propose an understanding of platforms as (a) public services and possessed of public service values, as proposed by Murdock (2005), Andrejevic (2013) and van Dijck et al. (2018); and (b) a symbol of the ways in which a transformed digital infrastructure has facilitated people's access to public or private services (Plantin et al. 2018) and without which their utilization of such services would be categorically different. Digital utility, defined in this way, is a subset of services like electricity and water but which are delivered by platforms. Digital utilities need to respond to public values such as affordability and they need to develop a sustainable infrastructure. For these reasons, they are in need of regulation by the state as opposed to being subject to self-regulation via the market.

There is, however, a crucial distinction between a digital utility and traditional utilities such as water and gas: the source of power for digital platforms – datafication – depends partly on algorithms, technology and non-human natural resources but, more crucially, on human labour, particularly couriers in the food delivery sector. Data extraction from the human elements of today's platform infrastructures – which we call 'digital utility labour' – is qualitatively different in its pivotal economic value as well as in its potential for panopticon-like control. No wonder that labour scholar Xiaoyi Wen (2018) contends that 'the real nature of the sharing economy is a labour-intensive economy'.

The rise of food delivery platforms necessitates state intervention because, as Srnicek notes, 'far from being owners of information, these companies are becoming owners of the infrastructures of society' (2016, p. 92). Digital utility platforms, like those for food delivery, have become critical urban service infrastructures. Their continuing operation requires the ongoing input of human labour from workers as much as capital investment and technological innovation. While food delivery in the past mostly generated jobs in the informal economy in Chinese cities, studies have shown that digital platforms now play the role of formalizing and standardizing service sectors that were previously informal (Ticona and Mateescu 2018). Thus, it would be desirable for local governments to have a sustainable platform economy that creates stable jobs and stable fiscal income. This would make local authorities potential allies for food delivery workers fighting precarity and unfair treatment, not to mention local branches of the official trade union (ACFTU, All China Federation of Trade Unions).

Like traditional utilities, food delivery giants such as Meituan and Ele.me should have the obligation of serving the public interest; in reality, however, they often act on the basis only of private interest, ignoring workers' rights and thus making digital utility labour a distinct source of systemic volatility. Couriers, their families and communities, and the functioning local state authorities share a common public interest in stabilizing this new platform economy, preventing the boom and bust cycle from wreaking havoc and disabling the essential urban infrastructure that food delivery has become. After all, in China as in other countries, food is delivered by humans not by algorithms.

NOTES

1. Authors' note: some of the material in this chapter comes from two published articles in the Chinese Journal of Communication: Sun (2019) and Chen and Qiu (2019).
2. Some of our research has benefited from the support of a local branch of ACFTU (All China Federation of Trade Unions) which has sought to improve employment relations in food delivery platforms to create better working conditions for couriers.
3. *Zhandian* in Chinese; literally meaning 'station points'.
4. At the same time, in our 2017 fieldwork, we found some platform-employed workers who did not have labour contracts.
5. The 'area' is the largest unit. Within each 'area', there would be several 'stations', while under each 'station' there are dozens of 'delivery groups'.
6. All the names of our interviewees have been anonymised.
7. The article, a long-form inquiry into the travails of food delivery work, was originally published in *Renwu* (People) magazine in China and then went viral online. It has been translated into English and thereafter published on the *Chuǎng* blog on 12 November 2020, at: http://chuangcn.org/2020/11/delivery-renwu-translation/.

REFERENCES

Andrejevic, M. (2013), 'Public service media utilities: rethinking search engines and social networking as public goods', *Media International Australia*, **146** (1), 123–132.

Arvidsson, A. (2019), *Changemakers: the industrious future of the digital economy*, Cambridge: Polity Press.

Beijing Statistics Bureau (2017), *Beijing statistical yearbook (2017)*, in Chinese, accessed 22 June 2021 at http://nj.tjj.beijing.gov.cn/nj/main/2017-tjnj/zk/indexch.htm.

Berg, J., M. Furrer, E. Harmon, U. Rani and M. S. Silberman (2018), *Digital labour platforms and the future of work: towards decent work in the online world*, Geneva: ILO.

Burawoy, M. (1979), *Manufacturing consent: changes in the labor process under monopoly capitalism*, Chicago, IL: University of Chicago Press.

Chen, J. Y. (2018), 'Thrown under the bus and outrunning it! The logic of Didi and taxi drivers' labour and activism in the on-demand economy', *New Media and Society*, **20** (8), 2691–2711.

Chen, J. Y. and J. L. Qiu (2019), 'Digital utility: datafication, regulation, labor, and DiDi's platformization of urban transport in China', *Chinese Journal of Communication*, **12** (3), 274–289.

CNNIC (China Internet Network Information Centre) (2020), *45th Statistical report on internet development in China*, Beijing: CNNIC.

Forward the Economist (2019), *Landscape of 2019 delivery industry in China: survey report*, March, in Chinese, accessed 21 June 2020 at www.qianzhan.com/analyst/detail/220/190308-8d14f0e9.html.

Gillespie, T. (2014), 'The relevance of algorithms', in T. Gillespie, P. J. Boczkowski and K. A. Foot (eds), *Media technologies: essays on communication, materiality, and society*, Cambridge, MA: MIT Press, pp. 167–194.

Hanser, A. (2007), 'Is the customer always right? Class, service and the production of distinction in Chinese department stores', *Theory and Society*, **36** (5), 415–435.

Hochschild, A. R. (1979), 'Emotion work, feeling rules, and social structure', *American Journal of Sociology*, **85** (3), 551–575.

Hochschild, A. R. (2003), *The managed heart: Commercialization of human feeling*, Berkeley, CA: University of California Press.

Jia, L. and D. Winseck (2018), 'The political economy of Chinese internet companies: financialization, concentration, and capitalization', *International Communication Gazette*, **80** (1), 30–59.

Mason, S. (2018), 'High score, low pay: why the gig economy loves gamification', *The Guardian*, 20 November, accessed 1 February 2021 at www.theguardian.com/business/2018/nov/20/high-score-low-pay-gamification-lyft-uber-drivers-ride-hailing-gig-economy.

Meituan-Dianping (2018), *Application proof of Meituan-Dianping*, accessed 1 February 2021 at www1.hkexnews.hk/listedco/listconews/sehk/2018/0920/a15846/emtd-20180622-05.pdf.

Murdock, G. (2005), 'Building the digital commons', in P. Jauert and G. F. Lowe (eds), *Cultural dilemmas in public service broadcasting*, Goteborg: Nordicom, pp. 213–230.

Plantin, J.-C., C. Lagoze, P. N. Edwards and C. Sandvig (2018), 'Infrastructure studies meet platform studies in the age of Google and Facebook', *New Media and Society*, **20** (1), 293–310.

Platt, M., B. S. Yeoh, K. A. Acedera, K. C. Yen, G. Baey and T. Lam (2016), 'Renegotiating migration experiences: Indonesian domestic workers in Singapore and use of information communication technologies', *New Media and Society*, **18** (10), 2207–2223.

Qiu, J. L., M. Gregg and K. Crawford (2014), 'Circuits of labour: a labour theory of the iPhone era', *TripleC: Communication, Capitalism, Critique*, **12** (2), 564–581.

Rosenblat, A. and L. Stark (2016), 'Algorithmic labor and information asymmetries: a case study of Uber's drivers', *International Journal of Communication*, **10**, 3758–3784.

Shapiro, A. (2018), 'Between autonomy and control: strategies of arbitrage in the "on-demand" economy', *New Media and Society*, **20** (8), 2954–2971.

Srnicek, N. (2016), *Platform capitalism*, Cambridge: Polity Press.

State Information Centre (2020), *China's sharing economy development report 2020*, Beijing: State Information Centre, accessed 18 February 2021 at www.sic.gov.cn/News/557/9904.htm.

Sun, P. (2019), 'Your order, their labor: an exploration of algorithms and laboring on food delivery platforms in China', *Chinese Journal of Communication*, **12** (3), 308–323.

Tencent.com (2020), 'Factories trapped in hard recruitment, while the number of delivery riders is growing', in Chinese, accessed 26 May 2020 at https://new.qq.com/omn/20200505/20200505A0F1O900.html.

Ticona, J. and A. Mateescu (2018), 'Trusted strangers: carework platforms' cultural entrepreneurship in the on-demand economy', *New Media and Society*, **20** (11), 4384–4404.

van Dijck, J., T. Poell and M. de Waal (2018), *The platform society: public values in a connective world*, Oxford: Oxford University Press.

van Doorn, N. (2014), 'The neoliberal subject of value: measuring human capital in information economies', *Cultural Politics*, **10** (3), 354–375.

van Doorn, N. (2017), 'Platform labor: on the gendered and racialized exploitation of low-income service work in the "on-demand" economy', *Information, Communication and Society*, **20** (6), 898–914.

Wang, H. (2016), 'ICTs, sharing economy and the transformation of labor politics in China: a case of Didi Dache', Paper presented at the International Workshop on ICT Development in East Asia, 29–30 August, Hallym University, Korea.

Wen, X. (2018), 'The real nature of the sharing economy is a labour-intensive economy', *The Paper*, 10 December, in Chinese, accessed 24 December 2018 at www.thepaper.cn/newsDetail_forward_2716522.

Wenger, E. (2000), *Communities of practice: learning, meaning, and identity*, Cambridge: Cambridge University Press.

19. Struggles over the power and meaning of digital labour platforms: A comparison of the Vienna, Berlin, New York and Los Angeles taxi markets

Hannah Johnston and Susanne Pernicka

19.1 INTRODUCTION

On-demand for-hire transportation platforms like Uber and Lyft are among the most popular and well-recognized digital labour platforms. Despite their global reach, the work performed on these platforms is highly localized and thus informed by local labour market conditions and the features of area transportation ecosystems. In much of the developed world, the platformization of the taxi industry has been the most recent attempt at sectoral deregulation (for example Dubal 2017; Rogers 2015); however, heterogeneous landscapes have engendered myriad responses by policy regulators, capital and labour vis-à-vis these new market actors. Unsurprisingly, these responses have varied geographically.

In this chapter we draw on a reconceptualized power resource perspective and present four case studies where workers' power resources vary along a power distribution continuum. At one end, we locate Austria, which exemplifies a strong position of neocorporatist industrial relations actors and a highly centralized workers' movement with significant institutional power. Germany is chosen as the second European case because of its strong neocorporatist tradition which has nevertheless had a declining influence on labour market structures and practices. In comparison to Austria, no collective agreement in respect of taxi and for-hire car drivers has been reached while institutionally weakened trade unions face a plurality of different business and employers' associations. As members of the European Union (EU) both Austria and Germany's market and industrial relations fields are at least indirectly influenced by a recent European Court of Justice (ECJ) ruling that classified Uber as a transportation company, requiring it to comply with rules governing traditional taxi companies. The Austrian and German cases contrast sharply with Los Angeles (LA) in the United States where industrial relations systems are comparatively weak and exclude (the often) self-employed taxi drivers. In LA the workers' movement is highly fragmented and pluralistic, and workers have struggled for recognition from transportation firms and policymakers at local and state levels. Between the European cases and LA, we highlight the example of New York City (NYC), where drivers' collective organizations and associational power have helped them to overcome weak institutional power and positioned them to engage in a symbolic struggle over what it means to work on digital labour platforms.

We examine the responses, strategic orientation, struggles and enacted resistance of on-demand transport workers and collective actors in the four cities mentioned above. Our reconceptualized approach is attentive to workers' structural, associational and institutional power resources – and to symbolic power which is understood as the perceived legitimacy of power resources. We view variations in the orientation of workers' and unions' responses to the introduction of digital labour platforms as contingent on workers' power positions in multi-scalar (local, national and transnational) political, economic and industrial relations spheres. We present these power positions as informative in identifying the array of key actors engaged in (re)shaping transportation markets and in establishing the subjective and scalar orientation of these actors' responses and struggles.

The chapter proceeds with an explanation of our theoretical perspective. We then present our four case studies which are followed by a brief comparative analysis as a conclusion. Our findings are based on semi-structured interviews with key stakeholders, including regulators, workers and their collective organizations and representatives of owners' groups in each of these cities. We have also gleaned important insight by monitoring public debates on transportation regulation that have taken place in the legislature and in government proceedings and via media outlets.

19.2 ORIENTING WORKERS' COLLECTIVE STRATEGIES: A RECONCEPTUALIZED POWER RESOURCE PERSPECTIVE

Our analysis is based on a reconceptualized power resource approach. Traditional power resource theory views structural and associational power as potential resources workers can draw upon to strengthen their position within market and political fields (Wright 2000; Silver 2003). Structural power consists of the power that accrues to workers 'simply from the location of workers within the economic system' (Wright 2000, p. 962). Tight labour markets or a strategic position within the production system, for instance, strengthen the bargaining position of labour in relation to employers. Associational power, in comparison, consists of 'the various forms of power that result from the formation of collective organisation of workers' (Wright 2000, p. 962), most importantly unions and political parties. A third and fourth type of power, introduced later into the debate, are institutional power, including as legal rules, norms and practices – a form of secondary social power in which past struggles crystallize (Brinkmann et al. 2007) – and societal power (Schmalz and Dörre 2018). Societal power is conceptualized as arising from the cooperative relations of workers and unions with other groups and organizations and the ability to recast the political project of unions as a political project of society generally (Schmalz and Dörre 2018). When successful, interests of workers and of society are aligned and broad coalitions are formed expanding workers' associational power resources.

We identify two limitations to the existing power resource approaches. The first is the (often implicit) assumption that the class interests of employers and labour are derived from their positions within the economic system which thus presents workers' and unions' interests as a fixed set of preferences. Such an approach neglects that workers' interests are socially constructed and an outcome of distinct struggles informed by the times, places and scales at which they take place. The second shortcoming relates to the prevailing understanding of institutional power as legal rules and norms, underestimating the informal and (often) unconscious symbolic dimension of power relations. Although power resource theory assumes that hegemonic narratives and worldviews play a role in determining the power of workers, the theory has limited capacity to explain the contested legitimacy of power resources and their effects on social inequality within and between groups of labour and employers. This is because traditional power resources theory remains focused on how workers employ society's prevailing power resources to advance their interests – and does not account for workers' and owners' symbolic struggles concerning the (il)legitimacy of such resources.

An analysis of the network of power positions in the taxi industry provides us with insight into the associational, structural and institutional power that is wielded by actors on both sides of the capital–labour relationship; symbolic struggles, meanwhile, occur over the recognition and legitimacy of power resources and the normalization of the stratified social order. Drawing inspiration from Bourdieu (1989, p. 17) we incorporate the concept of symbolic power into our analysis and understand it as a power resource rooted in the perception or recognition of power resources as legitimate. An example relates to the symbolic struggle over the classification of platforms as either technology networks or taxi companies. Unions and state agents have challenged new actors' claims that they are technology companies – contending instead that they are taxi companies who are denying drivers' wages and benefits by misclassifying them as independent contractors instead of employees. Independent contractors are rarely covered by labour law and lack collective bargaining rights, so the outcome of these struggles is decisive in shaping the legitimacy and effectiveness of their institutional and associational power resources. In this case, symbolic struggles (re)structure the social space within which the economic, political and industrial relations actors operate. They additionally determine the socio-spatial scale where these struggles manifest themselves. By institutionalizing rules, norms and practices on a local, regional or national scale the outcomes of these struggles shape transportation markets with an impact on present and future repertoires of contention.

To consider the multi-faceted struggles that have taken place in the taxi industry we have adopted a reconceptualized power resource approach that accounts for the multi-dimensional social space and social fields where workers' perceptions, interests and behaviours are constructed.[1] First, workers occupy different positions in a network of 'objective' power relations, being either economic, social or symbolic relations, giving them greater or lesser leverage in their struggles over the distribution of income or over the recognition or perceived legitimacy of their interests. Such power relations vary across scales; for example, some workers may receive recognition as powerful local actors but may be unknown regionally or nationally (or vice versa). Second, workers exhibit particular 'subjective' dispositions and position takings; these are understood as generative principles that correspond to, but are not determined by, the historical and current positions that actors occupy within social and geographic space. This reflects Bourdieu's *habitus* concept (Bourdieu 1990). Thus, in the case of on-demand transport, these features will shape the strategic orientation of workers' responses to the introduction of digital labour platforms and their strategies to influence their respective transportation markets. Workers who occupy, or have occupied, powerful institutional positions derived from their symbolic recognition as 'social partners' and from firmly established collective bargaining rights for taxi drivers are more likely to resume their strategic orientation based on a logic of influence. In contrast, drivers who have historically had less institutional power, or whose unions were not recognized or were even considered illegitimate, tend to orient their strategies towards building associational power; where successful, they are sometimes able to realize new institutional arrangements that better represent their interests.

The introduction of app-based transportation technologies has largely undermined workers' structural power by easing entry to the labour market and undermining occupational skill through the use of algorithms and navigation automation – we therefore consider their structural power to be low and do not discuss it here.

19.3 CASE STUDIES

19.3.1 Vienna

Austria boasts highly centralized and coordinated business and labour organizations and stable collective bargaining practices. The compulsory membership of business companies in the Chamber of Commerce contributes, in particular, to the country's exceptionally high level of collective bargaining coverage. Almost 100 per cent of private sector employees, including taxi and for-hire car drivers, as well as Uber driver partners, fall under the purview of sectoral collective agreements even though taxi and for-hire car drivers remain low-wage workers. Uber's cheapest service, UberPop, which operates with private vehicles, has been banned in many European cities and countries and has never been operational in Austria. Uber cooperates with for-hire car companies that have less strict regulations than taxis regarding their operational obligations, the qualifications requirements that apply to drivers and fare systems. In particular, for-hire cars must return to the place of business after each journey, a requirement Uber has been found to violate repeatedly.

Whereas institutional power resources in wage setting have remained intact over time, unions have lost some access to public policymaking and, hence, their highly regarded status as neocorporatist actors (Pernicka and Hefler 2015). This development is due to at least two factors. First, Austria's accession to the EU in 1995 shifted some competencies in legislative policymaking and jurisdiction from the national to the supranational level. European decisions such as the European Court of Justice ruling in 2017, which determined that Uber was a transportation company, rather than a digital technology firm, or the European Commission's urging of member states to open their markets to digital service providers and review existing national legislation 'to ensure that market access requirements continue to be justified by a legitimate objective' (European Commission 2016, p. 4), are likely to reshape the perceptions and evaluations of digital platforms in passenger transportation. However, taxi services are still considered part of the public transportation sector and thus, national and local actors maintain a strong interest in regulation. Second, Austrian governments since 1945 have been strongly committed to social partnership and a political culture oriented towards compromise, but this changed with the formation of coalition governments between the People's Party, ÖVP and the populist right-wing Freedom Party, FPÖ (and BZÖ, a splinter group from the Freedom Party), which came to power from 2000 to 2006 and from 2017 to 2019. These governments were hostile to tripartite concertation and denied labour organizations their previous informally granted role in legislative policymaking. Yet the unions and the Chamber of Labour still have close ties to the Social Democratic Party, SPÖ, currently the second strongest party in parliament, which lends political power to labour organizations and vice versa.

The most recent overhaul of taxi and for-hire car legislation can be partially attributed to the close relationship between the SPÖ and the unions. In February 2019 a former union official and social democratic member of parliament initiated legislation to bring all market players (taxis, for-hire car companies and Uber driver partners) under a common regulatory framework (Austrian Parliament 2019a). Interestingly, all major parties in parliament – ÖVP, SPÖ and FPÖ – eventually proposed legislation that was passed in June 2019 and which induced for-hire car companies, including Uber drivers, to follow the stricter rules of taxi companies (Austrian Parliament 2019b). Only two parties voted against the legislative proposal: the neoliberal Neos and Jetzt, a splinter group from the Green Party.

The policy actors thus created a level playing field that might nevertheless benefit the incumbents, including the two major taxi dispatch centres, over digital platform companies. Federal law has not yet been translated into local rules, however; this includes Vienna, Austria's capital and the only major city where Uber and a smaller Estonian company, Taxify (renamed Bolt in 2019), have operated so far (Kluge et al. 2020).

Since organized labour can still rely on institutional power, political backing and symbolic recognition, unions have tended to orient their strategy towards the logic of influence while engagement with their members and constituencies remains limited. This neglect of workers in the taxi and for-hire car sector has materialized in a very low union membership rate in Vida, the Transportation and Services Union, of less than 1 per cent (Vida, interview, 9 March 2018). The Chamber of Commerce representatives have also actively forged coalitions at national and transnational level (for example in the World Road Transport Organization) in order to influence the regulatory model reshaping the taxi and for-hire car markets (Pernicka 2019). The union relies on a voluntary membership model but the Chamber, however, benefits from compulsory membership and thus has more financial resources and staff.

19.3.2 Berlin

Despite its corporatist tradition, Germany exhibits a diffuse pattern of industrial relations, ranging from significant wage-bargaining coordination in core areas of manufacturing to more decentralized and fragmented industrial relations in the service sector. After the reunification of East and West Germany and, in particular, from the mid-1990s onwards, Germany's industrial relations system and collective bargaining practices began gradually to erode. Since the 1990s the collective bargaining coverage rate has declined from 72 per cent to 57 per cent in western Germany and from 56 per cent to 44 per cent in the eastern part (WSI 2020). Even though this decline has been less pronounced over the last 15 years, in 2015 unions welcomed the introduction of a general statutory minimum wage of €8.50 per hour. Industrial relations actors in the taxi and for-hire transportation sector, where a collectively agreed minimum wage had not previously existed, observed this development with mixed feelings.

As distinct from many other large German cities, including Munich, Düsseldorf and Frankfurt, the number of licensed taxi operators is not capped in Berlin. This situation has already exerted significant competitive pressures on both taxi companies and their drivers and has depressed taxi companies' profit margins. The introduction of global on-demand for-hire transportation platforms has fuelled greater industry competition and contributed to what a union activist has called 'predatory competition' (ver.di, interview, 11 November 2019). Like in Austria, Uber cooperates with for-hire car companies whose number has sharply increased in recent years (Linne und Krause, 2016, p. 54). Meanwhile regulatory authorities, in charge of granting new operating licences and of enforcing for-hire operating procedures, such as returning to the place of business after each journey, have faced notorious staff shortages (Taxi Innung Berlin, interview, 11 November 2019).

In contrast to Austria, Germany's United Services Trade Union, ver.di, which represents taxi and for-hire drivers, faces a plurality of voluntary and compulsory business association counterparts that operate in the sector. In Berlin there are no fewer than five voluntary business associations as well as the local and federal sections of the Association of German Chambers of Industry and Commerce (DIHK and IHK Berlin), which all taxi and for-hire car companies registered in Berlin are legally required to join. Despite some historical conflicts between the group of voluntary business associations, partly rooted in the previous bifurcation of Berlin into communist east and capitalist west, business representatives seem to share the view that taxi companies can only survive if politicians continue to regard them as a service in the public interest (*Daseinsvorsorge*) (Taxi Innung Berlin, interview, 11 November 2019; Taxiverband Berlin-Brandenburg, interview, 11 November 2019). A union official in charge of ver.di's transportation section in Berlin confirms this view (ver.di, interview, 12 November 2019).

Dependent employees, self-employed drivers and taxi owners all compete against each other. More recently, they have also faced increasing competition from for-hire drivers affiliated with Uber, adding to the difficulty of organizing and mobilizing members. Although a small group of active ver.di members in Berlin has been able to establish a union committee of lay officials, the total number of union members in the sector remains low. Nonetheless, ver.di maintains a strong neocorporatist position in that it remains confident that it will be able to reclaim the institutional power it has lost. The taxi and for-hire car sector – formally a part of the local public transportation sector – is not considered strategically important enough, however, either in material or in symbolic terms, to justify an increase in the union's organizing activities. This potentially casts doubt on the union's ability to cultivate associational power.

The most recent attempt by the Federal Minister of Transport to overhaul the main legislation applicable to taxis and for-hire cars at national (federal) level, the Carriage of Passengers Act (*Personenbeförderungsgesetz*), was met with harsh criticism on both sides of the capital–labour divide. The initial proposal to abolish for-hire cars' obligation to return to their place of business, in particular, was regarded as an unfair assault on the livelihood of taxi company owners and drivers who are already subject to stricter operating regulations. Both business associations and ver.di engaged in lobbying activities in protest of the proposal, but these efforts were spearheaded by business associations rather than by the union. This might be due to the business associations having comparatively more members (among taxi businesses and self-employed taxi drivers) than ver.di and also because they seem to be more aware of, and affected by, the material and symbolic consequences of Uber's strategic attempt to drive traditional taxis out of local transportation markets.

19.3.3 New York City

The United States is characterized as a liberal market economy with a labour market that is highly flexible and decentralized. While national legislation provides for freedom of association and collective bargaining, there is significant variability between individual states regarding how collective rights are enacted and the regulations regarding union formation and collective representation. Within the private sector, most bargaining agreements are negotiated at the enterprise level. This has led to a fragmented landscape and union membership and rights are continually under attack from legislative reforms which restrict, for example, unions' capacities to collect dues. Additionally, workers in a number of sectors, including agriculture and domestic work, have historically been excluded from the right to form unions and to engage in collective bargaining (these exclusions are rooted in a strong history of employment segregation and racial discrimination). Such exclusions also include self-employed workers, the employment classification of most taxi drivers throughout the United States.

While national union membership hovers around 10.5 per cent, NYC is unique in having among the highest rates of union membership nationally, rising to 22.3 per cent across the state (Bureau of Labor Statistics 2020a). This is attributed to NYC's enabling legal framework, high rates of employment in unionized industries and the state's legacy of pro-worker policies. For example, NYC has championed social policies including a $15 minimum wage and paid sick leave policies. Despite these progressive laws, for-hire drivers have rarely benefited because these are employment-related provisions and they are regarded as independent contractors.

The for-hire transport industry regulation falls under the purview of NYC's Taxi and Limousine Commission (TLC). This administrative body oversees the yellow cab industry and black car (limousine) services, the industry sub-sector through which Uber entered the market. The TLC maintains limits on the number of vehicles and regulations on minimum fares and lease prices for yellow cabs; for-hire vehicles, however, remained mostly un(der)regulated until 2018. This undermined drivers' structural power.

Among cab owners there are two distinct models for vehicle access. For years a significant portion of 'medallions'[2] were reserved for owner-operators who were also required to spend a certain number of days per year behind the wheel, although this latter requirement was lifted in 2017. Another category of owners typically owns fleets of medallions, leasing them out to drivers on a daily, weekly or monthly basis.

Fleet owners have historically been members of the Metropolitan Taxicab Board of Trade (MTBOT), a trade association which lobbies the city for favourable industry regulation and rules to facilitate lucrative leasing arrangements (Owner interview, November 2016). The interests of owner-drivers, meanwhile – and particularly since the introduction of Uber – are more closely aligned with lease drivers and are represented by the largest drivers' group, the New York Taxi Workers Alliance (NYTWA), a union without bargaining rights that represents workers in dialogue with the city. The relationship between these groups has historically been tense. Prominent MTBOT members have been publicly associated with worker overcharges and exploitation so, rather than engaging in bipartite dialogue, workers have directed their efforts at ensuring city-wide regulation (Johnston 2017). Since Uber entered the market, the Independent Drivers Guild, an association of app-based drivers, has been formed for the purpose of engaging in bipartite dialogue with Uber. This group, however, has had limited success in this forum and has thus adopted a strategy emulating that of NYTWA. Amidst this fragmented landscape there are also a small number of limousine drivers in select firms who have achieved union recognition, but these have not played a dominant role since the introduction of transportation apps.

The seven years of unbridled competition that transportation apps enjoyed has been attributed to the city's failure to regulate the limousine sector or to cap registration on the number of vehicles. Between 2014 and 2018, the number of black cars on the street increased by nearly 100,000 or over 300 per cent (De Blasio and Joshi 2016). Uber responded to mayoral discussions in 2011 about a prospective cap by successfully mobilizing its user base in opposition. Around the same time app-based transport companies attempted to implement state-wide legislation to permit widespread operation. In demonstrating their associational might and aiming to secure favourable institutional power resources, NYTWA organized repeated convoys to the state capital successfully to fend off these measures, citing the negative impact that state-wide regulation could have on the TLC's oversight powers and on working conditions for NYC drivers.

By 2018, following a rash of nine driver suicides, largely attributed to the toxic assets that drivers held in the form of medallions and an increasing inability to meet their financial obligations, the city introduced legislation to cap the number of vehicles and to institute a pay standard for limousine drivers. This regulation was pursued heavily by NYTWA, which oriented its efforts towards a broad-based and inclusive approach to membership with the goal of introducing city-wide sectoral regulation that would institutionalize the group's associational power. As part of a symbolic struggle, NYTWA championed a narrative that prioritized drivers' lives over app businesses while building alliances in city government and amongst the public. It also received support at rallies and on picket lines from service sector unions such as the Services Employees International Union. Finally, NYTWA led a dynamic media campaign that differentiated between the technology of the apps and the business model that these companies were perpetuating. A business model where standard practice included unilateral price cuts in the effort to attract more customers would lead app companies to a race to the bottom, NYTWA argued (Interview, November 2016). Discord between drivers would only fuel the erosion of working conditions and wages; in uniting drivers across the sector, however, drivers would be able to resist.

19.3.4 Los Angeles

California has been ground zero for ride-sharing applications and is an important site of struggle for workers and firms vying for associational and institutional power. Many digital labour platforms, including transportation networking companies Lyft and Uber, are headquartered in Silicon Valley which also renders the state an important site in the struggle to determine the narratives, and thus how symbolic power is distributed in the so-called sharing economy. Like NYC, LA is an important taxi market. Thus, the institutional regulatory arrangements established for transportation apps have significant bearing on the financial futures of ride-sharing platforms generally. Both cities, for example, are counted among Uber's five most lucrative markets globally (Uber Technologies 2019).

California boasts a fragmented field of associations representing business/employers and labour. Though the state has the largest absolute number of union members, this is due to California's sizeable population. Yet unionization has been in decline; after dropping four percentage points over the past decade, it now hovers below 15 per cent (Bureau of Labor Statistics 2020b). While decreasing unionization rates may be one catalyst for the development of coalitional organizing through worker centres, community and faith groups, unions have historically presented a centralized venue for the expression of workers' associational power which has also facilitated cohesion and the organization of businesses as a countervailing force. Thus, declines in unionism may also explain fragmentation and plurality amongst employer-based interest groups.

Like NYC, LA's taxi industry is regulated by municipalities and, as independent contractors and self-employed workers, drivers have been unable to join unions or to engage in collective bargaining since the 1970s. As a result, drivers have mostly organized through groups like the LA Taxi Workers Alliance (LATWA) which, along with NYTWA and affiliates from six cities around the country, is part of the National Alliance of Taxi Workers. Affiliates have adopted similar strategies by leveraging their associational power to create more favourable institutionalized regulation for taxi services. Fleet owners, meanwhile, are represented by the Taxicab and Paratransit Association of California (TPAC); while membership is open to individual owner-operators, larger member groups have greater influence over policy positions (Pernicka 2019). Rules and regulations overseeing limousine services, however, are determined at state-wide level by the Public Utilities Commission (PUC); under these, limousine drivers have long been considered employees and afforded key employment protections including the right to bargaining collectively, although this has never been achieved. Employers, meanwhile, have structured their associational membership at the state level through the Greater California Livery Association, a feature that is perhaps unsurprising given the state's approach to uniform (de)regulation.

In response to the proliferation of app-based transport services, state legislators created a new legal category for platform-based companies that absolved them of their compliance requirements at municipal and city level. This created an unequal regulatory landscape that placed fewer restrictions on platform operators relative to traditional taxis. TPAC responded by successfully lobbying the state to introduce another state-wide measure that provided for the deregulation of taxicab metered rates and, in particular, the elimination of the base fare rate. While this allowed taxi companies to compete more effectively with app-based services, for smaller taxi companies that could not afford to operate state-wide and drivers who bear much of the employment-related economic risk, this move symbolized, as in NYC, another race to the bottom (LATWA, interview, 18 October 2017). Deregulation eroded the institutional framework that was central to the organizing strategy of traditional taxi drivers in their efforts to achieve more favourable working conditions. Additionally, the trend to pre-empt local regulation with state-wide reform made it difficult for locally embedded groups like LATWA to respond because their struggles had traditionally been framed by their immediate geography.

Despite these challenges California legislators signed Assembly Bill 5 (AB5) into law in September 2019, which introduced a simplified 'ABC' employment classification test with the expected result of bringing hundreds of thousands – app-based drivers among them – into a formal employment relationship. This state-wide bill, codified rulings from *Dynamex* v. *Superior Court of LA*[3] and the legislative intervention that this necessitated had a bearing on the structural and institutional power of drivers and could have breathed new life into union organization. What is more, following the implementation of AB5, California's PUC issued a ruling that presumed app-based drivers to be employees, thus ensuring access to collective rights. Yet, these protections were not sufficiently institutionalized, and measures to exempt drivers from this legislation were rescinded via Proposition 22, a ballot initiative in the November 2020 election. While employment status represents a hopeful prospect to reinvigorate workers' struggles, fluctuations in state-wide regulation suggest that drivers' local strategic orientation may be strengthened by a more geographically expansive approach.

19.4 CONCLUSION

Our starting point was the observation that the emergence of digital platforms in taxi and limousine markets had engendered various responses by (organized) labour, capital and public policy actors. We drew on a reconceptualized power resource perspective to explain and comprehend differences in the strategic responses, struggles and enacted resistance of on-demand and traditional transport workers and their representatives. We find that these variations are contingent on organized workers' (historically evolved) material and symbolic power positions and their subjective dispositions in a multi-scalar and multi-dimensional social space.

Our empirical findings from our European case studies indicate that high and medium levels of institutional and symbolic power in (former) neocorporatist settings coincide with low levels of associational power in the taxi and limousine markets. Union actors are more likely to strategically rely on supporting institutions (collective or statutory minimum wages) and see no need to organize and mobilize members; a disposition and strategic orientation that reflects the historical trajectory of industrial relations and political fields in Germany and Austria where social partnership has been held in high esteem. In contrast to their Austrian counterparts, German unions have partly lost their strong institutional position and have been unable to establish the practice of collective bargaining in the taxi and for-hire sectors at federal and local/city levels. Ver.di has been neither able nor willing to pursue a coherent strategy on how to tackle the most recent liberalization attempts by the German government in favour of platform transportation companies, a response that could be explained in turn by the union's (traditionally) low membership density in the sector and because the sector was deemed to be of low organizing priority.

Scale also matters in highly institutionalized settings as the example of Austria exemplifies. The national regulation of the passenger transportation sector in 2019 established an equal playing field for digital platforms and traditional taxi and for-hire companies and their drivers. While this was well suited to union representatives' strategic orientation towards a national influence, it remains to be seen how these regulations will play out at the municipal level where authorities regulate fares. The Berlin case occupies a middle ground in terms of institutional power at national (statutory minimum wage) and sectoral levels where symbolic struggles are still taking place to establish a level playing field in passenger transportation; however, the union has chosen a wait-and-see strategy.

In comparison to Austria (Vienna) and Germany (Berlin) we positioned the United States, and especially California, at the opposite end of the power distribution continuum because of unions' weak position within industrial relations and political fields and their difficult symbolic struggles since the 1970s for legitimacy and recognition from transportation firms, policymakers and the general public. NYC, meanwhile, was chosen because it exemplifies how enacted associational power can help to overcome weak institutional power and position workers to engage in the symbolic struggle over what it means to work on digital labour platforms. A key finding relates to the spatial scale where power struggles actually take place. Workers are better able to achieve favourable outcomes when they demonstrate associational power at a similar scale to the regulation they seek and in places where they have greater symbolic power. Unions in NYC used locally embedded associational power to maintain control at that level, which helped to ensure more favourable institutional power arrangements that, in turn, have promoted local governance. This grassroots focus helped to unite the broader labour movement but NYTWA's success also depended on workers' greater symbolic power in NYC particularly.

California exemplifies a situation in which workers' collective organizing efforts neither mirrored the scale of TPAC nor of the PUC. Drivers were engaged locally while the TPAC and Uber had scaled up their power struggle to the state-wide level. This left drivers who were not well connected beyond their immediate geographies poorly positioned to respond and resist deregulation efforts. Workers had weak institutions during these events but, and in light of Proposition 22, are unlikely to be able to redistribute institutional and symbolic power to drivers in local jurisdictions throughout California.

NOTES

1. The same reconceptualized approach can be used to examine the power resources of business owners.
2. The transferable permits under which taxi services can be offered in New York.
3. Dynamex is a same-day courier and delivery service.

REFERENCES

Austrian Parliament (2019a), 'Neu im Verkehrsausschuss: Oppositionsparteien fordern verpflichtende Abbiegeassistenten für Lkw; SPÖ will Mietwagen- und Taxigewerbe zusammenlegen', Parlamentskorrespondenz Nr. 214, 05.03.2019, Vienna.

Austrian Parliament (2019b), 'Verkehrsausschuss beschließt einheitliches Gewerbe für Taxis und Mietwagen, Mehrere Gesetze mittels Initiativantrag von ÖVP, SPÖ und FPÖ auf den Weg gebracht', Parlamentskorrespondenz Nr. 709,19.06.2019, Vienna.

Bourdieu, P. (1989), 'Social space and symbolic power', *Sociological Theory*, **7** (1), 14–25.

Bourdieu, P. (1990), *The logic of practice*, Stanford, CA: Stanford University Press.

Brinkmann, U., H. L. Choi, R. Detje, K. Dörre, H. Holst, S. Karakayali and C. Schmalstieg (2007), *Strategic Unionism: Aus der Krise zur Erneuerung? Umrisse eines Forschungsprogramms*, Wiesbaden: Vs Verlag Fur Sozialwissenschaften.

Bureau of Labor Statistics (2020a), 'Union members 2019', News Release, 22 January, Washington, DC: Bureau of Labor Statistics, accessed 9 November 2020 at www.bls.gov/news.release/pdf/union2.pdf.

Bureau of Labor Statistics (2020b), 'Union members in California – 2019', News Release, 4 February, Washington, DC: Bureau of Labor Statistics, accessed 9 November 2020 at www.bls.gov/regions/west/news-release/pdf/unionmembership_california.pdf.

De Blasio, B. and M. Joshi (2016), '2016 TLC factbook', New York: Taxi and Limousine Commission.

Dubal, V. B. (2017), 'The drive to precarity: a political history of work, regulation, and labor advocacy in San Francisco's taxi and Uber economies', *Berkeley Journal of Employment and Labor Law*, **38** (1), 73–136.

European Commission (2016), 'Communication from the Commission to the European Parliament, the Council, the European Economic and Social Committee and the Committee of the Regions: a European agenda for the collaborative economy', COM (2016) 356 final, Brussels: European Commission.

Johnston, H. (2017), 'Workplace gains beyond the Wagner Act: The New York Taxi Workers Alliance and participation in administrative rulemaking', *Labor Studies Journal*, **43** (2), 141–165.

Kluge, J., M. G. Kocher, W. Müller and H. Zens (2020), *Empfehlungen für die Gestaltung eines Tarifs für die neue Konzessionsart „Personenbeförderungsgewerbe mit Pkw – Taxi" im Bundesland Wien*, Vienna: Institut für Höhere Studien.

Linne und Krause (2016), *Untersuchung zur Wirtschaftlichkeit des Taxigewerbes in der Bundeshauptstadt Berlin*, Erstellt für die Senatsverwaltung für Stadt und Umwelt, Hamburg: Linne und Krause.

Pernicka, S. (2019), 'The disruption of taxi and limousine markets by digital platform corporations in western Europe and the United States: responses of business associations, labor unions, and other interest groups', Working Paper, accessed 9 November 2020 at https://irle.ucla.edu/wpcontent/uploads/2019/06/Disruption-of-Taxi-and-Limousine-Markets.pdf.

Pernicka, S. and G. Hefler (2015), 'Austrian corporatism – erosion or resilience?' *Österreichische Zeitschrift für Politikwissenschaften*, **44** (3), 39–56.

Rogers, B. (2015), 'The social costs of Uber', *University of Chicago Law Review*, **82** (1), 85–102.

Schmalz, S. and K. Dörre (2018), *The power resources approach*, Bonn: Friedrich-Ebert-Stiftung.

Silver, B. (2003), *Forces of labor: worker's movements and globalization since 1870*, Cambridge: Cambridge University Press.

Uber Technologies (2019), 'Uber Technologies, Inc. Form S-1 registration statement', United States Securities and Exchange Commission, accessed 9 November 2020 at www.sec.gov/Archives/edgar/data/1543151/000119312519103850/d647752ds1.htm.

Wright, E. O. (2000), 'Working-class power, capitalist-class interests, and class compromise', *American Journal of Sociology*, **105** (4), 957–1002.

WSI (2020), 'Tarifbindung', accessed 11 December 2020 at www.wsi.de/de/tarifbindung-15329.htm.

20. Passenger transport in Australia: Injury compensation, public policy and the health pandemic

David Peetz

20.1 INTRODUCTION

This chapter examines a group of medium- to low-skilled, on-demand or location-specific workers – ride-share drivers in Queensland, Australia – and the policy response to a specific aspect of their disadvantage: injury compensation. 'Ride-share' is a term, popularized by the industry and its advocates, for passenger vehicle transport through apps operated by firms like Uber, Lyft and Sidecar. The main focus of the chapter is the efforts of a government review – led by the author of this chapter – into the injury compensation system (the Australian term is 'workers' compensation') in the state of Queensland, located in the context of international evidence on ride-share work, and the lessons this has for broader questions of the regulation of platform work (Peetz 2019).

Behind the superficially narrow focus of that review was the federal nature of industrial relations in Australia. Most states have passed their lawmaking responsibilities for industrial relations to the Commonwealth Parliament following the latter's use of the 'corporations power' in the 2000s to regulate industrial relations.[1] The ability of state governments to protect platform workers is therefore limited although one area where states can act is the injury compensation system.

This chapter contributes to our understanding of how concepts of social protection can be adapted to the status of independent contractors in Australia and elsewhere. The recent outbreak of the Covid-19 pandemic is also covered, illustrating how ride-share drivers are exposed to economic and social risks.

The chapter first examines labour and policy in the ride-share industry in the Australian context. It then considers the context of injury compensation and looks specifically at the 2018 workers' compensation review and the implications this had for platform workers, particularly in the ride-share sector. Finally, it considers developments after the review, including the Covid-19 pandemic, and the lessons for the regulation of platform workers more generally.

20.2 LABOUR AND POLICY IN THE PASSENGER VEHICLE TRANSPORT INDUSTRY

The platform economy is currently rather small as a proportion of the overall workforce (Katz and Krueger 2016) but it has considerable potential for growth. Its emergence reflects several factors. Managerial desire for greater flexibility has grown. New digital technologies have enabled 'algorithmic management' (Mohlmann and Zalmanson 2017) to substitute for control via the employment relationship. New models of organizational structure have also developed. Meanwhile, firms have sought to achieve 'apparent distancing' through what can be called 'not there employment' – a method of corporate organization whereby firms can avoid accountability for the misbehaviour of affiliates or subsidiaries by denying responsibility. The labour utilized by such organizations may be classed as 'employees' or as 'contractors', depending on the context (Peetz 2019). All these factors are relevant to the ride-share industry.

Ride-share work is not without its dangers. After 2015, when Uber started to make substantial inroads in the major capital cities, Australian road deaths appeared to be above what would have been expected given the downwards trend that had existed over the previous three decades. This trend was apparent in the national data and in data for the two largest states (Victoria and New South Wales); to a lesser extent, it was also there in the data for Queensland, the third largest (based on calculations from the Bureau of Infrastructure Transport and Regional Economics (2014, 2020)). These simple observations are consistent with a more thorough study by Barrios, Hochberg and Yi (2020) of cities in the United States, which found that the arrival of ride-share was associated with an increase of 2–3 per cent in the number of vehicle fatalities, along with increases in traffic, congestion, fuel consumption and new car registrations. While this is far from proving that Uber is more dangerous than alternative scenarios, it suggests that we should not downplay the safety implications of its emergence and that it is important to make sure that injury compensation is adequately dealt with in this sector.

Even without Uber, the vehicle passenger transport sector inherently shows above-average risk. WorkCover Queensland, the main insurer, estimates 'experience rated' premiums, meaning that premiums reflect the payouts the insurer has had to make in recent years. This is, in turn, a reflection of risk in the industry. It uses 560 industry classifications, each with its own premium. For road passenger transport (which includes buses and taxis) the premium (and, by implication, the risk) is 25 per cent higher than the all-industry median (WorkCover Queensland 2020). There is no disaggregation between ride-share and conventional taxi services, because ride-share workers have not been covered thus far.

From the foregoing, we can surmise that: the risk for ride-share drivers is significant and indeed higher than typical, but we do not know how it compares to taxi drivers; the death rate from motor vehicle accidents appears to have increased since the introduction of Uber but whether it is because of it cannot be determined; and injury compensation for ride-share drivers is an important and serious issue, even if we cannot state exactly how serious. As discussed later, the financial and health risks facing ride-share drivers have increased under Covid-19.

Ride-share workers are employed as contractors, not employees, which is why they have not been covered by injury compensation insurance up to now. The 'control' test is often applied in common law to establish whether an employment relationship exists. However, this is highly problematic. Large amounts of control could still be exercised by core capital (the central firm that organizes the work and directs the flows of money), even without the traditional indicators of control that the courts use to decide. Control is heavily there, yet 'not there'. The firm that attracts the most mentions in this chapter, Uber, says it is not a transport company hiring drivers; it is instead a technology company, acting as a client to drivers, but it is clearly in competition with transport companies and also celebrates its role in providing that competition (Castle 2017; Caspar 2015).[2]

Control of working time remains one of the indices used by the courts to determine whether someone is an employee or a contractor; yet amongst mainstream employers, organizational control of employees' working time has become less important over recent decades than organizational control of the product that employees generate for the employer (Fear 2011; Peetz et al. 2003). Uber can do many things that suggest it is an employer: terminate a driver's access to the app; deactivate a driver for cancelling trips, failing its background check policy, falling short of the required driver rating or soliciting payments outside of the app; make deductions against a driver's earnings; and limit the number of consecutive hours that a driver may work. However, when set against traditional measures, it is not an employer and its drivers are thereby open to being defined as independent contractors, as in the United States case of *Razak* v. *Uber Technologies Inc* (US District Court 2018).

The issue of whether platform economy workers are employees or independent contractors has been tested in the courts, tribunals and administrative bodies of a number of jurisdictions in Australia and overseas. The end result has been far from conclusive: some cases have led to platform workers being classed as employees; while others have led to them being treated as non-employees (BBC 2017; Hamilton 2016; Booth 2017; Grierson and Davies 2017; De Stefano 2018; Butler 2017), with developments in the California Supreme Court, legislature and political sphere taking on particular importance (Smith 2020; see also Chapter 8).

In Australia, the Fair Work Commission (FWC), in deciding an application for unfair dismissal, determined that a driver for Uber was not an employee. This outcome has been confirmed in subsequent cases but, in its initial judgment, the FWC (2017) commented that notions about what was necessary for an employment relationship to be established may be 'outmoded in some senses' and 'no longer reflective of our current economic circumstances', adding that perhaps one day 'the law of employment will evolve to catch pace with the evolving nature of the digital economy' (FWC 2017). Contradictory indications by courts and tribunals on whether location-based platform workers are employees do indeed arise, in part, from the way in which traditional legal conceptions of control and indicators of employment have failed to keep up with contemporary practices of corporate control and public understanding of what they mean.

Platform work has much in common with the Australian concept of 'labour hire', as the worker is paid through a party other than the person for whom they do the work. The biggest difference is that affected workers in the platform economy are mostly classified as independent contractors while most labour hire workers are employees of the labour hire firm. Employees paid by labour hire agencies are in uncertain 'triangular' relationships with the labour hire firm and the firm for whom they are doing the work, appearing to have inferior occupational health and safety outcomes (Underhill 2004; Quinlan 2015; Forsyth 2016) and little protection against unfair dismissal and redundancy.

As contractors, most ride-share workers are not unionized. The Transport Workers Union has sought to organize some delivery riders, mainly in New South Wales, but has little presence amongst ride-share drivers. Lacking unionization or employee status, many full-time workers in the platform economy are vulnerable. Expressed as an hourly rate, pay in the location-based platform economy is low and often below minimum wages while many of its workers have other jobs and a majority are underemployed (Chartered Institute of Personnel and Development 2017; https://twitter.com/FairGigForAll; Berg 2016; Unions NSW 2016; Smith 2016). They have low bargaining power. However, a financial transaction occurring between the client and another person, physically located in the same state, via an intermediary, does create an opportunity for state regulation.

20.3 WORKERS' INJURY COMPENSATION POLICY IN QUEENSLAND

The compensation of injured workers became government responsibility over a century ago when the inequity of allowing workers to be permanently disabled by workplace injuries without compensation became too politically odious while the cost of workers suing employers for such compensation became too burdensome both for workers and employers (Purse 2005). A universal, largely no-fault, system was recognized as the cheapest, most efficient and equitable approach.

Injury compensation is the constitutional responsibility of state governments in Australia rather than the Commonwealth. Systems vary between states, with private providers and competitive insurance markets having been established in some. In Queensland, with a population of 5 million, most workers' injury compensation insurance is handled by WorkCover Queensland, a single state agency. With employer premiums being 'experience rated', most firms pay for the cost of injuries in their workplaces in the long run. A small number of large organizations 'self-insure' but have to operate under largely the same rules.

20.3.1 The Queensland Review

In 2018, the Queensland government held a review of injury compensation, now required by statute every five years.[3] Over several weeks of preparation 28 organizations, including employer bodies, unions, insurers, lawyers and government departments, were consulted, mostly through formal meetings (Peetz 2019).

The review mainly looked at other issues to do with workers' compensation such as psychological injury, rehabilitation, access and compliance. However, the minister wanted action on the platform economy and one chapter of the report was devoted to this issue, identifying a number of problems facing platform workers and some of the constraints facing regulation in this area. It also made a number of key recommendations.

Under the existing system, the test of whether a worker was an employee was whether regular tax deductions should have been withheld by the organization. If a platform worker was found to be an independent contractor then they were not entitled to injury compensation and no premium was applied. This had been tested a few times when individuals applied for compensation, with platform workers having sometimes been found to be an employee and sometimes an independent contractor. Many injured workers would not have submitted claims because they believed they were not covered or it never occurred to them to do so. Furthermore, many platform workers did not have any alternative protection in place in case they were injured, with uncertainty having been created in part by the failure to modernize the law.

The distinction between workers who work in the online platform economy and those who work for location-based platforms or apps is important from both an analytical and policy perspective. Most of the options in connection with the compensation issue focused on extending coverage to those who work for location-based platforms, including ride-share apps. This is because of the greater difficulties in extending coverage to online platform work when much of this is undertaken across borders internationally. In contrast, for those working for location-based platforms, the worker and the client are located near each other, regardless of where the app is owned, and this occurs through multiple uses of the app. This makes it clear what geographic jurisdiction has responsibility for ensuring that minimum standards in employment, including in relation to workers' injury compensation, are set and complied with.

In seeking to regulate employment in the location-based platform economy, it is worth asking what are the points of leverage and what are the points of greatest resistance. The platform business model is based on minimizing costs while maintaining control through an app; this is critically threatened by any move to define platform workers as employees. Such firms might acquiesce in moves to ensure adequate coverage if these did not involve a redefinition of their workers as employees.

These particular matters were not canvassed during the review but are still relevant to any policy consideration. However, a key factor that was canvassed was the reduction of uncertainty – a term used 20 times in the report. As the report noted, 'Uncertainty is not satisfactory for administering a workers' compensation scheme' (2018, p. 99) while relying on the status quo was 'unlikely to remove uncertainty' (2018, p. 103). Furthermore, leaving matters to the courts would not reduce uncertainty. Several parties consulted during the review were keen that 'something' be done about workers in the platform economy but were unclear as to what that 'something' might be.

20.3.2 The Review's Findings on Platform Workers

The review identified several possible options for the coverage of platform economy workers in injury compensation systems. Most had a number of drawbacks, especially given the federal system of employment law dominated by the Commonwealth.

The review recommended redefining coverage to include those who work under agency arrangements and to require the payment of premiums by intermediaries or agencies. That is, if a platform economy firm supplied a worker that delivered a passenger or a meal, or undertook some other task for a third person, it would pay an injury compensation premium to cover that worker based on a percentage of their take. Such intermediary organizations gained their income by taking a proportion of the income paid by the client to the worker (a 'commission'). This commission varies between apps and over time. For example, in the case of Airtasker, the portion was, for a while, 15 per cent but was then reduced. Uber Eats set a commission of 35 per cent and then 30 per cent. In the ride-share market Uber's commission was 22 to 27 per cent, although the smaller firms charge less.

Consequently, premiums would be set as a proportion of the commission received by the agency. The net cost to insurers themselves would be zero as premiums in an experience-rated system would cover outlays. Basing payments on the revenue of the agent was technically a different concept to that used for conventional employees for whom wages were the basis. However, as commission was, in turn, calculated as a function of wages, and as premiums would be experience rated, premiums would in practice be calculated as a proportion of worker earnings, as they were for conventional employees.

Various exclusions would prevent the scheme from having unintended adverse effects while a programme to facilitate the return to work of injured workers was also recommended, along with an awareness campaign.

20.4 AFTERMATH OF THE REVIEW

To see what happened after the review reported, we need to refer to two aspects: the economic and health environment (briefly); and (in greater depth) the policy and political developments that happened within that environment.

20.4.1 Economic and Health Environment since the Pandemic

The main feature of the economic and health environment was, of course, the emergence of the Covid-19 pandemic in early 2020, which increased the financial and health risks facing ride-share drivers. The pandemic has influenced Australia less seriously than most of Europe and North America but has still had considerable effects. The first wave, initiated from inbound travellers, occurred across the country with the biggest effects in the largest states of New South Wales, Victoria and Queensland. The second, more serious, wave was concentrated in Victoria and started through poor infection control arising from the use in critical roles of casual workers and contracting – again, the influence of the 'not there' employment model (Sparrow 2020; Holden 2020).

The first impact of Covid-19 was on the ride-share business. Globally, ride-share usage collapsed where people were 'locked down' or encouraged to stay at home. In Australia Uber usage fell by 80 per cent in the first wave although the effects were partially reversed in Sydney, which has fared much better than Melbourne, the latter experiencing a second round of lockdowns from July (Rabe 2020). Economic effects were quickly transmitted to other workers, not just drivers. Globally, aside from the drivers who lost work, Uber sacked a quarter of all the staff it recognized as employees, in a three minute international Zoom meeting at which the unsuspecting invitees heard Uber's head of customer service, Ruffin Chaveleau, tell them: 'We are eliminating three thousand five hundred front line customer support roles. Your role is impacted and today will be your last working day with Uber. You will remain on payroll until the date [notified] in your severance package' (Ruffin Chaveleau in WION 2020).

From its inception, Uber has accumulated huge losses (Felton 2017). Yet Uber used the downturn to promote recognition of the role that ride-share work plays in supplementing the meagre incomes of workers financially stressed by the crisis – especially in the context of its campaign against the Californian legislation (Jennewein 2020), a campaign that ultimately succeeded with the passage of Proposition 22.

The second effect of the Covid-19 pandemic lay in the health risks faced by ride-share workers. Drivers in the United Kingdom and California have died from Covid-19 after exposure to coughing clients (Christmass 2020; Smith 2020). Uber did provide drivers with hygiene kits, sanitation bags and video messages and, in many countries, required drivers (and clients) to wear masks (Page 2020; Bellon 2020). Walking a tightrope between maintaining corporate legitimacy and not conceding its role as an employer, Uber also offered some drivers two weeks sick pay if they had to isolate due to Covid-19 (Canencia 2020). However, ride-share drivers did become exposed to occupational health risks greater than those they had experienced hitherto.

In short, Covid-19 has led to several important changes in the environment. Amongst drivers, it has seriously affected the health of some and threatened the health of many. Yet it has also reduced the incomes of many and made earnings more insecure for most. So, large numbers of people have taken on risks that they may not have done in the past as a result of that drop in income and in income security. Some responded by installing Perspex screens (Vida 2020) or using disinfectant supplied by the firm. Others, like those who subsequently died, have simply worked on – perhaps taking a calculated or not-so-calculated risk. Amongst ride-share companies, the pandemic has undermined their already weak financial viability but has also provided a political tool that might help them weather the policy storms.

20.4.2 Policy and Political Developments

Implementation of the recommendations of the Queensland inquiry required a 'business regulation review' and further consultation (Office of Industrial Relations 2019). Business regulation reviews were introduced in most states from the early 1990s as part of an Australia-wide 'competition' agenda driven by the national government, ostensibly to prevent the unnecessary regulation of business but with the effect of providing an additional avenue for the corporate sector to resist reforms it did not like.

The report covered many aspects of workers' injury compensation, some of which did not relate to platform workers or contractors. These other changes to workers' injury compensation law recommended in the report (which did not relate to platform workers or contractors) were introduced into the Parliament of Queensland in August 2019, a little over a year after the report was handed down. However, the changes to platforms were held over, initially until the regulation review was completed.

Reactions to the review's recommendations varied, not necessarily along predictable lines. Uber essentially accepted the report's recommendations, albeit half-heartedly. It is likely that the key thing for Uber was that the recommendations did not seek to redefine their drivers as employees. As the reaction to California's Assembly Bill 5 (AB5) legislation indicated, this is probably its 'line in the sand'. While workers remain defined as independent contractors, Uber can decide which protections it needs to offer its employees in which countries, subject to the norms and laws of those jurisdictions. In offering sick leave during the pandemic, the ride-share firm can act as a 'partial employer' – like Schrödinger's cat, it is simultaneously there and 'not there' as an employer. Ola, on the other hand, opposed the review's recommendations, arguing that the firm would 'encourage and support gig workers in providing for their own insurance' (Ola 2019, p. 7).

Perhaps the main opposition came from the Queensland Taxi Council, the body representing owners in the taxi industry. This was somewhat surprising as taxi drivers had been hardest hit by the rise of competition from ride-share, with the value of taxi licences falling by up to 78 per cent in Brisbane (Caldwell 2018). In some jurisdictions, such as Victoria, drivers were covered for workers' injury compensation through the taxi owner but this was not the case in Queensland, although drivers in Queensland were, for health and safety purposes, considered to be 'vulnerable' workers due to the danger of violence from passengers (Queensland Government 2011, p. 32; see also Queensland Government 2012). The Australian Taxi Federation (2016, p. 7) had previously argued that 'if workers' compensation cannot be changed to cover drivers, all owners, operators and partners must carry private insurance to cover drivers for accidents at work, and to and from work'. This implied support for the idea that workers' compensation should be changed to cover taxi drivers and highlighted the absence of insurance amongst drivers for Uber and other ride-share companies. In reality, requiring compulsory private injury insurance, outside the context of workers' compensation, for drivers of location-based platforms would not adequately address the problem, given that many do it on a part-time basis. It would be less efficient than applying coverage through WorkCover. However, the Taxi Council opposed the proposals. It already had an injury insurance scheme in place and might have suffered financially from its scheme's subversion by a replacement scheme that applied to taxis as well as ride-share.

Once the minister had made a decision, a submission would have to be drafted within the department and accepted by the minister before being presented to the Cabinet for discussion. Before that could happen, however, the Covid-19 pandemic came to Queensland. Within one day of New Zealand closing its borders Australia did the same, and then restrictions on movements and gatherings were quickly imposed in various Australian states including Queensland. Staff in government departments shifted to working from home and the focus of government quickly became handling the pandemic. The minister for industrial relations was also the minister for education and had to deal with the vexed issue of school closures. All matters unrelated to the pandemic disappeared from the government agenda. Amongst those that were put on hold was the response to those recommendations of the review dealing with the platform economy which, therefore, await a return to 'normal'. The government went to an election (which, in October 2020, it won) without making a public commitment on this issue.

In Queensland the Covid-19 pandemic has, in that sense, not only worsened the health and economic insecurity of ride-share workers, it has also considerably delayed the possibility of government amelioration of those problems.

20.5 CONCLUSION

What are the lessons from this experience for the regulation of the platform economy?

First, it is possible to find regulatory solutions to platform economy problems but these may be limited by the powers available to the relevant agency. In Australia's case, for example, it was possible for state governments to act. The issue of the changing nature of work and what it means for a legal understanding of 'employment' is, however, one that fundamentally requires national attention. It is only through legislation in the Commonwealth Parliament that reform to facilitate the regulation of the broader range of employment issues can happen.

Second, different aspects of the problems of workers in the location-based platform economy may require different, innovative solutions. Those solutions might not need to involve redefining contractors as employees but neither do they need to exclude their becoming employees – it may be a matter of finding a solution to a specific problem. In the case of Queensland, as the state parliament has no constitutional power to define employees and contractors, the review had to find a solution to the problems created in part by the classification of workers as contractors without being able to change that classification.

A lot of effort has been put into the question of whether and how workers should be defined as employees, and there are many arguments in favour of that although they are not always successful. Being unable to revise the contractor classification forced the review to be innovative about regulation and engage in some 'institutional experimentation' (Wright et al. 2019). The review had to find a way of creating a minimum standard for all workers, regardless of employee status.

Does it work if these solutions do not redefine contractors as employees? There may be some advantages. Some platforms will adapt if their core business model is not threatened – Uber will adapt to the minimum standards imposed upon it, presumably within limits, as long as its drivers are not redefined as employees. Others may resist resolutely, and we cannot assume that ride-share corporations will always act in unison since they are in competition with each other. In New York, for example, an innovative form of the regulation of minimum earnings in the ride-share and taxi industries was reluctantly accepted by Uber but immediately challenged in the courts by its competitor, Lyft (AFP 2019). Lyft was financially disadvantaged, relative to Uber, by the New York reforms because its drivers spent a higher proportion of their time waiting for fares and such differences fed through into the minimum fares that firms could charge. On the core issue of the contractor/employee definition affecting their common business model, however, Lyft and Uber united to oppose the AB5 reforms in California.

Understanding the political context is especially important. Opposition to reform can come from unusual and unexpected places. In Queensland it was not minimum earnings but workers' compensation that was being regulated, while an unexpected source of opposition was the Taxi Council. Taxi drivers are probably the group most disadvantaged of all by the emergence of ride-share, yet the body representing them had a financial interest in opposing reform. Policymakers have to be able to anticipate and adapt to such opposition.

A diversity of responses is appropriate for the broad range of situations and problems posed by the platform economy across various countries. One way to ensure adequate coordination of the various forms of institutional experimentation that are underway in a country is through a system of what can be called 'directed devolution'. This means that legal entitlements or obligations, for example minimum earnings, would be set at a higher, national level; subsequent to this a lower level would be required to determine the detailed application of the standard set by the higher-level policy and, in doing so, to protect the interests of the workers concerned. Crucially, such an approach must be designed to account for power (hence the 'directed' in 'directed devolution'). By tightly constraining the room to manoeuvre of those at the more decentralized level, this could minimize the likelihood of a power shift against the most vulnerable. At the same time, it would still allow the flexibility to account for differences in situation-specific circumstances.

During these events, the Covid-19 pandemic has highlighted many inadequacies in contemporary arrangements. The pandemic has also exposed gaps – for temporary/casual workers and contractors, and for platform economy workers in particular. In many respects, policy has ground to a halt and capital has used the opportunity to press its own agenda and maintain the power status quo. This was not, however, inevitable. The neoliberal dogma that allowed such gaps has been set aside by many countries in response to Covid-19 including, for a very short time, even in Australia. The ability of labour and progressive forces to organize in the pandemic, and the strategies they adopt, have a big influence on whether the pandemic ameliorates or worsens the situation for platform economy workers and others who are disadvantaged in labour markets. In Australia, as the second wave was traced to casual labour and contracting, it is far from clear that the conservative response to the pandemic's economic effects – to aim for more flexibility as per the platform economy – would gain support.

NOTES

1. The Australian Parliament is officially called the Federal Parliament but is also called the Commonwealth Parliament. The 'constitutions power' referred to here is set out in Article 51 of the Australian Constitution.
2. Lyft does not have significant operations in Australia although Didi, Bolt and Ola all commenced smaller-scale operations in 2018.
3. This was the second review since the Act was changed to require such a frequency.

REFERENCES

AFP (2019), 'Judge rules Lyft must follow New York rules for driver minimum wage', *Business Times*, accessed 18 December 2020 at www.businesstimes.com.sg/transport/judge-rules-lyft-must-follow-new-york-rules-for-driver-minimum-wage.

Australian Taxi Federation (2016), 'CTP premium system review', 20 March, accessed 18 December 2020 at www.sira.nsw.gov.au/__data/assets/pdf_file/0009/102330/Australian-Taxi-Federation.pdf.

Barrios, J. M., Y. V. Hochberg and L. Hanyi Yi (2020), 'The cost of convenience: ridesharing and traffic fatalities', Working Paper 2019-49, Chicago, IL: Becker Friedman Institute for Economics at the University of Chicago.

BBC (2017), 'Bike courier wins "gig" economy employment rights case', BBC News, 7 January.

Bellon, T. (2020), 'World's Uber drivers, riders to wear masks', AAP, 15 May, accessed 18 December 2020 at www.aap.com.au/worlds-uber-drivers-riders-to-wear-masks/.

Berg, J. (2016), 'Income security in the on-demand economy: findings and policy lessons from a survey of crowdworkers', Conditions of Work and Employment Series 74, Geneva: ILO.

Booth, R. (2017), 'Addison Lee wrongly classed drivers as self-employed, tribunal rules', *The Guardian*, 26 December.

Bureau of Infrastructure Transport and Regional Economics (2014), Road deaths Australia 2013 statistical summary, table 23 (spreadsheet tables from report), Canberra: Department of Infrastructure and Regional Development, p. 38.

Bureau of Infrastructure Transport and Regional Economics (2020), Road trauma Australia 2019 statistical summary, table 1.1 (spreadsheet tables from report), Canberra: Department of Infrastructure, Transport, Regional Development and Communication, p. 2.

Butler, S. (2017), 'Deliveroo wins right not to give riders minimum wage or holiday pay', *The Guardian*, 14 November.

Caldwell, F. (2018), 'Value of taxi licences plummets across Queensland, even without Uber', *Brisbane Times*, 20 February, accessed 7 January 2021 at www.brisbanetimes.com.au/politics/queensland/value-of-taxi-licences-plummets-across-queensland-even-without-uber-20180220-p4z0yo.html.

Canencia, G. (2020), 'Uber, Lyft halt ride-sharing in US and Canada amid coronavirus fears', Micky, 18 March, accessed 7 January 2021 at https://micky.com.au/uber-lyft-halt-ride-sharing-in-us-and-canada-amid-coronavirus-fears/.

Caspar (2015), 'Sydneysiders have spoken – and they choose ridesharing!', Uber Newsroom, 10 October, accessed 7 January 2021 at www.uber.com/en-AU/newsroom/sydney-has-spoken/.

Castle, J. (2017), 'UberX vs taxi – which is best?', Choice: Australian Consumers Association, accessed 7 January 2021 at www.choice.com.au/transport/cars/general/articles/uberx-vs-taxi-which-one-is-best.

Chartered Institute of Personnel and Development (2017), 'To gig or not to gig? Stories from the modern economy', London: CIPD.

Christmass, P. (2020), 'Coronavirus update: Uber driver dies after woman "coughed repeatedly" in his car', 7News.com.au, accessed 7 January 2021 at https://7news.com.au/lifestyle/health-wellbeing/coronavirus-update-uber-driver-dies-after-woman-coughed-repeatedly-in-his-car-c-969817.

De Stefano, V. (2018), 'Platform work and labour protection: flexibility is not enough', *Regulating for globalization: trade, labor and EU law perspectives*, Alphen aan den Rijn: Wolters Kluwer.

Fear, J. (2011), 'Polluted time: blurring the boundaries between work and life', Policy Brief 32, Canberra: The Australia Institute.

Felton, R. (2017), 'Uber is doomed', *Jalopnik*, 2 February.

Forsyth, A. (2016), 'Victorian inquiry into the labour hire industry and insecure work: final report', Melbourne: Industrial Relations Victoria, Department of Economic Development, Jobs, Transport and Resources.

FWC (2017), *Mr Michail Kaseris* v *Rasier Pacific V.O.F.*, Fair Work Commission. 6610.

Grierson, J. and R. Davies (2017), 'Pimlico Plumbers loses appeal against self-employed status', *The Guardian*, 11 February.

Hamilton, M. (2016), 'Judge finds NYC Uber drivers to be employees; upstate impact debated', *Timesunion*, 12 September, accessed 7 January 2021 at www.timesunion.com/allwcm/article/Judge-finds-NYC-Uber-drivers-to-be-employees-11220139.php.

Holden, R. (2020), 'Vital signs: Victoria's privatised quarantine arrangements were destined to fail', *The Conversation*, 23 July, accessed 7 January 2021 at https://theconversation.com/vital-signs-victorias-privatised-quarantine-arrangements-were-destined-to-fail-143169.

Jennewein, C. (2020), 'As California seeks injunction on gig workers, Uber says 158,000 will lose jobs', *Times of San Diego*, 24 June.

Katz, L. F. and A. B. Krueger (2016), 'The rise and nature of alternative work arrangements in the United States, 1995–2015', NBER Working paper 22667, Washington, DC: National Bureau of Economic Research, accessed 11 January 2021 at www.nber.org/papers/w22667.

Mohlmann, M. and L. Zalmanson (2017), 'Hands on the wheel: navigating algorithmic management and Uber drivers' autonomy', Proceedings of the International Conference on Information Systems, 10–13 December, Seoul.

Office of Industrial Relations (2019), 'Consultation Regulatory Impact Statement. Workers' compensation entitlements for workers in the gig economy and the taxi and limousine industry in Queensland: workers' compensation and rehabilitation Act 2003', Brisbane: OIR.

Ola (2019), 'Submission in response to the Consultation Regulatory Impact Statement (RIS): Workers' compensation entitlements for workers in the gig economy and the taxi and limousine industry in Queensland', Sydney.

Page, R. (2020), 'Uber and Dettol strike partnership', CMO from IDG, 9 June, accessed 11 January 2021 at www.cmo.com.au/article/680379/uber-dettol-announce-new-partnership/.

Peetz, D. (2018), *The Operation of the Queensland Workers' Compensation Scheme*, Nathan: Griffith Business School, accessed 2 August 2021 https://www.worksafe.qld.gov.au/__data/assets/pdf_file/0021/24087/workers-compensation-scheme-5-year-review-report.pdf.

Peetz, D. (2019), *The Realities and Futures of Work*, Canberra: ANU Press.

Peetz, D., K. Townsend, R. Russell, C. Houghton, A. Fox and C. Allan (2003), 'Race against time: extended hours in Australia', *Australian Bulletin of Labour*, **29** (2), 126–142.

Purse, K. (2005), 'The evolution of workers' compensation policy in Australia', *Health Sociology Review*, **14** (1), 8–20.

Queensland Government (2011), 'Queensland government response to report on investigation into the taxi industry in Queensland by the Queensland workplace ombudsman', Brisbane: Queensland Parliament.

Queensland Government (2012), 'Work health and safety for taxi drivers and operators', Brisbane: Queensland Government, accessed 11 January 2021 at www.worksafe.qld.gov.au/__data/assets/pdf_file/0027/15786/whs-taxi-drivers.pdf.

Quinlan, M. (2015), 'The effects of non-standard forms of employment on worker health and safety: inclusive labour markets, labour relations and working conditions branch', Conditions of Work and Employment Series 67, Geneva: ILO.

Rabe, T. (2020), '"Tale of two cities": Uber data sheds light on Sydney, Melbourne amid COVID-19', *Sydney Morning Herald*, 14 July, accessed 11 January 2021 at www.smh.com.au/national/tale-of-two-cities-uber-data-sheds-light-on-sydney-melbourne-amid-covid-19-20200713-p55bnp.html.

Smith, A. (2016), 'Gig work, online selling and home sharing', Pew Research Center, 17 November, accessed 11 January 2021 at www.pewresearch.org/internet/2016/11/17/gig-work-online-selling-and-home-sharing/.

Smith, J. E. (2020), 'A Covid-19 death renews questions of responsibility of Uber and Lyft to drivers', *Los Angeles Times*, 24 July, accessed 11 January 2021 at www.latimes.com/california/story/2020-07-25/covid-19-death-uber-lyft-drivers.

Sparrow, J. (2020), 'Workplace insecurity pervades the whole economy, just when every job is under threat', *The Guardian*, 25 July, accessed 11 January 2021 at www.theguardian.com/commentisfree/2020/jul/25/workplace-insecurity-pervades-the-whole-economy-just-when-every-job-is-under-threat.

Underhill, E. (2004), 'Changing work and OHS: the challenge of labour hire employment', in M. Barry and P. Brosnan (eds), *New economies: new industrial relations: proceedings of the 18th AIRAANZ Conference: AIRAANZ*, 3–6 February, Noosa: Association of Industrial Relations Academics of Australia and New Zealand.

Unions NSW (2016), *Innovation or exploitation: busting the Airtasker myth*, Sydney: Unions NSW.

US District Court (2018), *Razak* v. *Uber Technologies Inc.*, US District Court for the Eastern District of Pennsylvania. Case No. 2:16-cv-00573, 25.

Vida, S. (2020), 'Lyft is installing coronavirus sneeze guards in some cars', *Wall Street Nation*, 21 July, accessed 11 January 2021 at https://wallstreetnation.com/breaking-news/lyft-is-installing-coronavirus-sneeze-guards-in-some-cars.html.

WION (2020), 'Uber lays off 3500 employees in 3 minutes, *Coronavirus News*, accessed 11 January 2021 at www.youtube.com/watch?v=1yAoxqYUps0.

WorkCover Queensland (2020), 'Gazette notice: workers' compensation and rehabilitation Act 2003 (Q)', WorkCover Queensland Notice (No. 1) of 2020, Brisbane: WorkCover Queensland.

Wright, C. F., A. J. Wood, J. Trevor, C. McLaughlin, W. Huang, B. Harney, T. Geelan, B. Colfer, C. Chang and W. Brown (2019), 'Towards a new web of rules: an international review of institutional experimentation to strengthen employment protections', *Employee Relations*, **41** (2), 313–330.

PART V

Closing thoughts

21. Institutional experimentation and the challenges of platform labour

Maria Figueroa

21.1 INTRODUCTION

This volume has discussed the regulatory and organizational strategies that social actors have been implementing with regard to digital labour platforms over the past few years. As changes in digital technology disrupt industries and the geography of labour markets, new forms of work arrangements are being introduced, causing confusion to the existing regulatory frameworks for labour protection and collective representation and challenging labour organizations to adapt and to advance innovative policy and organizing solutions. The analysis reveals significant challenges but also promising opportunities for those labour organizations that are able to advance the rights of platform workers.

21.2 OUTLINING THE CONTOURS AND CHALLENGES OF PLATFORM LABOUR

Jan Drahokoupil and Gérard Valenduc both discussed the contours of the platform economy and how it has shaped platform labour markets. According to Drahokoupil, the platform economy comprises platforms that use data as a direct source of revenue, including the big technology companies (Google, Apple, Facebook and Amazon (GAFA); and Baidu, Alibaba, Tencent and Xiaomi (BATX)) and their relatively smaller competitors; as well as the labour platforms which use data to improve their services and their labour management functions. As Drahokoupil noted, these two segments of the platform economy follow two 'partially overlapping revenue models'. The first model involves value added and income generated through the structuring of interactions 'in a way that facilitates the proliferation of transactions' and mainly aims at monetizing user-generated data and network effects. This first strategy is mainly applied by the big tech companies. The business model followed by the labour platforms, on the other hand, involves increased efficiency of transactions and the algorithmic management of labour, but user-generated data do not 'represent a major source of revenue' for them.

In developing a typology of platform labour, Drahokoupil distinguished between location-based platform work and online and largely global platform work. Location-based platform labour involves a range of service sectors such as transportation, delivery, maintenance and housecleaning, and personal care, in which platforms like Uber, Deliveroo, Handy.com and others match workers with consumers on an on-demand basis. Online platform labour includes microtasking or macrotasking with the former involving less skilled work, such as tasks performed by Amazon Mechanical Turk, and the latter comprising creative and higher-skilled jobs.

Compared to the technology giants, labour platforms, particularly location-based ones, are at the low end of the platform economy. They are typically not profitable and, despite recent growth in their market capitalizations, they are dwarfed by the GAFA and BATX capitalizations, which reached a combined $4.5 trillion in 2019 (Chevré 2019). Drahokoupil's and Valenduc's contributions serve not only to situate platform labour in its economic and industrial context but also help us to better understand value creation and appropriation in the global platform supply chain. In this value creation process, platform workers as well as other users or consumers are shortchanged as they lack control over the data they generate. Drahokoupil's example of Deliveroo workers, for whom the performance metrics and ranking data collected by the platform represent valuable economic assets, and indeed societal power, is key to this part of the analysis. Workers accrue this capital but they can neither control nor leverage it as it cannot be transferred between platforms. This was also a recent issue with food delivery workers in Argentina who worked for Glovo, acquired by Delivery Hero in September 2020 (Senen 2020).

Closely linked to platforms' control over user-generated data are issues related to measuring the extent of platform labour. In her discussion of these issues, Agnieszka Piasna pointed to the conundrum of the lack of data on the true scale of platform labour despite, unlike in traditional informal sectors, 'all transactions mediated by online platforms [being] digitally recorded'. This lack of hard data on labour platforms makes greater sense when viewed in the context of the largely unregulated environment in which such platforms operate. In the United States (US), for instance, the only reliable source of data on the pay, mileage driven and hours worked by Uber and Lyft drivers is the New York City Taxi and Limousine Commission which regulates ride-sharing companies. In the absence of a regulatory agency with data collection powers, there is no mechanism for public access to the data controlled by the platforms.

The lack of data on labour platforms is both the result of and a barrier to regulatory action in this sector; as Piasna pointed out, 'a patchwork of statistics from singular surveys is certainly insufficient to inform and guide policy'. Indeed, advocates in the US currently rely on local survey efforts to collect the worker data needed to inform legislation directed towards labour protection. This was the case for the recent minimum pay and other benefits achieved in Seattle for Uber and Lyft drivers (Parrott and Reich 2020), as well as the ongoing organizing initiatives in food delivery in New York City (Irizarry Aponte and Velasquez 2020). With these problems in mind, Piasna offered a number of recommendations for possible ways of tackling this challenge, including the development of harmonized instruments for use in official, regularly repeated labour force surveys and the measurement of platform work within the broader framework of precarious and casual work.

Failure to regulate labour platforms is leaving workers across the platform economy without basic protection, including with regard to occupational health and safety that has become critical in the face of the Covid-19 pandemic, particularly for location-based platform workers. Pierre Bérastégui and Sacha Garben highlighted that platform workers across online and location-based work face a wide range of workplace health and safety issues. Microtaskers may be especially at risk of cognitive overload and stress related to managing the queue of assignments; macrotask platform workers face high pressure to work long hours; and location-based platform workers face hectic and potentially hazardous conditions due to platforms' widespread use of surveillance technology. The case study of Australia, which David Peetz discussed in this volume, illustrates how the vulnerability of location-based platform workers in ride-sharing has been exacerbated because of the outbreak of the Covid-19 pandemic. As reported by Peetz, the pandemic has revealed the inadequacies of the current work arrangements for temporary or casual workers and independent contractors, including platform workers, who are exposed to the health, social and economic risks of the crisis.

According to Bérastégui and Garben, the stress caused by heightened digital surveillance, job instability, isolation and workplace fragmentation not only has psychological and physical impacts on workers but also hampers their collective action via their ability to defend their rights and interests and to engage in social dialogue. The chapter by Benjamin Herr, Philip Schörpf and Jörg Flecker discussed these challenges and explored the opportunities for labour organizing and regulatory action. They highlighted that employers and labour activists in the platform economy both engage in politics of scale, 'trying to shift [the] scales in order to challenge power relations', and aggregated spatial scales across the platform economy into three: workplace/local; national; and transnational. Unlike platform workers who do not share a common national jurisdiction, 'location-based platform workers are subject to a common regulatory framework' which determines the appropriate scale for regulation. These spatial scales are constantly contested, however: Herr et al. offered the example of the Dutch group Takeaway.com which sued the works council in its Liferando branch in Vienna, Austria, claiming that this branch was not part of their operations, thereby arguing that the founding of the works council was illegal.

The chapter by Pamela Meil and Mehtap Akgüç examined the challenges for regulating online labour platforms, focusing on business-to-business platforms. This chapter explored the question of whether the outsourcing of work to platforms contributes to new forms of value chains and identified three modalities for the reintegration of online platform labour into such new forms. These modalities include anonymous microtasking by human microworkers; the isolation and modularization of tasks requiring little reintegration (for example information technology, software development and design tasks); and a task management and coordination role 'similar to that of tier one suppliers in conventional outsourcing models'. The authors concluded that these new trends in the externalization of work to business-to-business platforms erode labour standards for workers mainly through deskilling and low pay.

Bérastégui and Garben, Herr et al. and Meil and Akgüç all agreed that the challenges for collective action are more severe for online platform labour than for location-based platform workers. For these authors physical distance, workplace fragmentation and the resulting isolation are key barriers to remote global 'virtual' workers organizing and promoting regulatory measures in their sector. The case studies in this volume substantiate this argument as they illustrate how labour organizing has concentrated on location-based platform work which also offers more favourable spatial scales for regulatory action. As Valenduc indicated, keeping a historical perspective will be critical for labour advocates to formulate proactive organizing, policy and legal strategies.

21.3 REGULATING PLATFORM WORK

Regulating labour platforms and establishing a framework for labour relations in the platform economy requires the exploration of opportunities across spatial scales at the sub-national, national and supra-national levels. Valerio De Stefano and Mathias Wouters offered us three possible approaches. The first builds on the idea of Personal Work Relations and the ABC tests used in the US, with an initial presumption of employee status at its core. An example of this approach is set out in the Action Programme 2019–2023 of the European Trade Union Confederation, which 'advocates for any person engaged in a "personal work relationship" to have collective rights, including the right to strike, regardless of competition rules'. An example of the application of the ABC test model is the adoption by the Federation of Dutch Trade Unions (*Federatie Nederlandse Vakbeweging*) of the Dutch Commission's proposal for an 'employee unless' model. The second approach involves the application of the existing conventions of the International Labour Organization that can be used to regulate labour platforms as intermediaries of contracts for personal services (Convention No. 181) or as employers of crowdworkers who are legally classified as homeworkers (Convention No. 177). The third, but perhaps less feasible, approach would involve the formulation of a Crowdworkers' Bill of Rights, similar to the Seafarers' Bill of Rights established in the Maritime Labour Convention 2006.

Focusing on the experience within Europe of contending with the challenges of regulating the platform economy, Sacha Garben advanced three possible methods. First, there is the simple application (enforcement) of current legal provisions, although she noted that this has resulted in a policy patchwork with varying outcomes within and between jurisdictions. An example of this type of result involves Uber drivers who have been reclassified as workers (instead of independent contractors) in many national courts while a lower-level court ruled in 2019 that UberX drivers in Brussels were self-employed. The second method would involve narrowing the group of persons that could be considered self-employed by adding an intermediate or independent worker category (a formula first proposed in the US). The third method would consist of providing specific protection for platform workers regardless of their employment status. In this case, Garben referred to the example of France, where a law enacted in 2016 provides that 'independent workers in an economically and technically dependent relationship with a platform' can receive benefits such as accident insurance, professional training and union representation rights. This is also the approach taken in Queensland, Australia, to provide workers' compensation benefits to ride-sharing drivers. At the level of the European Union, Garben praised the new Directive on Predictable and Transparent Working Conditions for addressing, at least partially, the precarity and lack of occupational health and safety, and social, protections that prevail in non-standard work arrangements including platform labour.

Sai Englert and his colleagues examined the ongoing struggles for platform regulation by focusing firstly on the platform companies' own approach based on self-regulation and secondly on the grassroot alternatives. Platforms indeed go as far as to claim that they can 'develop better standards than the state'. These standards, platforms claim, can involve effective ways of addressing issues related to business transactions as well as measures for disciplining and promoting efficiency among consumers and workers (for example, negative ratings and the deactivation of workers' accounts). Englert et al. identified three categories of grassroot alternatives to platform self-regulation: industrial action; legal challenges; and extra-legal standard-setting initiatives including the creation of networks of allies and stakeholders. Grassroot industrial action involves a range of possible approaches from the organizing of Turkernation, enabling Amazon Mechanical Turk workers to rate clients, to the development of unions for platform workers. Although these worker-led regulatory efforts have achieved some victories, Englert et al. recognized that 'the balance of forces remains everywhere to the advantage of the platforms'.

The contribution by Simon Joyce and Mark Stuart assessed trade unions' responses to platform work and examined the factors explaining the uneven nature of the responses from established unions as well as the drivers for the rapid growth of grassroot unions. Their analysis uncovered two main sources for the limitations of traditional trade union responses: their historical approach of focusing on their 'core membership at the expense of emerging industries' and new groups of workers; and that the practices and forms of platform work 'do not fit established, institutionalized organizing models'. In contrast, grassroot unions have developed direct action methods from which mainstream unions could learn. The implications of this approach for the current state of organizing is that mainstream unions have been 'reticent' although 'some progress' has been made '[w]here platform workers have legal employee status and can be incorporated into existing collective bargaining arrangements'. This has been the case for delivery riders with Just Eat in Denmark, Foodora in Sweden and Norway and Deliveroo in Austria and Italy.

Joyce and Stuart pointed out that the 'gains from orthodox collective bargaining have, so far, been minimal' but, even so, they are able to cite examples of successful union intervention where location-based platform workers are not employees. This includes the efforts of Teamsters Local 117 in the ride-sharing industry in Seattle, where the union focused on lobbying local government for labour protection and greater regulation; while another example is the successful campaign of Service Employees International Union and Gig Workers Rising for the reclassification of platform workers as employees through the enactment of Assembly Bill 5. This was a short-lived victory, however, as the platforms gained exemption from the new law after running a $200 million campaign in 2020. A third example would be the experience of the Canadian Union of Postal Workers (CUPW) with Foodora in Toronto, which Raoul Gebert discussed in this volume.

Joyce and Stuart attributed the rapid growth of grassroot unions in platform work to two broad factors: the emergence of new groups of workers resulting from changes in digital technology and the consequent reorganization of work in a number of industries; and the rise of a current of self-organizing among platform workers aided by the widespread utilization of digital technology as a tool for creating communities of workers and a launching platform for collective action. This includes international industrial action (for example, the Uber drivers' international strike in 2019). While acknowledging the need for further research, the authors found sufficient evidence to support the claim that grassroot unions' strategic approach is focused on 'mobilizing as many workers as possible rather than … narrowly in terms of organizational membership'; something that is key to understanding the differential offer, and indeed the experimentation, of grassroot unions.

21.4 VIRTUAL CONNECTIONS IN DISLOCATED LABOUR MARKETS: ONLINE PLATFORM LABOUR CASES

As noted by Mariya Aleksynska and her colleagues, online labour platforms match workers from any location to global pools of work opportunities without requiring workers' physical mobility, helping to overcome distances and constraints of local demand through virtual migration. In this process, however, online labour platforms also create challenges for workers as well as for local and national economies. The chapters led by Mark Graham and by Janine Berg also underscored the limitations of online platform work for human and economic development at the global margins. Graham et al. investigated the potential for this labour engagement modality to benefit human development in south-east Asia and sub-Saharan Africa and found that online platform work in particular offers improved earning opportunities, stimulating work and increased autonomy for workers. At the same time platform work also involves health risks and erodes workers' well-being because of social isolation, overwork and the lack of job security. These findings are consistent with those of Bérastégui and Garben who noted that threats to workers' physical and psychological health pervade the online platform labour sector. Additionally, the human development benefits are not spread evenly across online platform labour markets due to the high levels of inequality inherent in the industry's work organization, discriminatory practices on the part of clients and the proliferation of predatory intermediaries.

Both Graham et al. and Berg et al. agreed that, as online labour platforms operate outside geographically defined regulatory and normative frameworks, and are disconnected from local economies, they have limited income multiplier effects and do not generate tax revenues to fund human and economic development. Aleksynska et al. make the same point from the perspective of platforms that are plugged into local and regional economies. Focusing on the shift from outsourcing to crowdsourcing in India, Berg et al. examined how this results in the further precarization of online work and the deterioration of local economies. While online labour, previously engaged by the business processing outsourcing industry in India, has represented a key source of foreign exchange, jobs and technology transfer for the past few decades, the entrance of labour platforms has resulted in some of the outsourcing being channelled directly to individual workers through crowdsourcing. Berg et al. also noted that, unlike workers engaged by business processing outsourcing firms who are generally classified as employees, crowdworkers are considered independent contractors which further contributes to precarity in this sector. The key issues that Berg et al. identified also encompass the lack of mechanisms for workers to communicate with clients, algorithmic management, difficulty in leveraging skills for increased job opportunities, few learning opportunities and, very importantly, deskilling as a result of microtasking.

Focusing on the examples of Russia's and Ukraine's online platform labour markets, Aleksynska et al. discussed how digitalization and globalization are creating new economic geographies and argued that the presence of geographic and linguistic variety in online labour markets is linked to 'the socio-economic benefits of operating in the context of a common language, similar culture and proximate geography'. Additionally, they found that geographic and linguistic variety influences skill transferability and skills recognition as 'skills can be valued differently in different markets'. The emergence of domestic and regional online markets which value skills differently has policy implications including the need for an international governance system for digital labour platforms, the emergence of issues related to the diversification of skill or workforce development programmes and the high degree of informality in online work. Aleksynska et al. argued that the ideas set out in the International Labour Organization's 2019 Global Commission on the Future of Work for an international governance system setting minimum rights and protections for platform workers, along with a representative board to adjudicate disputes between platforms, clients and workers, takes us in an important direction but that there remains additional scope for the domestic regulation of local online markets.

21.5 INSTITUTIONAL EXPERIMENTATION: LOCATION-BASED PLATFORM CASES

The location-based platform cases discussed in this volume serve to tease out elements of institutional experimentation and the innovative organizing strategies required by the emergence of new economic sectors and new groups of workers. In his study of the Foodora case in Toronto, Raoul Gebert noted that new economic sectors are disrupting existing institutional and legal frameworks, driving creative actors to become institutional entrepreneurs who, in their turn, engage existing resources to develop 'new institutions, networks and alliances' in a process of institutional experimentation.

In their case studies of taxi markets in four cities in Europe and the US, Hannah Johnston and Susanne Pernicka incorporated the concept of symbolic power to reconceptualize power resource theory. With this new theoretical framework, they examined four case studies 'where workers' power resources vary along a power distribution continuum' from Vienna, with a strong position of neocorporatist industrial relations actors and a highly centralized workers' movement with significant institutional power, to Berlin with its strong neocorporatist tradition but declining influence on labour market structures and practices, to Los Angeles 'where industrial relations systems are comparatively weak and exclude (the often) self-employed taxi drivers'. Los Angeles is placed at the other end of the continuum because of its highly fragmented and pluralistic workers' movement which has struggled to confront the power that platform companies wield in California. On this continuum, New York City lies between the European cities and Los Angeles; here, 'collective organizations and associational power have helped [drivers] overcome weak institutional power and positioned them to engage in a symbolic struggle over what it means to work on digital labour platforms.' It should be noted, however, that drivers in New York City have not challenged the platforms on the issue of independent contractor status, which constitutes the keystone of these companies' business model, whereas the Los Angeles labour movement was able to mount defiance with the passage of Assembly Bill 5. Johnston and Pernicka found that workers' ability to achieve favourable outcomes increases 'when they demonstrate associational power at a similar scale to the regulation they seek and in places where they have greater symbolic power'.

The experience of the CUPW with Foodora in Canada and of ride-sharing drivers in Australia are cases of institutional experimentation. As Gebert pointed out, the case of Foodora is an example of institutional experimentation in two ways. First, the union did not ask for salaried employee status but helped to create the new legal status of dependent contractor. Second, Foodora's departure from the Canadian market 'led to the establishment of a CUPW-funded community union without legal bargaining rights'. This case is also an example of how the resources of traditional labour unions 'can be redeployed in dynamic, new sectors to gain collective bargaining rights for platform workers even when the institutional framework is not favourable'.

Developing the theme of institutional experimentation, the study of the mobilization process of platform-based food delivery couriers in Köln, Oslo and Paris by Kristin Jesnes, Denis Neumann, Vera Trappmann and Pauline de Becdelièvre discussed the variety in the couriers' collective (proto-)organization, coalition-building with trade unions and the effectiveness of social protest. A critical mass of couriers has been instrumental in driving the mobilization process to address the negative effects of algorithmic management, non-standard contractual arrangements and low pay. Outcomes have varied, much in line with the dominant logic of the national industrial relations system.

The chapter by Jack Linchuan Qiu, Ping Sun and Julie Chen also analysed food delivery platforms. Focusing on Beijing, it showed how algorithmic management is manipulating employment terms and conditions. In fact, it led to a deflexibilization as the incentive structures and algorithmic management encouraged full-time engagement with very little working time flexibility. At the same time, couriers resisted algorithmic management collectively, through social protest, and, in particular, individually, through deceiving and subverting the algorithms.

In his analysis of the ride-sharing industry in Australia, David Peetz focused on the efforts of a government review into the injury compensation system in Queensland which, similarly to the experience of Foodora in Toronto but with a narrower scope, involved achieving protections for platform workers without dealing with employment status or the need for reclassification. According to Peetz, there may even be some advantages from not reclassifying contractors as employees since '[s]ome platforms will adapt if their core business model is not threatened'. As Peetz warned, however, it cannot always be assumed that ride-sharing companies 'will always act in unison' as they are in competition with each other and not equally positioned to absorb the increased costs resulting from changes in market conditions.

The cases of Alia in the US and Aliada in Mexico, discussed by Andrea Santiago Páramo and Carlos Piñeyro Nelson, involve a case of institutional experimentation with the establishment of Alia as a platform for employer contributions to a portable benefits fund for domestic workers. Alia emerged as a worker response to the lack of social protections that domestic workers have historically experienced in their situation as excluded workers. This chapter contrasted Alia with labour platforms such as Aliada, Care.com and Handy.com which engage domestic workers in north America; focusing in detail on the case of Aliada in Mexico, Santiago and Piñeyro found that this platform has, in some ways, helped to formalize the domestic work sector in Mexico. This formalization comes at a cost to worker organizing potential, however, as it has created perceived distinctions between cleaners who 'see themselves as "professional" and "modern" and those that do not fit into these categories'.

The cases of Alia and Aliada illustrate the complexities resulting from the entrance of digital platforms into domestic work. Digital technology in domestic work, as in most industries examined in this volume, is likely to have a dual impact on labour, creating new challenges and disrupting work arrangements and institutional frameworks while also enabling workers to self-organize and experiment with network-supported and infrastructure solutions to their lack of protection.

21.6 THE TAKEAWAYS

This volume offers lessons that can be helpful for drafting new roadmaps for efforts to advance workers' rights. Adopting a historical perspective is critical to the development of proactive responses to the disruptive aspects of platforms, including their economic models and the new industrial and power structures that have emerged. The varying modalities for engaging workers indicate that 'one size fits all' solutions would not be effective in providing protections for workers in the segmented labour markets of the platform economy. International regulatory frameworks might be appropriate for online labour platforms while location-based platform labour might necessitate sub-national policy solutions. Consideration of the appropriate spatial scale is critical to the identification of agencies with regulatory powers in the relevant jurisdictions and is key to the viability and ultimate success of policy interventions.

The analysis presented in this volume indicates that worker organizing and regulatory initiatives have, perhaps inevitably, concentrated on the location-based platform economy as opposed to online platforms. The emergence and diffusion of digital platforms have created new battlegrounds for worker organizations to alter power relations in favour of labour. Technological change is not a new challenge for unions and workers. In the face of labour platforms' disruptive effects across industries and institutional frameworks for worker protections, however, labour advocates must confront the following questions: is it enough simply to enforce existing rules and strengthen our institutions? Is digital technology just an old challenge dressed in new clothes? Alternatively, are we facing an irreversible shift in the paradigm for work organization and labour relations? This debate is unlikely to be resolved in the near future but the analysis presented in this volume suggests that worker organizations need to be nimble and to seek out new formulas to represent and gain protection for the new groups of workers that have emerged with the spread of digital platform technologies. As the conceptual frameworks and case studies presented in this volume indicate, worker organizations will need to engage in institutional experimentation if they are to confront effectively the challenges of the digital economy.

REFERENCES

Chevré, C. (2019), 'GAFA vs BATX: to rule them all – leaders league', accessed 26 February 2021 at www.leadersleague.com/en/news/gafa-vs-batx-to-rule-them-all.

Irizarry Aponte, C. and J. Velasquez (2020), 'NYC food delivery workers band to demand better treatment. Will New York listen to Los Deliveristas Unidos?', accessed 26 February 2021 at www.thecity.nyc/work/2020/12/6/22157730/nyc-food-delivery-workers-demand-better-treatment.

Parrott, J. and M. Reich (2020), *A minimum compensation standard for Seattle TNC drivers*, Report for the City of Seattle, New York: New School, Center for New York City Affairs and Michael Reich Center for New York City Affairs, accessed 26 February 2021 at www.centernyc.org/s/Parrott_Report_July22020-f876.pdf.

Senen, C. (2020), *Incertidumbres de Los Repartidores Autónomos*, Presented at 'Advancing Decent Work in the Platform Economy', Reshaping Work Onward, OSF, Ithaca, NY: Cornell Worker Institute.

Index

Printed and bound by CPI Group (UK) Ltd, Croydon, CR0 4YY

16/04/2025

14658485-0005